WAVES AND STABILITY IN CONTINUOUS MEDIA

Series on Advances in
Mathematics for Applied Sciences — Vol. 4

Proceedings of the Vth International Meeting

WAVES AND STABILITY IN CONTINUOUS MEDIA

Sorrento, Italy 9 – 14 October 1989

Editor

S. Rionero

Università degli Studi di Napoli
Italia

World Scientific

Singapore • New Jersey • London • Hong Kong

Published by

World Scientific Publishing Co. Pte. Ltd.
P O Box 128, Farrer Road, Singapore 9128
USA office: 687 Hartwell Street, Teaneck, NJ 07666
UK office: 73 Lynton Mead, Totteridge, London N20 8DH

WAVES AND STABILITY IN CONTINUOUS MEDIA

ISBN 981-02-0554-6

Printed in Singapore by Utopia Press.

Scientific Committee

G. Carini	(Univ. of Messina)
G. P. Galdi	(Univ. of Ferrara)
A. Greco	(Univ. of Palermo)
P. Renno	(Univ. of Naples)
S. Rionero	(Univ. of Naples)
A. Romano	(Univ. of Naples)
T. Ruggeri	(Univ. of Bologna)
G. Toscani	(Univ. of Ferrara)

Organizing Committee

A. D'Anna
R. Grassini
P. Renno
S. Rionero (Chairman)
A. Romano

Organizing Secretariat

F. Dell'Isola
G. Guerriero
A. Maio

UNDER THE AUSPICES OF
COMITATO PER LA MATEMATICA OF ITALIAN C.N.R.;
ISTITUTO ITALIANO PER GLI STUDI FILOSOFICI;
DIPARTIMENTO DI MATEMATICA E APPLICAZIONI
"R. CACCIOPPOLI" - UNIVERSITA "FEDERICO II"
NAPLES ITALY.

PREFACE

The 5th Biennial Conference on *"Waves and Stability in Continuous Media"* was held in Sorrento (Italy) from October 9 to 14, 1989. Same as the previous four conferences this was a gathering of researchers who are interested in discussing and presenting current problems on Waves and Stability in Continuous Media. The scientific program included about twenty invited speakers and short contributed papers.

The warm atmosphere created during the Conference has produced a deep interchange of ideas and new common projects in this area.

The success of the Conference was due to the generous financial support of several agencies that are listed in p. viii.

I wish to thank the members of both the Scientific and Organizing Committee as well as the members of the Organizing Secretariat for their help.

Last, but not least, I must acknowledge the participation of so many registrants from different countries. Their collaboration was essential for the success of the Conference.

Salvatore Rionero

CONFERENCE SPONSORS

The organization of the 5th International Conference on Waves and Stability in Continuous Media was made possible by the financial support of:

Mathematical Committee of the Italian National Research Council (C. N. R.)

Gruppo Nazionale per la Fisica Matematica (G. N. F. M.–C. N. R.)

Gruppo Nazionale M. P. I. 40% Problemi di Evoluzione nei fluidi e nei solidi

Universita' di Napoli "Federico II"

Dipartimento di Matematica ed Applicazioni "R. Caccioppoli"

Istituto Italiano per gli Studi Filosofici

Istituto per le Applicazioni della Matematica del C. N. R. (Naples)

CONTENTS

ON STABILITY AND LONG TERM BEHAVIOUR FOR RAREFIED GASES

LEIF ARKERYD

Department of Mathematics
Chalmers University of Technology
and the University of Göteborg
S-41296 Göteborg Sweden

ABSTRACT

The talk discusses the stability and long term behaviour of rarefied gases modeled by the Boltzmann equation. A review of results for initial values far from equilibrium is given.

A Formal Discussion

Let us consider a rarefied gas described through its one particle density F in phase space. Back in 1872 Boltzmann gave the evolution of such a gas through an equation which expresses the material derivative of F in terms of the gain from molecules colliding into a region by binary collisions minus the loss from those colliding out of the region. In the absence of exterior forces his equation is

$$\partial_t F + v\nabla_x F = \text{gain} - \text{loss} = QF.$$

Here ∇_x is the gradient with respect to the space variable x, v is velocity and t is time. Finally Q is the so called collision operator,

$$QF(x,v_1)=\int_{R^3\times B}(F(x,v_1')F(x,v_2') - F(x,v_1)F(x,v_2))k(v_1,v_2,u)dv_2du.$$

Given two molecules of initial velocities (v_1,v_2) and initially separated in space, (v_1',v_2') denotes the velocities after collision. The details of the collision process are described with the help of the parameter set B, which is often chosen as angular coordinates for a sphere. Finally k is a weight function depending on the interaction between the molecules.

On physical grounds, the gas left to itself should settle down towards some equilibrium, and that in a stable way, i.e. gases with a similar initial preparation should approach similar equilibria in a similar way. And of course, this ought to be mirrored by the solution F of the Boltzmann equation model. Surprisingly , after the equation has been around for almost 120 years, the mathematical understanding of these questions is still far from complete. For the great variety of more complicated phenomena such as shocks, bifurcations and waves, which in many cases are fairly well understood on the gas dynamic level, any more fundamental kinetic studies have hardly even started, and for a reasonable picture above the perturbation level, we can only put our hopes to the future.

Before turning to the mathematical results on equilibrium and stability, let us see what the Boltzmann equation seems to tell us formally about the long time behaviour. Consider the entropy

$$HF(t) = \int F(t)\log F(t)\,dxdv.$$

We have

$$(\partial_t + v\nabla_x)(F\log F) = (\partial_t + v\nabla_x)F + \log F(\partial_t + v\nabla_x)F = QF + \log F\cdot QF,$$

and (by a change of variables argument)

$$\int QF\,dxdv = 0.$$

Hence, under suitable boundary conditions

(1) $$\partial_t \int F\log F\, dxdv = \int(\partial_t + v\nabla_x)F\log F dxdv = \int QF dxdv + \int \log F\cdot QF dxdv$$

$$= \int \log F\cdot QF dxdv = \text{(change of variables)}$$

$$= \frac{1}{4}\int (F_1'F_2' - F_1F_2)(\log F_1F_2 - \log F_1'F_2')k\, dxdv \le 0.$$

And so the entropy is decreasing. By a similar change of variables, mass, v-moments, and energy can be seen to be conserved. Finally a lemma by Gibbs tells us that the entropy is bounded from below by its value for a certain maxwellian $E_o = a\exp(-b|v|^2 + (c\cdot v))$ depending on the initial value.

Integrating (1) with respect to time we formally get

$$\int E_o \log E_o dxdv \le \int F(t)\log F(t) dxdv = \int F_o \log F_o dxdv -$$

$$- \frac{1}{4}\int_0^t \int (F_1'F_2' - F_1F_2)(\log F_1'F_2' - \log F_1F_2)k\, dxdvds,$$

and so

$$F_1'F_2' - F_1F_2 \to 0 \text{ as } t\to\infty.$$

As mentioned above we also expect F to converge to some equilibrium w, which should then satisfy $w_1'w_2' - w_1w_2 = 0$. This gives for f=logw that

$$f(v_1') + f(v_2') = f(v_1) + f(v_2).$$

Essentially by arguments going back to Cauchy, it follows that

$$f = \alpha + (v_o\cdot v) - \beta|v|^2, \quad w = \exp(f) = a\exp(-b|v|^2 + (v_o\cdot v))$$

with coefficients possibly depending on other variables such as x and t. Hence the equilibrium is a (possibly local) maxwellian.

In the case of the forces between the molecules being inversely proportional to the fifth power of the distance, the v-moments formally satisfy a system of ordinary differential equations which can easily be analysed having exponential convergence to equilibrium and stability.

Strict Results

The formal arguments indicate the presence of convergence to equilibrium, even exponential convergence, and stability. When we next turn to the strict mathematical results and starting far from equilibrium, we will see that the strength of the known results steadily decreases, when progressing from a) the spacehomogeneous case via b) the spacedependent case close to spacehomogeneous to c) the spacedependent case in general.

In the spacehomogeneous case a) the understanding is good for hard forces, i.e. those of inverse power >5 under suitable cutoffs in the u-variable, which means playing down the grazing collisions. Let

$$k(v_1,v_2,u) = h(u)|v_1-v_2|^{\beta}\ ,\ 0<\beta\leq 1,\ \text{with } h \text{ bounded},$$

$$L^1_r = \{\text{measurable functions } f \text{ with } (1+|v|^r)f \in L^1(R^3)\},$$

$$L^1_{r\gamma} = \{f:R_+ \to L^1_r \text{ with } \sup_{t>0} \exp(\gamma t)\ ||f(t)||_r < \infty\}.$$

Given a maxwellian E_o the following result holds.

Theorem[1]) *There exists* $r_o\geq 2+\beta$, *and* $\mu>0$, *such that if* $r_1\geq r_o$ *and*

$$F_o \varepsilon L^1_{r_1},\ \int(1,v,v^2)F_o dv = \int(1,v,v^2)E_o dv,\ F_o \log F_o \varepsilon L^1,$$

then

$$||F - E_o||_{r,\gamma} < \infty \quad \text{for } r<r_1,\ 0\leq\gamma<\mu.$$

Moreover, if $2\leq r\leq r_1-\beta$, *then given* $\varepsilon>0$, *and a bounded set* $B\subset L^1_{r_1}$, *there is a* $\delta>0$ *such that (stability)*

$$\sup_{t>0}||F-\tilde{F}||_{r,0} < \varepsilon \quad \text{if} \quad ||F_o - \tilde{F}_o||_r < \delta \text{ and } \tilde{F}_o \varepsilon B.$$

The proof uses the spectral properties of the linearized collision operator and a powerful apriori estimate for Q. It is fairly technical and the interested reader is referred to the article[1]).

In the spacedependent case with initial values close to spacehomogeneous b), we let $u = (\theta,\phi)$ be an angular variable and take

$$h(\theta) \sim \sin 2\theta, \quad 0 \leq \theta \leq \pi/2.$$

The following function spaces are used;

$$B_r = \{\text{continuous functions } f \text{ on } R^3, \text{ with } ||f||_r < \infty\}$$

where

$$||f||_r = \sup_{R^3} (1+|v|^r)|f(v)|,$$

and

$$B_{p,r} = \{\text{continuous functions } f \text{ on } R^3/Z^3 \times R^3 \text{ with } ||f||_{p,r} < \infty\}$$

where

$$||f||_{p,r} = || \sum_{k\varepsilon 2\pi Z^3} (1+|k|^2)|Ff(k,v)|^2)^{1/2}||_r .$$

The following stability result holds.

Theorem[2)] *Let* ρ, $\sigma > 0$ *be given and let the initial value* F_o *satisfy*

$$F_o(x,v) = g_o(v) + u_o(x,v)$$

with

$$g_o(v) = \int F_o(x,v)dx, \int g_o(v)dv = \rho, \int |v|^2 g_o(v)dv = \sigma.$$

Then there are p_o, $r_o > 0$, *such that for* $r > r_o$, $p > p_o$, *there exists* $a > 0$ *such that if* $||u_o|| < a$, *then there is exactly one solution* F *of the Boltzmann equation with initial value* F_o *with* $F(t) \varepsilon B_{p,r}$ *and*

$$||F(t) - E_o||_{p,r} \to 0 \text{ when } t \to \infty.$$

Given $\eta > 0$, *there is a* $\delta > 0$, *such that (stability)*

$$\sup_{t>0} ||F(t) - \tilde{F}(t)||_{p,r} < \eta \text{ if } ||F_o - \tilde{F}_o||_{p,r} < \delta.$$

The theorem tells us that for hard forces, if the gas starts close enough to a spacehomogeneous initial value, then the solution converges to the relevant global maxwellian. In particular any detailed memory of the initial value is lost in the limit. Moreover, the solution is stable in $B_{p,r}$. The proof is similar but less technical than the previous L^1-case.

If the torus R^3/Z^3 is replaced by full space R^3, then the integral $\int_0^t \int QF \log F \, dxdvds$ is not always strong enough to force the gas to equilibrium. Instead a memory of the initial value may be retained into the limit. Physically that is so because the expansion of the rarefied gas may separate the molecules too quickly for them to undergo enough collisions. But the situation is still stable with respect to nearby initial values. This was discovered by G.Toscani.

Set $f^{\#}(x,v,t) = f(x+tv,v,t)$, and introduce the following function spaces;

$$C_b = \{\text{continuous functions } f \text{ on } R^3\times R^3 \text{ with } ||f||_{p,r} < \infty\}$$

where

$$||f||_{p,r} = \sup_{x,v} (1+|x|)^p (1+|v|^2)^r |f(x,v)|,$$

$$B_r^p = \{\text{continuous functions } f \text{ on } R^3\times R^3\times R_+ \text{ with } \sup_{t>0} ||f^{\#}(t)||_{p,r} < \infty\}.$$

Theorem[7)] *Let* $p>1$, *and* $r>3/2$ *be given. There exists* $a > 0$, *such that if* $||F_o||_{p,r} < a$, *then there exists exactly one solution* F *with* $F(t) \in B_r^p$ *and* $\sup_{t>0} ||F(t)||_{p,r} < 2a$. *If* $||\tilde{F}_o||_{p,r} < a$, *and* $||F_o - \tilde{F}_o||_{p,r}$ *is small enough, then (stability)* $\sup_{t>0} ||F - \tilde{F}||_{p,r} \leq 2||F_o - \tilde{F}_o||_{p,r}$.

The proof uses a contracting mapping argument via good a priori estimates in the p,r-norm. The same kind of arguments gives that $f^{\#}$ has a limit when time tends to infinity, and by the exponential form of the equation that this limit is $\geq F_o/2$ for small F_o not necessarily of exponential type. Hence the limit cannot be a local maxwellian in general.

Finally in a bounded domain with reasonable boundary conditions, there is always convergence to some equilibrium even for spacedependent data far from spacehomogeneous ones. The proofs depend on weak L^1 compactness arguments of the DiPerna Lions type[6], which cannot be used to study stability questions in the sense of this talk. Such results have been obtained by DiPerna and Lions[7], Arkeryd[3], Desvillettes[5], and Cercignani[4]. Here is one example.

<u>Theorem</u>[3] *Suppose that* $k \in L^1_{loc}$ *and that* $\int_{R^3/Z^3} F_o(1+v^2+|\log F_o|)dxdv < \infty$. *Given a solution in* L^1 *of DiPernaLions type and a sequence* $(t_n)_N$ *with* $t_n \to \infty$ *with* n, *then there is a subsequence* $(t_{n'})$ *and a global maxwellian* E *such that for* $T>0$, $f(.+t_{n'}) \to E$ *weakly in* $L^1(R^3/Z^3 \times R^3 \times [0,T])$, *and for* $t>0, f(.,t+t_{n'}) \to E$ *weakly in* $L^1(R^3/Z^3 \times R^3)$.

References

1. Arkeryd L.,"Stability in L^1 for the Spatially Homogeneous Boltzmann equation",*Arch.Rat.Mechs.Anal.* <u>103</u>, 151-167 (1988).

2. Arkeryd L., Esposito R. and Pulvirenti M.,"The Boltzmann Equation for Weakly Inhomogeneous Data", *Comm.Math.Phys.* <u>111</u>, 393-407 (1987).

3. Arkeryd L.,"On the long term behaviour of the Boltzmann Equation in a periodic box", *Tech.Rep.*Chalmers Univ. of Tech.Göteborg, <u>23</u>, 1988.

4. Cercignani C., "Equilibrium states and trend to equilibrium in a gas according to the Boltzmann equation",*Preprint* Politecnico di Milano, 1989.

5. Desvillettes L.,"Convergence to equilibrium in large time for the Boltzmann and B.G.K. equations",*Arch.Rat.Mechs.Anal.* to appear 1990.

6. DiPerna R.J. and Lions P.L., On the Cauchy problem for Boltzmann equations: Global existence and weak stability, *Ann.of Math.* 130, 321-366 (1989).

7.Lions P.L. personal communication Oct.1988.

8. Toscani G., "H-theorem and Asymptotic Trend of the Solution for a Rarefied Gas in the Vacuum", *Arch.Rat.Mechs.Anal.* 100, 1-12 (1987).

A macroscopic model for the electromagnetic pulse

F. Bampi
Dipartimento di Matematica e Applicazioni "R. Caccioppoli"
Università di Napoli, Via Mezzocannone 8, 80134 Napoli, Italy

C. Zordan
Dipartimento di Ingegneria Biofisica ed Elettronica
Università di Genova, Via all'Opera Pia 11a, 16145 Genova, Italy

1. EMP phenomenology

Nuclear explosions, as well as natural lightning, produce electromagnetic effects, in the form of electromagnetic pulses (EMP). Schematically, the nuclear explosion generates an intense pulse of gamma-rays with a rise time of the order of a few nanoseconds. The gammas, while traveling in the air, undergo Compton scattering with atomic electrons. This process creates a flux of Compton recoil electrons whose angular distribution peaks in the forward direction so that a net electric current density results. Such a current is usually called Compton current density J_c.

The magnitude of J_c is obviously related to the flux Φ_e of recoil electrons through the formula

$$J_c = -e\Phi_e, \tag{1}$$

$e \approx 1.60 \cdot 10^{-19}$C being the absolute value of the electron charge. Longmire [1] observes that the electron flux Φ_e in a given direction is generated by a steady flux Φ_γ of collimated gammas in the same direction; precisely, on denoting by λ the scattering mean free path of the gammas and by R the mean forward range of the recoil electron we have

$$\Phi_e = \frac{R}{\lambda}\Phi_\gamma \approx 0.007\Phi_\gamma. \tag{2}$$

It is convenient to express Φ_γ in terms of the so-called *dose rate* $\dot{D}$ [2, p. 77]. Since a dose rate of 1 rad/s corresponds to a flux of about $2 \cdot 10^9$ gamma-MeV/(cm^2 s), we can write

$$\Phi_\gamma = 2 \cdot 10^9 \dot{D}. \tag{3}$$

From (1), (2), and (3) we conclude that

$$J_c\left(\frac{\mathrm{A}}{\mathrm{m}^2}\right) \approx 2 \cdot 10^{-8}\dot{D}\left(\frac{\mathrm{rad}}{\mathrm{s}}\right). \tag{4}$$

An important factor in limiting the amplitude and determining how EMP propagates is the electric conductivity of the air. Indeed the Compton recoil electrons collide with atomic electrons which, in turn, are dislodged from their atoms thereby producing ionization. In these collisions, the Compton electrons gradually lose their energy and are stopped in a suitable distance called the track length. A typical approximation is that the electrons maintain their initial velocity $\mathbf{v}$ throughout the track length and then abruptly stop [3].

The literature bears evidence of two different procedures for evaluating the conductivity of the air due to ionization. A first approach [1] relies on a microscopic picture of the recombination chemistry relative to Compton electrons and ions and on approximate solutions of the equations involved. Alternatively, Karzas and Latter [3] adopt the viewpoint that collisions are adequately modeled by an equivalent viscosity ν_e. Both approaches arrive at determining the air conductivity σ in accordance with the formula [1]

$$\sigma = eN_e\mu_e, \tag{5}$$

where N_e is the density of the electrons generated by ionization and μ_e is their mobility. The mobility of an electron is defined as the ratio of its drift speed to the electric field causing the drift; in formula [3]

$$\mu_e = \frac{e}{m\nu_e}$$

where m is the electron mass. At sea level the value of μ_e is

$$\mu_e\left(\frac{\mathrm{m}}{\mathrm{s}}\right) \Big/ \left(\frac{\mathrm{V}}{\mathrm{m}}\right) \approx 0.3.$$

It is possible to evaluate the magnitude of σ in terms of the dose rate $\dot{D}$. To this purpose, notice that, at sea level, the ionization source S is related to the dose rate by the formula

$$S\left(\frac{\text{electrons}}{\mathrm{m}^3\ \mathrm{s}}\right) = 2.2 \cdot 10^{15}\dot{D}\left(\frac{\mathrm{rad}}{\mathrm{s}}\right).$$

On assuming that the ionization source S rises exponentially, $S \sim \exp(\alpha t)$, as for fission chain reactions, the electron density N_e is given by

$$N_e = \frac{S}{\alpha + k_1};$$

a typical value for the parameter k_1 is 10^8/s at sea level. According to the previous estimates, the air conductivity σ can be given the expression

$$\sigma\left(\frac{\mathrm{mho}}{\mathrm{m}}\right) \approx \frac{1 \cdot 10^{-4}}{\alpha + k_1}\dot{D}\left(\frac{\mathrm{rad}}{\mathrm{s}}\right). \tag{6}$$

We turn now our attention to an application of the previous results in the case when $\mathbf{J}_c = J_c\mathbf{n}$, $\mathbf{n}$ being a fixed direction; meanwhile we assume that both J_c and σ depend on the single space variable $x = \mathbf{n} \cdot \mathbf{x}$ and on time t. Accordingly, the Compton current generates an electric field $\mathbf{E} = E(x,t)\mathbf{n}$ which is trivially irrotational. Therefore no magnetic field $\mathbf{B}$ is produced and the only nonvanishing Maxwell equation is

$$\epsilon_0 \frac{\partial E}{\partial t} + \sigma E = -J_c, \tag{7}$$

where ϵ_0 is the dielectric constant of vacuum. Owing to the dependence of σ on $\dot{D}$, and consequently on J_c, eq. (7) turns out to be nonlinear. Nonetheless the approximate behavior of the electric field is easily deduced. Following Longmire [1], at early times J_c and σ take arbitrarily small values, and then increase. Thus the term σE is negligible and, in this regime, the solution of eq. (7) is

$$E \approx \frac{1}{\epsilon_0}\int_{-\infty}^{t} J_c\, d\hat{t}.$$

Later, the term σE becomes comparable with, and then greater than the displacement current, making this latter unimportant; hence

$$E \approx -\frac{J_c}{\sigma} \equiv E_s.$$

From (4) and (6) we realize that E_s is independent of $\dot{D}$; consequently E becomes constant at the *saturated* value E_s, thus proving that neglecting the displacement current is indeed a good approximation. On appealing to the estimates (4) and (6) we find [1]

$$E_s\left(\frac{\mathrm{V}}{\mathrm{m}}\right) \approx 2 \cdot 10^{-4}(\alpha + k_1)(s^{-1}).$$

A flavor on the magnitude of E_s can be obtained by assuming that α takes the typical value $2 \cdot 10^8$; in this instance we get $E_s = 6 \cdot 10^4 \mathrm{V/m}$ at sea level.

2. Minkowski model

According to the model illustrated in the preceding section, calculation relative to EMP generation and propagation is carried out by looking at the Compton current as a known source, driving the EMP. In particular, this knowledge requires a microscopic picture of the gamma-ray sources and of the recombination chemistry involving Compton electrons and ions, in order to estimate the time behavior of the Compton current.

As usual in Continuum Mechanics, a macroscopic approach provides an appealing, although less detailed, alternative to microscopic theories. In the present case, a model for describing the behavior of EMP can be elaborated by having recourse to the Minkowski approach to moving elecromagnetic continua (see [4] for a review). Indeed, Compton electrons can consistently be modeled by a charged fluid, where the Compton current coincides with the convection current due to the electron motion. At a macroscopic level, the Compton current need be characterized by an appropriate constitutive relation. Thus, the behavior of the charged fluid is governed by the Maxwell equations and by pertinent mechanical equations, supplemented by suitable constitutive assumptions. In this framework, EMP phenomenon is mathematically described by a discontinuity wave, possibly a shock.

Let us now enter into details. According to the Minkowski model, in low speed limit the constitutive equations, relating to each other the electromagnetic fields **E**, **D**, **B**, and **H**, read [5]

$$\mathbf{D} = \epsilon\mathbf{E} + (\epsilon\mu - \epsilon_0\mu_0)\mathbf{v} \times \mathbf{H}, \tag{8}$$

$$\mathbf{B} = \mu\mathbf{H} - (\epsilon\mu - \epsilon_0\mu_0)\mathbf{v} \times \mathbf{E}, \tag{9}$$

$$\mathbf{J} = \sigma(\mathbf{E} + \mathbf{v} \times \mathbf{B}) + \rho\mathbf{v}, \tag{10}$$

where $\mathbf{v}$ denotes the velocity of the fluid; in the vacuum, $\epsilon = \epsilon_0$ and $\mu = \mu_0$. In agreement with the previous section, we specialize the Minkowski model to the one-dimensional case. In this instance the constitutive equations (8)–(9) takes the usual form

$$D = \epsilon E, \qquad B = \mu H, \tag{11}$$

while (10) becomes

$$J = \sigma E + \rho v. \tag{12}$$

In conclusion, the Minkowski scheme for a charged fluid consists in the following Maxwell equations

$$\frac{\partial}{\partial t}(\epsilon E) + J_c = -\sigma E, \tag{13}$$

$$\frac{\partial}{\partial x}(\epsilon E) = \rho, \tag{14}$$

along with suitable mechanical equations; in (13) $J_c = \rho v$ denotes the convection current. Our task is now that of deducing a constitutive equation for J_c, thus making eqs (13) and (14) into a consistent differential system.

Two comments seem to be in order. First, as well known, Minkowski theory serves as the framework for magnetofluid-dynamics (MFD), a theory which mainly describes low-frequency phenomena. Consistently, the displacement current is always considered negligible. In the present model it is just the displacement current that drives the EMP (see eq. (7)). In this sense the present approach could be viewed as standing at the opposite side of MFD.

Second, starting from (2.11), (2.12) a standard procedure leads to the continuity equation for the electric charge in the form

$$\frac{\partial \rho}{\partial t} + \frac{\partial J_c}{\partial x} + \frac{\partial}{\partial x}(\sigma E) = 0. \tag{15}$$

On letting n_e be the electron number density, we have $\rho = -en_e$; hence eq. (15) implies

$$\frac{\partial n_e}{\partial t} + \frac{\partial}{\partial x}(n_e v) = \frac{1}{e}\frac{\partial}{\partial x}(\sigma E), \tag{16}$$

whereby the electron number density is not conserved. That is why, as usual in MFD, one-component fluid is assumed to be neutral [6,7]. In our case such a circumstance appears to be a profitable way of accounting for electron creation due to Compton scattering.

3. Application to EMP

Application of the Minkowski model to EMP theory requires that we first identify the convection current $\mathbf{J}_c$ with the Compton current. Then we state some assumptions which follow directly from the discussion of Sect. 1. Precisely, we make the following hypotheses:

(i) the dielectric constant and magnetic permeability of the fluid coincide with those of vacuum, i.e. $\epsilon = \epsilon_0$, $\mu = \mu_0$;

(ii) Compton electrons move at a constant velocity $\mathbf{v}$ [3], i.e. $\dot{\mathbf{v}} = \mathbf{0}$;

(iii) the electric conductivity depends linearly on J_c [1,3], i.e.

$$\sigma = k\, J_c, \tag{17}$$

k being a constant.

The task of deducing an evolution equation for the Compton current—which make our scheme into a self-consistent macroscopic model for EMP—is accomplished by observing that the irrotationality of $\mathbf{J}_c$ implies the existence of a scalar field, say Q, such that

$$\mathbf{J}_c = \nabla Q.$$

As proved in [8], taking the material time derivative of $\mathbf{J}_c = \rho\mathbf{v}$ and exploiting the previous hypotheses and equations enable us to arrive at the relation

$$\frac{\partial}{\partial x}\left(\frac{\partial Q}{\partial t} + 2v\frac{\partial Q}{\partial x} + v\sigma E - v^2\rho\right) = 0,$$

whence

$$\frac{\partial Q}{\partial t} + 2v\frac{\partial Q}{\partial x} + v\sigma E - v^2\rho = s(t),$$

for a suitable function $s(t)$; finally the charge density ρ can be eliminated by using eq. (14). Recalling eq. (17), we conclude that the model for one-dimensional EMP is described by the following differential system

$$\frac{\partial E}{\partial t} + \frac{1 + kE}{\epsilon_0}\frac{\partial Q}{\partial x} = 0, \tag{18}$$

$$\frac{\partial Q}{\partial t} + v(2 + kE)\frac{\partial Q}{\partial x} - v^2\epsilon_0\frac{\partial E}{\partial x} = s(t). \tag{19}$$

4. Discontinuity waves

As is well known, in EMP generation the field variations occurs in a few nanoseconds. Therefore, discontinuity waves, and shocks, prove to be a reliable mathematical model for the description of EMP phenomenology. In this section we consider the discontinuity waves; comments on shocks are deferred to next section.

Equations (18), (19) take the form of a quasilinear system

$$\frac{\partial \mathbf{u}}{\partial t} + A(\mathbf{u})\frac{\partial \mathbf{u}}{\partial x} = \mathbf{f}, \tag{20}$$

where the column vector $\mathbf{u}$ represent the unknown field, i.e.

$$\mathbf{u} = \begin{pmatrix} E \\ Q \end{pmatrix},$$

while the 2×2 matrix A and the column vector $\mathbf{f}$ are given by

$$A = \begin{pmatrix} 0 & \dfrac{1+kE}{\epsilon_0} \\ -v^2\epsilon_0 & v(2+kE) \end{pmatrix}, \qquad \mathbf{f} = \begin{pmatrix} 0 \\ s(t) \end{pmatrix}. \tag{21}$$

We are interested in determining the weak discontinuity waves (i.e. acceleration waves) compatible with the system (20); hence we let the field $\mathbf{u}$ suffer jump discontinuities in its first derivative across the wave surface $\varphi(x,t) = 0$. According to the standard theory [9,10], the discontinuity waves can be determined by formally substituting

$$\frac{\partial}{\partial t} \to -\lambda\delta, \qquad \frac{\partial}{\partial x} \to \delta$$

into the system (20); here $\delta\mathbf{u}$ denotes the jump of $\partial\mathbf{u}/\partial x$ across $\varphi(x,t) = 0$ and λ is the characteristic speed.

Calculation shows that the characteristic speeds are

$$\lambda^{(1)} = v, \tag{22}$$

$$\lambda^{(2)} = v(1+kE). \tag{23}$$

If $E \neq 0$ the eigenvalues are distinct and hence the system is strictly hyperbolic; we assume that this is the case.

It is worth noticing that $\lambda^{(1)}$ is the speed of an exceptional wave, while $\lambda^{(2)}$ corresponds to a wave that can degenerate in a shock wave. To determine the evolution law for the discontinuity $\delta\mathbf{u}$ along the ray corresponding to the speed $\lambda^{(2)}$ we set $s(t) = 0$ and consider a wave propagating into a constant state $\mathbf{u}^*$. On denoting by $\mathbf{d}^{(2)}$ the right eigenvalue associated to $\lambda^{(2)}$ we have

$$\delta\mathbf{u} = \Pi(\tau)\mathbf{d}^{(2)} = \Pi(\tau)\begin{pmatrix} 1/(vk) \\ \epsilon_0/k \end{pmatrix}, \tag{24}$$

the amplitude Π depending on the time τ evaluated along the ray. In the present case the evolution law for Π takes the form [8]

$$\Pi(\tau) = \frac{\Pi(0)}{1+\tau\Pi(0)}. \tag{25}$$

Therefore, the quantity Π tends to infinity as τ tends to the so-called critical time

$$\tau^* = -\frac{1}{\Pi(0)}.$$

In order for the wave to break we must have $\tau^* > 0$, or $\Pi(0) < 0$. In view of the positivity of σ this condition can never be met. Indeed, from (24) we find that

$$\delta Q = \Pi\epsilon_0/k;$$

now, since in the constant state $J_c = 0$, it turns out that δQ coincides with the current J_c behind the wave front. As a consequence, the previous relation and eq. (17) imply that

$$\Pi\epsilon_0 = kJ_c = \sigma,$$

whereby condition $\Pi < 0$ is physically inadmissible.

Let us briefly comment on the condition $E \neq 0$. It is readily shown that, if $E = 0$, there is only one eigenvector associated to the multiple eigenvalue v; consequently the system loses its hyperbolicity. As this circumstance is not atypical in the literature, Boillat and Pluchino [11] suggest that the presence of constraints could restore hyperbolicity when a system of conservative equations with a convex extention is dealt with. In this case, it is possible to make the system symmetric in the sense of Friedrichs and Lax [12], which is always hyperbolic. Of course, constraints are vital for this procedure to work; therefore we cannot employ this technique in our case.

5. Shock waves

The properties of shock waves can be investigated provided the system we are considering is a conservative system, namely it can be written in the form

$$\frac{\partial \mathbf{f}^0}{\partial t} + \frac{\partial \mathbf{f}^i}{\partial x^i} = \mathbf{f},$$

where $\mathbf{f}^0$, $\mathbf{f}^i$, and $\mathbf{f}$ depend on the field $\mathbf{u}$. As we shall prove, the system (18), (19) is a conservative one, thereby allowing for shock-wave calculation.

It is a straightforward matter to derive the conservative form of the system (18), (19), where, for definiteness, we choose $s(t) = 0$. Indeed, multiplying (18) by v and subtracting (19) yield

$$\frac{\partial}{\partial t}(v\epsilon_0 E - Q) + \frac{\partial}{\partial x}(v^2\epsilon_0 E - vQ) = 0, \tag{26}$$

while dividing (18) by $1 + kE$ and integrating give

$$\frac{\partial}{\partial t}\left(\frac{\epsilon_0}{k}\ln(1 + kE)\right) + \frac{\partial Q}{\partial x} = 0. \tag{27}$$

Apparently the system (26), (27), equivalent to the original differential system, is in a conservative form. A more convenient expression for the system (26), (27) can be achived by introducing the following new fields

$$F = v\epsilon_0 E - Q, \qquad S = \frac{\epsilon_0}{k}\ln(1 + kE), \tag{28}$$

whereby

$$E = \frac{1}{k}(e^{kS/\epsilon_0} - 1), \qquad Q = \frac{v\epsilon_0}{k}(e^{kS/\epsilon_0} - 1) - F.$$

The system (26), (27) takes eventually the form

$$\frac{\partial F}{\partial t} + v\frac{\partial F}{\partial x} = 0, \tag{29}$$

$$\frac{\partial S}{\partial t} + \frac{\partial}{\partial x}(ve^{kS/\epsilon_0}) - \frac{\partial F}{\partial x} = 0. \tag{30}$$

The Rankine-Hugoniot conditions can be deduced by the formal substitution

$$\frac{\partial}{\partial t} \to -s[\;], \qquad \frac{\partial}{\partial x} \to [\;],$$

where s denotes the shock speed; in our case we have

$$(v - s)[F] = 0,$$
$$-s[S] + v[e^{kS/\epsilon_0}] - [F] = 0.$$

It should come as no surprise that the condition $[F] \neq 0$ implies $s = v$ in that v is an exceptional speed and hence a shock is associated with it. In the other instance $[F] = 0$ the Rankine-Hugoniot conditions simplify to

$$-s[S] + v[e^{kS/\epsilon_0}] = 0$$

which cannot explicitly be solved.

As a final remark, we note that the condition $[Q] \neq 0$ implies that the Compton current take the form of a delta distribution defined on the shock surface. It has been proved recently that this is not so atypical [13].

References

[1] C. L. Longmire, *On the electromagnetic pulse produced by nuclear explosions*, IEEE Trans. EMC **EMC-20**, 3 (1978).

[2] M. Fazio, "Manuale delle unità di misura", ISEDI, Milano (1973).

[3] W. J. Karzas and R. Latter, *Detection of the electromagnetic radiation from nuclear explosions in space*, Phys. Rev. **137B**, 1369 (1965).

[4] G. Paria, *Magneto-elasticity and magneto thermo elasticity*, Adv. Appl. Mech. **10**, 73 (1967).

[5] Y. H. Pao, *Electromagnetic forces in deformable continua*, in "Mechanics Today", vol. 4, Pergamon Press, New York (1978).

[6] H. Cabannes, "Theoretical Magnetofluiddynamics", Academic Press, New York (1970).

[7] J. D. Jackson, "Classical Electrodynamics", J. Wiley, New York (1962).

[8] F. Bampi and C. Zordan, *A hyperbolic model for the one-dimensional electromagnetic pulse* (submitted for publication).

[9] G. Boillat, "La Propagation des Ondes", Gauthier-Villars, Paris (1965).

[10] C. Cattaneo, "Elementi di Teoria della Propagazione Ondosa", Quaderno UMI n. 20, Pitagora, Bologna (1981).

[11] G. Boillat and S. Pluchino, *Sopra l'iperbolicità dei sistemi con vincoli e considerazioni sul superfluido e la magnetoidrodinamica*, ZAMP **36**, 893 (1985).

[12] K. O. Friedrichs and P. D. Lax, *Systems of conservation equations with a convex extention*, Proc. Nat. Acad. Sci. USA **68**, 1686 (1971).

[13] F. Bampi and C. Zordan, *Higher order shock waves*, ZAMP **41** (1990).

Fourier Spectra in the Supercritical Regime

M. Bartuccelli

Department of Mathematics
Imperial College
University of London
Queens Gate
London SW7 2BZ, England

1. ABSTRACT
We calculate analytically the Fourier spectrum for a class of area-preserving maps, namely the sawtooth maps. Because these maps are completely chaotic, our analysis concerns only the supercritical regime. We emphasize the relevance that these spectra may have for the understanding of other dynamical systems which show a supercritical regime, such as the standard map.

2. INTRODUCTION

Over the past twenty years, the properties of nonintegrable maps have been used for describing real dynamical systems, such as solid state systems, fluids, charge density waves, adsorbed monolayers, incommensurate alloy structures and structural properties of condensed matter. Two of the most interesting maps which have been extensively studied are the Chirikov-Taylor map[1)] and the sawtooth map[2)]. The Chirikov-Taylor map describes, among other things, a one-dimensional collection of atoms connected by Hooke's-law springs between nearest neighbours in a sinusoidal potential (this system is also described by the discrete sine-Gordon equation and was introduced more than fifty years ago by Frenkel and Kontorova[3)]). It corresponds to a periodic crystal which can be modelled by the set of equations[4),5)]

$$\begin{aligned} p_{j+1} &= p_j + K sin(x_j) \qquad (\text{mod } 1) \\ x_{j+1} &= x_j + p_{j+1} \qquad (\text{mod } 1), \end{aligned} \tag{1}$$

where $K \geq 0$.

Here x_j defines the abscissa of the atom j in the periodic crystal and $p_j = x_j - x_{j-1}$. It is now well known that (1) shows a very rich non-chaotic and chaotic behaviour (mixed phase space)[6),7),8)]. It has been studied in great detail especially in the sub-critical and critical regime. However, very few papers have been devoted to understanding its behaviour in the supercritical regime. This is because it is very difficult to make general statements on the physical properties of dynamical systems in the chaotic regime. The aim of this paper is to give a small contribution to the understanding of dynamical systems in the supercritical regime by studying a completely chaotic system, the sawtooth map[2),9),10)], which is defined as

$$\begin{aligned} p_{t+1} &= p_t + Kx_t \qquad \pmod 1 \\ x_{t+1} &= x_t + p_{t+1} \qquad \pmod 1, \end{aligned} \tag{2}$$

where $K \geq 0$ and the toral coordinates are reduced so as to lie in the unit interval [-1/2,1/2). This map is important because it is one of the simplest systems which is completely chaotic. The understanding of its properties gives insight and useful information on the behaviour of other chaotic systems. It is the limit of the Chirikov-Taylor map for large K, that is in the supercritical regime. In fact for large K the configuration points of map (1) are very similar to the corresponding configuration points of map (2)[11)].

In what follows we quote without proofs, a number of results which are true for map (2) (for proofs see[9)]). These results are expected to be approximately true for the standard map for $K \gg 1$. When $K = 0$, equation (2) describes a linear twist all of whose orbits are ordered. When $K > 0$ all tori are destroyed and replaced by cantori (Cantor sets on which the system has a quasi-periodic motion). A cantorus of frequency ν may be obtained as the limit of periodic ordered orbits of frequency $\nu_n = p_n/q_n$, $\nu_n \to \nu$. The rational number ν_n is usually taken as the nth convergent in the continued fraction expansion of ν[12)]. In order to construct the periodic approximants to a cantorus, we study its gaps, that is, the intervals between neighbouring points . The gap ξ_t between neighbouring points with respect to temporal ordering is defined as

$$\xi_t = x_t - x_{q_n - q_{n-1} + t}. \tag{3}$$

The analytical expression for the gap is given by[2)]

$$\xi_t = -\delta_{0,t} + \left(\frac{\lambda - 1}{\lambda + 1}\right) \frac{\lambda^{-t} + \lambda^{-(q_n - t)}}{1 - \lambda^{-q_n}}, \qquad 0 \leq t \leq q_n - 1, \tag{4}$$

where $\delta_{0,t}$ is the Kronecker delta, $\lambda = (K + 2 + \sqrt{D})/2$ and $D = K^2 + 4K$.

3. FOURIER ANALYSIS

Let $\{x_t\}_{t=0}^{q_n-1}$ be an ordered periodic orbit of the sawtooth map (2). We consider the function which represents deviation from integrable behaviour, that is

$$y_t = x_t - x_0 - \frac{p_n}{q_n} t; \tag{5}$$

which is periodic with period q_n.

From (3) it follows that

$$y_t - y_{q_n - q_{n-1} + t} = \xi_t^\star, \tag{6}$$

where $\xi_t^\star = \xi_t - \frac{(-1)^n}{q_n} + p_n$.

Taking the Fourier transform of both sides of (6) we get

$$\begin{aligned} F^{\nu_n}(u) &= \frac{1}{q_n}\left(\frac{1}{1-\exp(2\pi i u/q_n)}\right)\left(\frac{\lambda-1}{\lambda+1}\right) \times \\ &\left[\frac{1}{1-\lambda^{-1}\exp(-2\pi i u \frac{p_n}{q_n})} - \frac{1}{1-\lambda\exp(-2\pi i u \frac{p_n}{q_n})}\right]. \end{aligned} \tag{7}$$

where $1 \leq u \leq q_n - 1$.

Formula (7) is the Fourier transform of an ordered periodic orbit $\{x_t\}_{t=0}^{q_n-1}$ of period q_n and frequency p_n/q_n. The asymptotic behaviour as q_n tends to infinity gives the spectrum for cantori. In this case (7) becomes

$$F^{\nu}(u) = \frac{1}{2\pi i u}\left(\frac{\lambda-1}{\lambda+1}\right)\left[\frac{\lambda}{\lambda - \exp^{-2\pi i u \nu}} - \frac{1}{1-\lambda\exp^{-2\pi i u \nu}}\right], \tag{8}$$

provided $0 < u \ll \infty$. With $F^{\nu}(u)$ we denote the Fourier transform for cantori, and ν is the irrational frequency of the cantorus.

4. DISCUSSION

We shall now investigate formula (8), which expresses the Fourier transform of an ordered periodic solution for the sawtooth map. In what follows we restrict ourselves to quadratic irrationals.

Let ν be a quadratic irrational, then its continued fraction expansion

$$[b_1, \ldots, b_k, \overline{a_0, a_1, a_2, \ldots, a_{T-1}}].$$

is T-periodic with $k \geq 0$ non-periodic leading coefficients. We are interested in the renormalization properties of the spectrum $|F^\nu(u)|$ between any two consecutive convergents, say p_{n-1}/q_{n-1} and p_n/q_n. From (8) it follows that

$$\frac{|F^\nu(u)|}{|F^\nu(l)|} = \left(\frac{l}{u}\right) \frac{1+\lambda^2 - 2\lambda\cos(2\pi l \nu)}{1+\lambda^2 - 2\lambda\cos(2\pi u\nu)}. \tag{9}$$

for $u, l \gg 1$. If $u = q_n$ and $l = q_{n+1}$ then (9) becomes

$$\frac{|F^\nu(q_n)|}{|F^\nu(q_{n+1})|} = \left(\frac{q_{n+1}}{q_n}\right) \frac{1+\lambda^2 - 2\lambda\cos(2\pi q_{n+1}\nu)}{1+\lambda^2 - 2\lambda\cos(2\pi q_n\nu)}. \tag{10}$$

If $\lambda \gg 1$ then (10) reduces to

$$|F^\nu(q_n)| \simeq \frac{q_{n+1}}{q_n} |F^\nu(q_{n+1})|. \tag{11}$$

Formula (11) gives the scaling of the heights of the spectrum computed on the set of q_n's of a quadratic irrational, for large λ.

We shall now investigate formula (8) in the two limiting cases when λ is very close to the critical regime, and when λ is very large. Let us consider first the supercritical case. Inserting $\lambda = 1 + \epsilon$ in (8), with ϵ small, we get

$$|F^\nu(u)| = \frac{\epsilon^2}{4\pi u} \frac{1}{1 - \cos(2\pi u\nu)}. \tag{12}$$

So when u assumes values which are q_n's, then the Fourier spectrum grows exponentially (see fig. 1). Exponential growth in the supercritical regime has also been found in the Chirikov-Taylor map[13)].

We now consider the case when λ is very large, that is $\lambda = 1/\mu$, with $\mu \ll 1$. Then (8) becomes

$$|F^\nu(u)| = \frac{1}{2\pi u} \frac{1}{1+\mu^2 - 2\mu\cos(2\pi u\nu)}. \tag{13}$$

Formula (13) tells us that the Fourier spectrum decays algebraically like $1/u$, so that the peaks of $u|F^\nu(u)|$ lie on lines (see fig. 2).

These Fourier spectra may be relevant for the understanding of the properties of any dynamical system which shows a supercritical regime.

References

[1] B. V. Chirikov, A Universal Instability of Many Dimensional Oscillator Systems, Phys. Rep. 52, 263 (1979).

[2] I. C. Percival, F. Vivaldi, A Linear Code for the Sawtooth and the Cat Maps. Physica 27D, 373 (1987).

[3] Y. I. Frenkel, T. Kontorova, Zh. Eksp. Teor. Fiz. 8, 1340 (1938).

[4] S. N. Coppersmith, D. S. Fisher, Pinning Transition of the Discrete Sine-Gordon Equation, Phys. Rev. B, 28, Numb. 5, (September 1983).

[5] S. Aubry, P. Y. Le Daeron, The Discrete Frenkel-Kontorova Model and its Extensions I, Physica 8D, 381 (1983).

[6] J. M. Greene, A Method for Determining a Stochastic Transition , J. Math. Phys. 20 , 1183 (1979).

[7] R. S. Mackay, J. D. Meiss, I. Percival, Transport in Hamiltonian Systems, Physica 13D (1984) 55-81; R. S. Mackay, J. D. Meiss, I. Percival, Resonances in Area-Preserving Maps, Physica 27D (1987) 1-20.

[8] S. J. Shenker, L. D. Kadanoff, Critical Behaviour of a KAM Surface: I. Empirical Results. J. Stat. Phys. 27, 631 (1982).

[9] M. Bartuccelli, Cantori Spectra for the Sawtooth Map, Preprint (Submitted to "Nonlinearity").

[10] N. Bird, F. Vivaldi, Periodic Orbits of the Sawtooth Maps, Physica 30D (1988) 164-176.

[11] Q. Chen, J. D. Meiss, I. C. Percival, Orbit Extension Method For Finding Unstable Orbits, Physica 29D (1987), 143-154.

[12] I. Niven, H.S. Zucherman. An Introduction to the Theory of Numbers. (Wiley, 1980).

[13] B. V. Chirikov, D. L. Shepelyansky, Chaos Border and Statistical Anomalies. Report at the International Workshop "Renormgroup-86" Dubna, 1986.

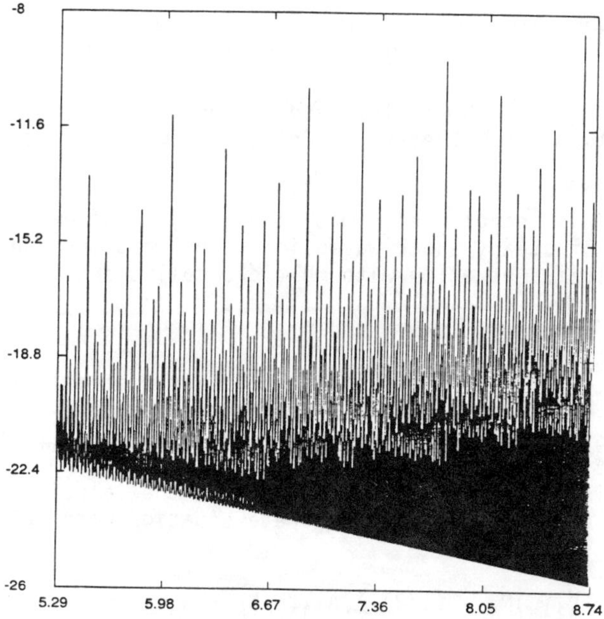

Figure 1. The Fourier coefficients of formula (12) for the irrational frequency $\sqrt{2}$ on a log-log plot. $\epsilon = 0.001$, $200 < u < 6000$.

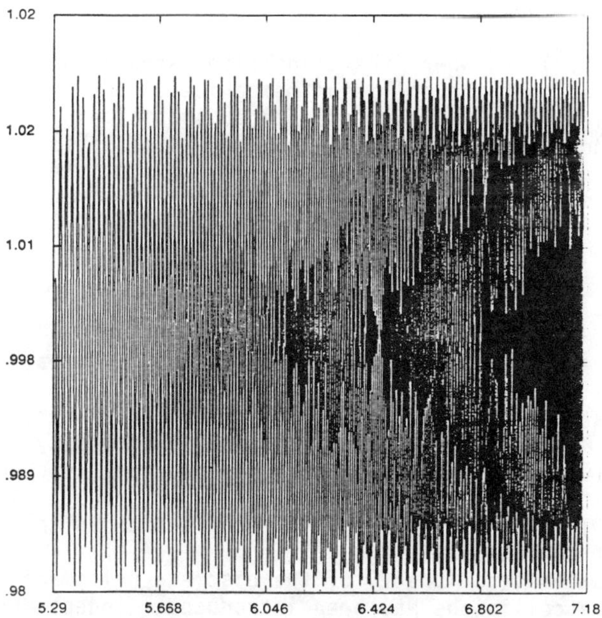

Figure 2. The Fourier coefficients of formula (13) for the irrational frequency $\sqrt{5}$ multiplied by the integer variable u. The abscissa is on a log scale. $\mu = 0.01$, $200 < u < 1100$.

SHOCK WAVE PROFILES IN ONE DIMENSION BY THE DISCRETE BOLTZMANN EQUATION WITH MULTIPLE COLLISIONS

N. Bellomo and E. Longo
Department of Mathematics, Politecnico, Torino Italy

ABSTRACT: This paper deals with the analysis of existence and computing of steady shock waves in one space dimension by the discrete Boltzman equation with multiple collisions.

1. Introduction

The discrete Boltzmann equation is a mathematical model, in nonlinear kinetic theory, which describes the time-space evolution of a gas particles which can attain only a finite number of velocities $\vec{v}_i$, for $i=1,\ldots,n$.

If N_i is the number density joined to the velocity $\vec{v}_i$, then the formal expression of the discrete equation with binary and triple collisions is

$$(\partial/\partial t + \vec{v}_i \underline{\nabla}_x)N_i = (1/2)\sum_{jhk} A_{ij}^{hk}(N_h N_k - N_j N_i) +$$

$$+ \ (1/3!)\sum_{jlghk} A_{ijl}^{ghk}(N_g N_h N_k - N_j N_l N_i) \qquad (1.1)$$

where $N_i=N_i(t,x)$ and where the transition rates A can be modelled taking into account the collision mechanics of the gas particles in a fashion that, the "indistinguishability" of the gas particles undergoing "reversible" collisions is satisfied.

The transition rates A are related to the transition probability densities through the relations

$$A_{ij}^{hk} = S|\vec{v}_i - \vec{v}_j| a_{ij}^{hk} \qquad (1.2)$$

and

$$A_{ijl}^{ghk} = (c/3)S\ S^{3/2}(|\vec{v}_i - \vec{v}_j|+|\vec{v}_j - \vec{v}_l|+|\vec{v}_i - \vec{v}_l|)a_{ijl}^{ghk} \qquad (1.3)$$

where c is a constant, S the cross sectional area of the particles, a_{ij}^{hk} defines the probability density that the encounter $(\vec{v}_i,\vec{v}_j)$ gives the product $(\vec{v}_h,\vec{v}_k)$ and viceversa and a_{ijl}^{ghk} refers to the encounters $(\vec{v}_i,\vec{v}_j,\vec{v}_l)$ with product $(\vec{v}_g,\vec{v}_h,\vec{v}_k)$. Formula (1.3) holds only for special models such as the one dealt with in what follows.

The mathematical theory of the discrete Boltzmann equation with multiple collisions has been introduced, independently, in papers [1, 2] and [3]. The first two papers provide a quite general and formal framework, whereas paper [3] is more oriented to provide a specific model and applications to nonlinear fluid-dynamics. Moreover, paper [4] deals with the analysis of the stability properties, with respect to perturbations of equilibrium, of discrete velocity models with multiple collisions. This paper shows indeed, by a suitable stability theorem, that specific models which are unstable when binary collisions only are taken into account may become stable when also triple collisions are included.

It is also worth mentioning that the study of models with triple collisions is relevant towards the study of lattice gas dynamics [5] also taking into account that introducing multiple collisions consists in increasing the gas viscosity and makes the mathematical model more consistent with fluid-dynamic applications.

This paper deals with the specific 6-velocity model proposed in [3] and studies the steady shock wave problem in one space dimension. In particular, after the mathematical description of the model and of its properties dealt with in the second section, with particular attention to the characterization of the maxwellian state, the third section deals with the computing of the sound and shock wave propagation velocities in one dimension. The steady shock wave problem is then studied in the fourth section.

2. The Mathematical Model and Maxwellian State

The six velocity model of the discrete Boltzmann equation describes the time and space evolution of the densities

$$N_i = N_i(t,x) : [0,T] \times R^2 \to R^+ \quad , \quad i=1,\dots,6 \tag{2.1}$$

joined to the velocities $\vec{v}_i = c\,\vec{e}_i$ where

$$\vec{e}_i = \cos((i-1)\pi/3)\vec{i} + \sin((i-1)\pi/3)\vec{j} \tag{2.2}$$

and where $\vec{i}$ and $\vec{j}$ are the unit vectors of a plane orthogonal frame.

Such a model, proposed in [3], can be written as follows

$$(\partial/\partial t + \vec{v}_i \cdot \underline{\nabla}_x)N_i = J_i[N] = J_i^{(2)}[N] + J_i^{(3)}[N] \tag{2.3}$$

where $J_i^{(2)}$ is the classical binary collision operator

$$J_i^{(2)} = cS \sum_{h=1}^{6}(N_h N_{h+3} - N_i N_{i+3}) \tag{2.4}$$

and $J_i^{(3)}$ is the triple collision operator

$$J_i^{(3)} = (\sqrt{6}/(\pi/3))(cS)S^{3/2}\sum_{h=1}^{6}(N_h N_{h+2} N_{h+4} - N_i N_{i+2} N_{i+4}) \qquad (2.5)$$

We have now to define and characterize the Maxwellian state for such a model. This aim can be achieved making use of the following definitions and properties already stated elsewhere [4]

Definition 1: A vector $\Phi = \{\phi_i\} \in R^6$ is defined "collision invariant" if

$$< \Phi, J^{(p)}[N]> = 0 \quad \text{for} \quad p=2,3 \qquad (2.6)$$

where <.,.> denotes the standard inner product and $J^{(p)} = \{J_i^{(p)}[N]\} \in R^6$. The space of collision invariants will be denoted, in what follows, by $\mathcal{M}$.

Definition 2: A vector $N > 0$, $N=\{N_i\} \in R^n$, is a Maxwellian if

$$J^{(p)}[N] = 0 \quad \text{for} \quad p = 2,3 \qquad (2.7)$$

The standard analysis applied to this class of mathematical models (it is not worth repeating the proof) yields to the following

PROPOSITION 2.1

The following relations are equivalent

i) $\Phi \in \mathcal{M}$,

ii) $< \Phi, J^{(p)}[N]> = 0$ for $p=2,3$, $\forall N \in R^n$,

iii) $\forall i,j=1,\ldots,6$: $\phi_i + \phi_{i+3} = \phi_j + \phi_{j+3}$

$\phi_i + \phi_{i+2} + \phi_{i+4} = \phi_{i+1} + \phi_{i+3} + \phi_{i+5}$

PROPOSITION 2.2

Let $0 < N \quad R^n$, then the following relations are equivalent

i) N is a Maxwellian,

ii) $\ln(N) \in \mathcal{M}$,

iii) $J^{(p)}[N] = 0$ for $p=2,3$,

iv) $J[N] = J^{(2)}[N] + J^{(3)}[N] = 0$.

We have now all tools to characterize the Maxwellian state according to

Remark 1: The space $\mathcal{M}$ of the collision invariants is the linear subspace of R^6 spanned by the orthogonal basis $\{\psi^{(\alpha)}\}_{=0,1,2}$

defined by

$$\psi^{(0)} = (1,\dots,1)^T \in R^6$$
$$\psi^{(1)} = (\cos\theta_1,\dots,\cos\theta_6)^T \in R^6 \qquad (2.8)$$
$$\psi^{(2)} = (\sin\theta_1,\dots,\sin\theta_6)^T \in R^6$$

where $\theta_i=((i-1)/3)\pi$.

Remark 2: The moments of N with respect to the basis defined in Remark 1, namely the so called "hydrodynamical moments" are given by

$$\alpha = 0,1,2: \qquad W_\alpha = W_\alpha(t,x) = \langle \psi^{(\alpha)}, N(t,x)\rangle \qquad (2.9)$$

where $n=W_0$ is the density and $U_1=(c/n)W_1$, $U_2=(c/n)W_2$ are the mass velocity components along the coordinate axes, then

$$W_0 = n = \sum_{i=1}^{6} N_i$$

$$W_1 = nU_1/c = \sum_{i=1}^{6} \cos\theta_i N_i \qquad (2.10)$$

$$W_2 = nU_2/c = \sum_{i=1}^{6} \sin\theta_i N_i$$

Remark 3: The Maxwellian densities have, according to Proposition 2.2, the following expression

$$i=1,\dots,6: \quad N_i = N_i(b(t,x)) = \exp\{b_0(t,x) + b_1(t,x)\cos\theta_i + b_2(t,x)\sin\theta_i\} \qquad (2.11)$$

where $b=\{b_0,b_1,b_2\}^T \in R^3$ is the vector of the parameters characterizing the Maxwellian equilibrium state.

It can be easily shown [9] that the mapping

$$b \longrightarrow W = \{W_0, W_1, W_2\} \qquad (2.12)$$

is a one to one mapping. In fact, since the $N_i(b) > 0$ for all $i=1,\dots,6$, then the Jacobian matrix referred to (2.12)

$$[J(b)]=[\partial W/\partial b]=[\langle \psi^{(\alpha)}, \mathrm{diag.}(N_1,\dots,N_6)\,\psi^{(\beta)}]_{\alpha,\beta=0,1,2} \qquad (2.13)$$

is positive defined for each $b \in R^3$.

On the other hand, the explicit inversion of the map (2.12) in order to define the Maxwellian state through the hydrodynamical moments (2.9) cannot be, in general, performed in terms of elementary functions [3].

3. Sound and Steady Shock Wave Propagation Velocities

Shock wave phenomena have been studied by various authors who have dealt with, using the discrete Boltzmann equation, the following problems:

- Search for analytic solutions of the shock profiles [6,7,8] -.
- Analysis of stability problems for steady shock waves [9, 10, 11].
- Analysis of shock wave problems for gas mixtures [12,13,14]-.

Some of the papers which have been cited above also deal with the computing of the sound propagation velocity.

This present section deals with the calculation of the sound and steady shock wave propagation velocity according to the mathematical model of the discrete Boltzmann equation described in the preceding section.

Referring to the shock wave problem, it will be shown that such a problem is well formulated only when the triple term is included. This kind of result is somewhat analogous to the one found in [4], where it has been shown that good stability properties are found in the perturbation of equilibrium only when the triple collision term is included.

Consider then the mathematical model of the discrete Boltzmann equation defined in eqs.(2.3-5) in the one dimensional case

$$\begin{gathered} i = 1,\dots,6: \quad N_i = N_i(t,x), \quad x \in R^1 \\ N_2(t,x) = N_6(t,x), \quad N_3(t,x) = N_5(t,x), \quad \forall t,x \end{gathered} \tag{3.1}$$

In this case, the kinetic equations can be rewritten as

$$\begin{aligned} \partial N_1/\partial t + c\,\partial N_1/\partial x &= c(2B[N] + T[N]) \\ \partial N_2/\partial t + (c/2)\,\partial N_2/\partial x &= -c(B[N] + T[N]) \\ \partial N_4/\partial t - c\,\partial N_4/\partial x &= c(2B[N] - T[N]) \\ \partial N_5/\partial t - (c/2)\,\partial N_5/\partial x &= -c(B[N] - T[N]) \end{aligned} \tag{3.2}$$

where

$$B = (2/3)S(N_2N_5 - N_1N_4)$$

$$T = (3\sqrt{6/\pi})S^{5/2}(N_2^2N_4 - N_5^2N_1)$$

It is now usefull introducing the terms P_q defined as

$$q=0,1,2,3: \quad P_q = P_q(t,x) = \sum_{i=1}^{6} \cos^q\theta_i N_i(t,x) \tag{3.3}$$

where P_0 identifies the number density, i.e. $P_0 = n$, P_1 the mass velocity, i.e. $P_1 = nU_1/c$, and the other two quantities higher order moments.

The following conservation equations

$$\begin{aligned} \partial P_0/\partial t + c\,\partial P_1/\partial x &= 0 \\ \partial P_1/\partial t + c\,\partial P_2/\partial x &= 0 \end{aligned} \tag{3.4}$$

can be recovered from (3.2).

The velocity of propagation of sound waves can be classically determined looking for the properties of the solutions of the conservation equations referred, as we shall see, to the Maxwellian state.

Keeping this in mind, recall that the Maxwellian one-dimensional flow can be defined through eq.(2.11) which, with obvious meaning of symbols, can be rewritten (in one space dimension) as follows

$$i=1,\ldots,6: \quad N_i(t,x)=(1/2)a(t,x)\exp\{b(t,x)\cos\theta_i\} \tag{3.5}$$

with $a > 0$.

Substituting (3.5) into (3.3) shows

$$\begin{aligned} q=0: \quad P_0 &= P_0(a,b) = a[\mathrm{ch}(b) + 2\mathrm{sh}(b/2)] \\ q>0: \quad P_q &= \partial P_{q-1}/\partial b \end{aligned} \tag{3.6}$$

So that one obtaines a symmetric hyperbolic evolution equation for the parameters

$$C(a,b)\frac{\partial}{\partial t}\begin{pmatrix} a \\ b \end{pmatrix} + D(a,b)\frac{\partial}{\partial t}\begin{pmatrix} a \\ b \end{pmatrix} = 0 \tag{3.7}$$

where

$$C = \begin{pmatrix} P_0 & P_1 \\ P_1 & P_2 \end{pmatrix} \; ; \; D = \begin{pmatrix} P_1 & P_2 \\ P_2 & P_3 \end{pmatrix} \tag{3.8}$$

System (3.7), according to the fact that C and D are real and symmetric and that C is positive defined, admittes the solution

$$a = a_0 \exp\{x - \lambda ct\}, \qquad b = b_0 \exp\{x - \lambda ct\} \tag{3.9}$$

corresponding to the propagation of sound waves with velocity λc where λ is the solution of the characteristic equation

$$(P_0 P_2 - P_1^2)\lambda^2 - (P_0 P_3 - P_1 P_2)\lambda - (P_2^2 - P_1 P_3) = 0 \tag{3.10}$$

Solving eq.(3.10) yields the expression of the propagation of sound waves in the directions of the x-axis

$$c_s/c = \lambda = \{U + (1/2)(d^2U/db^2)/(dU/db) +$$
$$+[((1/2)(d^2U/db^2)/(dU/db))^2 + dU/db]^{1/2}\} \tag{3.11}$$

where the term b can be related to the expression of the mass velocity in the direction of the x-axis

$$U = U_1/c = (sh(b) + sh(b/2))/(ch(b) + 2ch(b/2)) \tag{3.12}$$

In particular, when $U = 0$, and therefore $dU/db = 1/2$ and $d^2/db^2 = 0$, one has $c_s = c(\sqrt{2}/2)$.

Consider now the problem of computing the propagation velocity (when it exists) of shock waves in the direction of the x-axis. This problem has to be studied joining to the evolution equations (3.2) suitable boundary conditions at $x \to -\infty$ and $x \to +\infty$, corresponding to Maxwellian states.

In particular, the problem can be studied with the following conditions:

$$i=1,\ldots,6, \; \forall t > 0 : \; N_i(t,-\infty) = (a/2)\exp\{b \cos\theta_i\} \tag{3.13}$$

where $a > 0$ and $b > 0$ and

$$i=1,\ldots,6, \; \forall t > 0 : \; N_i(t, x \to +\infty) = N_\infty \tag{3.14}$$

Then one has to look for solutions invariant with respect to the translation

$$z = x + \beta ct \, , \quad \beta > 0 \tag{3.15}$$

where 'βc' is the velocity of propagation of the shock wave in the direction of the x-axis.

Two shock wave problems, which will be dealt with in what follows, can be mathematically formulated,

Problem 1: Let $a > 0$ and $b > 0$ be given, then one has to find the propagation velocity 'βc' and the asymptotic density N_∞ consistent with (3.2, 3.13, 3.14, 3.15).

Problem 2: Let a,b (and consequently β and N_∞ according to the solution of Problem 1) be given then one has to find the profiles $N_i = N_i(z)$ consistent with the equations already mentioned in Problem 1.

The first problem will be dealt with in this section, the second one in the last section of this present paper.

Let us now deal with Problem 1. Applying the change of variables (3.15) to the conservation equations (3.4) yields

$$\partial(P_1 + \beta P_2)/\partial z = 0 , \qquad \partial(P_2 + \beta P_1)/\partial z = 0 \tag{3.16}$$

which are the only linearly independent conservation equations which can be integrated referring the boundary conditions (3.13, 3.14) with respect to the new variables P_i

$$\begin{aligned} P_0(-\infty) &= a[ch(b) + 2ch(b/2)] \\ P_1(-\infty) &= a[sh(b) + sh(b/2)] \\ P_2(-\infty) &= a[ch(b) + (1/2)ch(b/2)] \\ P_0(\infty) &= 6N_\infty \\ P_1(\infty) &= 0 \\ P_2(\infty) &= 3N_\infty \end{aligned} \tag{3.17}$$

therefore combining eqs.(3.16) and (3.17) yields

$$\begin{aligned} 6N_\infty/P_0(-\infty) &= \{P_2/P_0+1/2+[(P_2/P_0-1/2)^2+2(P_1/P_0)^2]^{1/2}\}(-\infty) \\ \beta &= (P_1/P_0)/(6N\ /P_0-1)(-\infty) \end{aligned} \tag{3.18}$$

where the terms P_0, P_1 and P_2 at $x \to \infty$ are given as by eqs.(3.17). N_∞ and β are then computed.

Remark 4: The same problem formulated for the six velocity discrete Boltzmann equatiuon with binary collision only is not well formulated. In fact, it can easily be verified that one has, in this case, one additional linearly independent conservation equation. Consequently, Problem 1 cannot be solved for the whole range of Mach numbers with three conservation equations. In fact one can only find some special combinations of particular values of β and N_∞ solving Problem 1.

4. Steady Shock Wave Density Profiles

Consider now Problem 2 as it has been defined in the preceding section. Such a problem deals with the integration of eqs.(3.2) joined to the boundary conditions (3.13-14) where the constants a and b are given.

Equations (3.2) can be rewrittwen, after the change of variables (3.15), as

$$\begin{aligned}(1 + \beta)dN_1/dz &= 2B[N] - T[N]\\ (1 + 2\beta)dN_2/dz &= -2(B[N] + T[N])\\ (1 - \beta)dN_4/dz &= (2B[N] + T[N])\\ (1 - 2\beta)dN_5/dz &= -2(B[N] - T[N])\end{aligned} \tag{4.1}$$

where the terms B and N have already been defined in Sec.3.

If a and b are given, then and N can be computed by eqs.(3.18). This result can be re-enterpreted according to the following:

THEOREM 1

Suppose that the system (4.1) of ordinary differential equations admittes a smooth solution $N_i^* = N_i^*(z)$ consistent with the boundary conditions (3.13-3.14), then the initial value problem for eqs.(3.2) with initial conditions $N_i^*(x)$ admittes a travelling wave solution propagating with constant velocity β defined by eq.(3.18) joining the Maxwellian states defined by (3.13, 3.14).

Remaining at a technical level, the density profiles N_i can be found, following [14], solving the initial value problem for eqs.(3.2) with initial conditions

$$\begin{aligned}t=0, \quad x \leq 0 &: N_i = (a/2)\exp\{b\cdot\cos\theta_i\}\\ t=0, \quad x > 0 &: N_i = N_\infty\end{aligned} \tag{4.4}$$

Paper [14] has show that the initial value problem defined

by eqs.(3.2, 4.4) provides the transient shock-wave formation and the asymptotic steady travelling wave solution for $t \to \infty$. We shall concentrate our attention to the initial value problem (3.2, 4.4) as it gives additonal information on the transient behaviour.

As known, after the paper by Tartar [15], Beale [16], Bony [17], Cabannes and Kawashima [18] and Toscani [19], the initial value problem for the discrete Boltzmann equation in one space dimension always has a solution for a gas with finite mass. In particular papers [15, 16, 17, 18] need initial conditions bounded and integrable, whereas Toscani [19] was able to remouve the L restriction adding a very weak decay at infinity. In all cases smallness assumptions are not needed to ensure global existence of the solutions. In addition, Lachowicz and Monaco [20] were able to show that one can obtain classical solutions under suitable differentiability conditions to be imposed on the initial data.

Unfortunately, no-one of the mathematical results which have been mentioned above can be used for the shock wave problem. This is essentially due to two main reasons: The initial data (4.4) do not belong to L_1 (this means, in a physical language, that the mass of the gas is not finite). The proofs of papers [15-20] hold only when the equations are characterized by the binary collision term only. On the other hand, if $J^{(3)} \neq 0$ then the mathematical proof cannot be straightforwardly extended.

As a matter of fact, the only global existence proof for the discrete Boltzmann equation with triple collisions which is available in the literature is the one of paper [4] which holds for perturbation of equilibrium.

Therefore one can use the local existence theorem proposed in the paper in progress [21] in order to go through the two following steps:

a) Defining a local existence theorem in a suitable time interval $[0, T]$ where T is computed in details in terms of the norm $\|N_0\|$ of the initial data N_0.

b) Problem (3.2, 4.4) is then numerically integrated for $t \in [0,T]$ in a fashion that if T is sufficiently large, i.e. $\|N_0\|$ is sufficiently small, then the behaviour of the asymptotic travelling wave is reached.

Existence and uniqueness of the solution to the initial value problem (3.2, 4.3) is stated by

<u>THEOREM 2</u> [21]:
Let $N_0 \in X_+$ and $T < (c_1\|N_0\| + 2c_2 \|N_0\|)$ where c_1 and c_2 are constants and where X is the Banach space of the L_∞-functions endowed with the norm

$$\|N\| = \max_{1<i<6,\ x\in R} \text{ess. sup} |N_i|, \tag{4.4}$$

then a unique positive mild solution exists in the time-interval [0,T].

The result of the computing obtained by classical finite difference schemes, is shown in the figure below where two different asymptotic profiles are compared at different Knudsen numbers Kn. The figure shows that the profile at small Knudsen number is characterized by a steepest behaviour with respect to the one at larger Knudsen number.

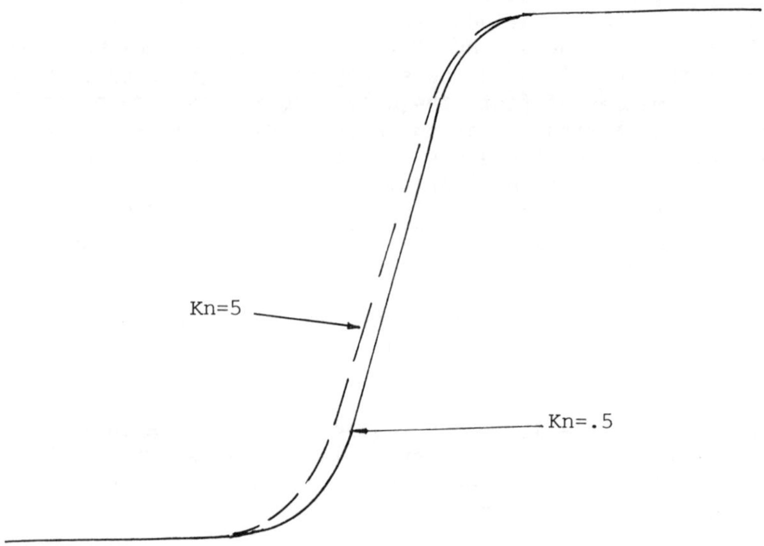

Fig.1 - Asympotic density profiles at different Knudsen numbers -

References

[1] Gatignol R & Coulouvrat F., Comp. Acad. Sci. Paris, I, 306, (1988), 169-174.

[2] Coulouvrat F. & Gatignol R., Comp. Acad. Sci. Paris, I, 306, (1988), 393-398.

[3] Monaco R. & Longo E., in Rarefied Gas Dynamics, Progress in Aeron. and Astron. n.118,(1989),118-130.

[4] Bellomo N. & Kawashima S., J. Math. Phys.,31,(1989),245-253.

[5] Monaco R. Ed., Discrete Kinetic Theory, Lattice Gas Dynamics and Foundation of Hydrodynamics, World Scientific, London, Singapore, (1989).

[6] Broadwell J., The Phys. of Fluids, 7, (1964), 1243-1247.

[7] Caflisch R., Comm. Pure Appl. Math., 32, (1979), 521-554.

[8] Platkowski T., in Discrete Kinetic Theory, Lattice Gas Dynamics and Foundation of Hydrodynamics, Monaco R. Ed., World Scientific, London, Singapore, (1989), 248-257.

[9] Kawashima S. & Matsumura A., Comm. Math. Phys., 101, (1985), 97-127.

[10] Caflisch R. & Tai Ping Liu, (1986), Internal Report.

[11] Caflisch R., in Rarefied Gas Dynamics, Boffi V. & Cercignani C. Eds., Teubner, Stuttgart, Vol 1, (1986), 3-9.

[12] Monaco R., Acta Mechanica, 55, (1985), 239-251.

[13] Monaco R., in Rarefied Gas Dynamics, Boffi V. & Cercignani C. Eds, Teubner, Stuttgart, Vol.1, (1986), 245-254.

[14] Monaco R., Pandolfi Bianchi M. & Platkowski T., Acta Mechanica, 84, (1990), 175-184.

[15] Tartar L., Int. Report University Wisconsin Madison, (1980).

[16] Beale J. T., Comm. Math. Phys., 102, (1986), 659-678.

[17] Bony J., Equationes aux derivees partielles, Centr. Mat. Ecole Polyt. Paris, (1987).

[18] Cabannes H. & Kawashima S., Comp. Rend. Acad. Sci. Paris, 304, (1988), 29-34.

[19] Toscani G. Comm. Math. Phys., 121, (1989), 121-142.

[20] Lachowicz M. & Monaco R., Siam J. Appl. Math., 4, (1989), 1231-1241.

[21] Bellomo N. and Toscani G., Stab. Meth. Appl. Anal., (1990), To Appear.

THE SEMIDISCRETE ENSKOG EQUATION

G. Borgioli[1], R. Monaco[1] and G. Toscani[2]

[1] Dipartimento di Matematica - Politecnico di Torino
[2] Dipartimento di Matematica - Universita' di Ferrara

1. Introduction

In 1922 D.Enskog [1] has proposed a model for the evolution of a moderately dense gas of hard spheres. The structure of the equation, which from his pioneer work has been called Enskog equation, differs from the Boltzmann one in two points: first, the radius σ of the spheres is taken into account, so that the change in flux due to collisions in a point $\underline{x}$ of the physical space depends on the presence of the particles whose centers lie in a sphere of radius 2σ and center $\underline{x}$; second, the collision frequency is increased by a factor Y which represents the influence of the pair correlation function. According to these assumptions the Enskog equation is written as follows (for details see [2])

$$\partial f/\partial t + \underline{v}\cdot\underline{\nabla}_x f = E(f;f,f)$$

where $E(f;f,f) = E^+(f;f,f) - E^-(f;f,f)$ and

$$E^+(\cdot)(t,\underline{x},\underline{v}) = 4\sigma^2 \int_{R^3}\int_{S^2} Y^+(f;\sigma) f(t,\underline{x},\underline{v}') f(t,\underline{x}+2\sigma\underline{\nu},\underline{v}_1') \cdot \Phi((\underline{v}_1-\underline{v})\cdot\underline{\nu})\, d\underline{\nu}\, d\underline{v}_1$$

$$E^-(\cdot)(t,\underline{x},\underline{v}) = 4\sigma^2 f(t,\underline{x},\underline{v}) \int_{R^3}\int_{S^2} Y^-(f;\sigma) f(t,\underline{x}-2\sigma\underline{\nu},\underline{v}_1) \cdot \Phi((\underline{v}_1-\underline{v})\cdot\underline{\nu})\, d\underline{\nu}\, d\underline{v}_1$$

The Enskog equation has been neglected for a long time by mathematicians. Only recently, after the progresses on the Cauchy problem for the Boltzmann equation, a new interest has been devoted to it. One of the reasons, in our opinion, of this interest, concerns the fact that the presence in the collisional operator of a product of densities evaluated in different points seems to add some further properties in order to solve the Cauchy problem for large L_1 initial data.

The first studies on the Cauchy problem have been performed by Lachowicz [3]; afterwards Bellomo and Toscani [4], Cercignani [5] and Arkeryd [6] have obtained global existence theorems for particular initial data and after suitable modifications of the collisional operator. The proof by Bellomo and Toscani, which holds for a general Y function, is given for initial data which vanish at infinity. The proof by Arkeryd, which holds in 2D and for L_1 initial values, assumes Y=1 in order to have an usual H-theorem. In the last two years, two lines of research have been developed. From one side the interesting connection between the Boltzmann and Enskog equations, when the molecule radius tends to zero, has been investigated by Bellomo and Lachowicz [7]; on the other side the recent result by Di Perna and Lions [8] on the existence of solutions (but without uniqueness) for the Boltzmann equation for L_1 data brings to a new interest the study, with the same methods, of the Enskog equation. In this context Arkeryd [9] and Polewczak [10] have given successful existence results for small L_1 initial data; moreover Polewczak [11] has proved an H-theorem for the Enskog equation with almost general Y function with some simmetry properties. In the same last years great interest has been devoted to discrete and semidiscrete models of the Boltzmann equation (see [12] and its related bibliography). The reason of such an interest is essentially due to the fact that the mathematical stucture of these models is relatively simple with respect to the full Boltzmann equation and suitable to obtain quantitative results in fluid dynamics of rarefied gases. Nevertheless, in spite of their simplicity in the collisions rules, the macroscopic behaviour of the gas represented by these models is rather accurate and in general fits experimental results.

Accordingly we think that the extension of discrete kinetic theory to the Enskog equation may be interesting and promising especially towards the application of such an equation in fluid dynamic problems. At this end, in the present paper, we derive a semidiscrete model of the Enskog equation in 2D and we prove a formal H-theorem which gives some insight to the thermodynamics described by this model.

Let us finally recall that a semidiscrete model (for the Boltzmann version of this model see ref.[13]) is a model where the discretization is assumed only on the moduli of the velocities which, conversely, can attain any direction in the plane or in the space.

2. The model

Let us consider a monoatomic gas with molecules which collide according to the following rules:

a) Each molecule moves on the (x,y)-plane and can be viewed as a rigid elastic disc of mass m and diameter 2σ;

b) All the molecules have velocity (with same modulus c)
$$\underline{c} = c(\cos\theta\ \underline{i} + \sin\theta\ \underline{j})$$
where θ is the angle between the x-axis and $\underline{c}$;

c) All the admissible collisions are those which mantain, before and after the shock, the velocity modulus c;

d) The collision frequency is increased by a factor which is function of the local mass-densities computed where the molecules are located during the shock itself.

Under the above hypotheses, the collisions which mantain momentum, angular momentum and energy are

"Exchange" collisions: $[\underline{c}(\theta),\underline{c}_1(\theta_1)] \leftrightarrow [\underline{c}'(\theta_1),\underline{c}_1'(\theta)]$
(see Fig.1).

Such a class of collisions is defined $\forall\psi \in (-\pi/2,\pi/2)$ and assures obviously all the conservation laws.

Let us remark that these collisions are not considered within the semidiscrete Boltzmann models [2,13] because they do not modify the incoming and outcoming fluxes, thanks to the fact that the molecules are treated as mass points. Conversely in the Enskog scheme the fluxes are modified, since the actual position on the plane of each molecule during a collision is taken into account in the kinetic equation.

If $\theta_1=\theta+\pi$ it is interesting to remark that there exists another admissible collision beside the one already considered in the class of "exchange" collisions. We define such a collision as:

"head-on collision": $[\underline{c}(\theta),\underline{c}_1(\theta_1)] \leftrightarrow [\underline{c}'(\theta'),\underline{c}_1'(\theta_1')]$
(see Fig.2).

In both Figures we call ψ the "impact angle" between the directions of $\underline{c}$ and of the unit vector $\underline{\nu} = \cos(\theta+\psi)\underline{i}+\sin(\theta+\psi)\underline{j}$ which links the two molecules centers; in Fig.2 ϕ is the "scattering angle" between the direction of $\underline{\nu}$ and the direction of the velocity $\underline{c}_1'(\theta_1')$, $\theta_1'=\theta+\psi+\phi$, after the collision. Such a class of collisions is defined $\forall\psi \in (-\pi/2,\pi/2)$; conservation of momentum and energy is obviously assured for any pair θ, θ'. Conservation of angular momentum determines uniquely the scattering angle ϕ: in fact if we write the angular momenta, before and after the shock, with respect to the center of mass G, we have

$$|\underline{c}|\sigma \sin\psi + |\underline{c}_1|\sigma \sin\psi = |\underline{c}'|\sigma \sin\phi + |\underline{c}_1'|\sigma \sin\phi$$

which implies $\phi=\psi$ and therefore $\theta_1'=\theta+2\psi$.

After this analysis we can write the time-space evolution equation on the distribution function N of molecules with velocity $\underline{c}(\theta)$

$$N(t,\underline{x};\theta): R_+\times R^2\times[0,2\pi) \rightarrow R_+ \ , \ t\in R_+ \ , \ \underline{x}=\{x,y\}\in R^2 \ , \ \theta\in[0,2\pi)$$

We will have:

$$\partial N/\partial t + \underline{c}\cdot\underline{\nabla}_x N = E(N,N) \qquad (1)$$

where the right-hand-side term represents the so-called collision operator. The latter can be split, as usual, in two terms, "gain" and "loss", given respectively by

$$E(N,N) = E^+(N,N) - E^-(N,N) \qquad (2)$$

The two terms E^+ and E^- can be derived according to the classical arguments [13] of semidiscrete velocity models taking into account, as we have outlined, the particles position during collisions. Thus we have

$$E^+ = \int Y^+(N,N)N(t,\underline{x};\underline{c}')N(t,\underline{x}+2\sigma\underline{\nu};\underline{c}_1')A(\underline{c}',\underline{c}_1'\rightarrow\underline{c},\underline{c}_1)d\psi d\theta_1 \qquad (3)$$

$$E^- = N(t,\underline{x};\underline{c})\int Y^-(N,N)N(t,\underline{x}-2\sigma\underline{\nu};\underline{c}_1)A(\underline{c},\underline{c}_1\rightarrow\underline{c}',\underline{c}_1')d\psi d\theta_1 \qquad (4)$$

where $A(\underline{c},\underline{c}_1\rightarrow\underline{c}',\underline{c}_1')$ is the transition probability rate that two particles with velocities $\underline{c}$ and $\underline{c}_1$ will have, after the collision, velocities $\underline{c}'$ and $\underline{c}_1'$, respectively; such a rate is defined by

$$A(\underline{c},\underline{c}_1\rightarrow\underline{c}',\underline{c}_1') = 2\sigma V^r(\underline{c},\underline{c}_1)a(\underline{c},\underline{c}_1\rightarrow\underline{c}',\underline{c}_1') \qquad (5)$$

where V^r is the relative velocity before the shock and "a" is the probability density function of the transition $(\underline{c},\underline{c}_1)\rightarrow(\underline{c}',\underline{c}_1')$. The rate $A(\underline{c}',\underline{c}_1'\rightarrow\underline{c},\underline{c}_1)$ appearing in the gain term (3a) is defined in the same fashion but is referred to the inverse transition.

The integrals appearing in Eqs.(3-4) are extended to all angles ψ and θ_1 characterizing particles with velocity $\underline{c}_1$ which collide with the "reference particle" of velocity $\underline{c}$. In Eq.(4) the distribution function of particles with velocity $\underline{c}(\theta)$ is settled outside the integral because it does not depend either on ψ either on θ_1.

The interaction factor $Y^{\pm}$ is assumed as function of the mass-densities in the particles position during collisions and will be expressed by

$$Y^{\pm}(N,N) = Y(\rho(t,\underline{x}),\rho(t,\underline{x}\pm 2\sigma\underline{\nu})) = Y^{\pm}(t,\underline{x},\psi) \tag{6}$$

$$\rho(t,\underline{x}) = m\int_0^{2\pi} N(t,\underline{x};\theta)d\theta \tag{7}$$

Such hypothesis is coherent with the so-called "revised Enskog equation" [7]. The fact that in general $Y^+ \neq Y^-$ assures to this model the capability to describe phenomena of irreversible thermodynamics.

The computation of the transition rate A will be here made by taking into account the usual hypothesis of equal probable shocks towards any direction in the (x,y)-plane.
The relative velocity is given by

$$V^r(\theta,\theta_1) = c\{2[1-\cos(\theta-\theta_1)]\}^{1/2} \tag{8}$$

The transition rate has only the following indistinguishability property

$$A(\underline{c},\underline{c}_1\rightarrow\underline{c}',\underline{c}_1') = A(\underline{c}_1,\underline{c}\rightarrow\underline{c}_1',\underline{c}') \tag{9}$$

because the scattering angle ϕ is restricted to $(-\pi/2,\pi/2)$ as well as ψ. Moreover the transition rate has also the reversibility property

$$A(\underline{c},\underline{c}_1\rightarrow\underline{c}',\underline{c}_1') = A(\underline{c}',\underline{c}_1'\rightarrow\underline{c},\underline{c}_1) \tag{10}$$

These properties follow immediately from the dynamics of the collisions. The transition rate is then given by

$$A(\underline{c},\underline{c}_1\rightarrow\underline{c}',\underline{c}_1')=2\sigma V^r a(\theta,\theta_1\rightarrow\theta_1,\theta)=(c\sigma/\pi)\{2[1-\cos(\theta-\theta_1)]\}^{1/2} \tag{11}$$

In Eq.(11) the probability density, according to the uniform probability hypothesis of shocks, is equal to $1/2\pi$.
On the other hand the "head on" collisions occur for a unique angle within the domain of variation of θ_1 in the range $[0,2\pi]$; therefore they provide a set of measure zero which does not give any contribution to the collisional integral.

By using expressions (3-11) we can finally re-write Eq.(1) in a more detailled form

$$\partial N/\partial t + c\cos\theta\, \partial N/\partial x + c\sin\theta\, \partial N/\partial y = \tag{12}$$

$$= (c\sigma/\pi)\int_0^{2\pi}\int_{-\pi/2}^{\pi/2} Y^+(t,\underline{x},\psi)N(t,\underline{x};\theta_1)N(t,\underline{x}+2\sigma\underline{\nu}(\theta,\psi);\theta)\cdot$$
$$\cdot\{2[1-\cos(\theta-\theta_1)]\}^{1/2}d\psi d\theta_1 -$$

$$- (c\sigma/\pi)\, N(t,\underline{x};\theta)\int_0^{2\pi}\int_{-\pi/2}^{\pi/2} Y^-(t,\underline{x},\psi)N(t,\underline{x}-2\sigma\underline{\nu}(\theta,\psi);\theta_1)\cdot$$
$$\cdot\{2[1-\cos(\theta-\theta_1)]\}^{1/2}d\psi d\theta_1$$

which represents the Enskog semidiscrete equation.

3. The H-theorem

A formal H-theorem can be proved for the semidiscrete Enskog equation. The proof is essentially based on the arguments by Polewczak [11] who has proved such a theorem for the full equation in R^3.

The proof will be based on a further hypothesis on the interaction factor Y which will be considered as a symmetric function of its two arguments, i.e.

$$Y(\rho(t,\underline{x}),\rho(t,\underline{x}\pm 2\sigma\underline{\nu})) = Y(\rho(t,\underline{x}\pm 2\sigma\underline{\nu}),\rho(t,\underline{x}))$$

This hypothesis is physically reasonable and is justified by particles indistinguishability.

Moreover let us rewrite the collision integral (3-4) in a more compact form

$$E(t,\underline{x},\theta) = \int_0^{2\pi}\int_{-\pi/2}^{\pi/2}[Y(\rho,\rho^+)N(\theta')N^+(\theta_1')-Y(\rho,\rho^-)N(\theta)N^-(\theta_1)]\cdot A(\theta,\theta_1)d\theta_1 d\psi \tag{13}$$

In Eq.(13) the superscript ± refers to a quantity calculated in the points $\underline{x}+2\sigma\underline{\nu}$ and $\underline{x}-2\sigma\underline{\nu}$, respectively. Without superscripts the quantity is calculated in the reference point $\underline{x}$ of the physical plane. All the quantities which appear in Eq.(13), except the rates A, are functions of time: for sake of simplicity we will omit t in the arguments of the functions.

Therefore, following [11], it is possible to prove the H-theorem according to the steps:

1) Multiply the collision integral by a function $\gamma(t,\underline{x},\theta)$ and integrate with respect to both $\underline{x}\in R^2$ and $\theta\in[0,2\pi]$. It is easy to show the following property

$$\int_{R^2}\int_0^{2\pi}\gamma(t,\underline{x},\theta)E(t,\underline{x},\theta)d\underline{x}d\theta = \tag{14}$$

$$= (1/2)\int_{R^2}\int_0^{2\pi}\int_0^{2\pi}\int_{-\pi/2}^{\pi/2}[\gamma(\theta')+\gamma^-(\theta_1')-\gamma(\theta)-\gamma^-(\theta_1)]\cdot Y(\rho,\rho^-)N(\theta)N^-(\theta_1)A(\theta,\theta_1)d\underline{x}d\theta d\theta_1 d\psi$$

Such an equality follows from the fact that the left-hand-side integral of Eq.(14) is invariant with respect to the transformations

$$\text{i)}\quad \theta\leftrightarrow\theta' \quad\text{and}\quad \theta_1\leftrightarrow\theta_1'$$
$$\text{ii)}\quad \theta\leftrightarrow\theta_1$$

thanks to the properties of reversibility (10), indistinguashibility (9) and symmetry of the factor Y. More in details equality (14) has been obtained by applying both transformations (i) and (ii) to the gain term and only transformation (ii) to the loss term.

2) Introduce the function H(t)

$$H(t) = \int_{R^2}\int_0^{2\pi}N(t,\underline{x},\theta)\log[N(t,\underline{x},\theta)]d\underline{x}d\theta - \int_0^t I(s)ds$$

where

$$I(t) = (1/2)\int_{R^2}\int_0^{2\pi}\int_0^{2\pi}\int_{-\pi/2}^{\pi/2} [N^+(\theta_1)Y(\rho,\rho^+)-N^-(\theta_1)Y(\rho,\rho^-)] \cdot \cdot N(\theta)A(\theta,\theta_1)d\underline{x}d\theta d\theta_1 d\psi$$

3) Multiply both terms of Eq.(1) by the function (1+log N) and integrate over $\underline{x}$ and θ, using the standard differentiation properties of the logarithmic function. We obtain

$$dH/dt = \int_{R^2}\int_0^{2\pi} E(t,\underline{x},\theta)\log[N(t,\underline{x},\theta)]d\underline{x}d\theta - I(t) \qquad (15)$$

4) Taking into account property (14), Eq.(15) becomes

$$dH/dt=(1/2)\int_{R^2}\int_0^{2\pi}\int_0^{2\pi}\int_{-\pi/2}^{\pi/2}\{[\log[N(\theta')N^-(\theta_1')]-\log[N(\theta)N^-(\theta_1)]]\cdot \cdot Y(\rho,\rho^-)N(\theta)N^-(\theta_1) - \qquad (16)$$
$$- [Y(\rho,\rho^+)N(\theta)N^+(\theta_1)-Y(\rho,\rho^-)N(\theta)N^-(\theta_1)]\}A(\theta,\theta_1)d\underline{x}d\theta d\theta_1 d\psi$$

5) Exchanging again $\theta \to \theta'$ and $\theta_1 \to \theta_1'$, the term $Y(\rho,\rho^+)N(\theta)N^+(\theta_1)$ is transformed into $Y(\rho,\rho^-)N(\theta')N^-(\theta_1')$. Taking into account the last result in Eq.(16) and recalling the inequality $\forall y,z>0$: $(\log y - \log z)z \leq y - z$ we finally obtain

$$dH/dt \leq 0 \qquad (17)$$

and the H-theorem holds.

From the H-theorem one can recover some informations about the equilibrium thermodynamical state (dH/dt = 0) of the semidiscrete Enskog gas. In fact, conversely to the semidiscrete Boltzmann equation, where the H-functional depends on the space $\underline{x}$ (see ref.[13]), here H is a function of the time only. Therefore the equilibrium condition, given by the equality in (17), holds if and only if y=z, which in our case corresponds to

$$N(\theta')N^-(\theta_1') = N(\theta)N^-(\theta_1) \qquad (18)$$

Such a condition can be satisfied only by an equilibrium distribution function which is constant with respect to space, namely by an "absolute Maxwellian". Let us note that this result is in agreement with the one obtained by Resibois [14] for the full Enskog equation in R^3.

Acknowledgements
This research has been partially supported by the funds of the Italian Ministery of University and Scientific Research, and by the project MMAI of the Gruppo Nazionale per la Fisica Matematica of National Council for Research.

References

1. D. Enskog, Kinetische Theorie der Waermeleitung, Reibung und Selbstdiffusion in gewissen verdichteten Gasen und Flussigkeiten, Kungl. Sv. Vetenskapsakademiens Handl. 63 no.4, 3-44 (1922). English Transl. in S.G. Brush, Kinetic Theory, vol.3, Pergamon, New York (1972).
2. N. Bellomo, A. Palczewski and G. Toscani, Mathematical Topics in Nonlinear Kinetic Theory, World Scientific, Singapore (1988).
3. M. Lachowicz, On the local existence and uniqueness of solution of the initial value problem for the Enskog equation,

Bull. Polish Acad. Sci., 31, 89-97 (1983).
4. G. Toscani and N. Bellomo, The Enskog-Boltzmann equation in the whole space R^3: some global existence and stability results, in Advances in Hyperbolic Equations, M. Witten Ed., Comput. Math. Appl., 13, 851-859 (1987).
5. C. Cercignani, Existence of global solutions for the space inhomogeneous Enskog equation, Transport Theory Stat. Phys., 16, 213-231 (1987).
6. L. Arkeryd, On the Enskog equation in two space variables, Transport Theory Stat. Phys., 15, 673-683 (1986).
7. N. Bellomo and M. Lachowicz, On the asymptotic equivalence between the Enskog and Boltzmann equations, J. Stat. Phys., 51, 233-247 (1988).
8. R.L. Di Perna and P.L. Lions, On the Cauchy problem for the Boltzmann equation: global existence and weak stability, to appear in Ann. of Math. (1989).
9. L. Arkeryd, On the Enskog equation with large initial data, preprint (1989).
10. J. Polewczak, Global existence in L^1 for the nonlinear Enskog equation, SIAM J. of Appl. Math., 49, 952-959 (1989).
11. J. Polewczak, Global existence in L^1 for the modified nonlinear Enskog equation in R^3, J. Stat. Phys., 56, 159-173 (1989).
12. T. Platkowski and R. Illner, Discrete velocity models of the Boltzmann equation: a survey on the mathematical aspects of the theory, SIAM Review, 30, 213-255 (1988).
13. E. Longo, R. Monaco and T. Platkowski, Sound and shock wave propagations by the semidiscrete Boltzmann equation, J. de Mecan. Theor. et Appl., vol.17, No.3, 233-243 (1988).
14. P. Resibois, H-theorem for the (modified) nonlinear Enskog equation, J. of Stat. Phys., vol.19, 593-609 (1978).

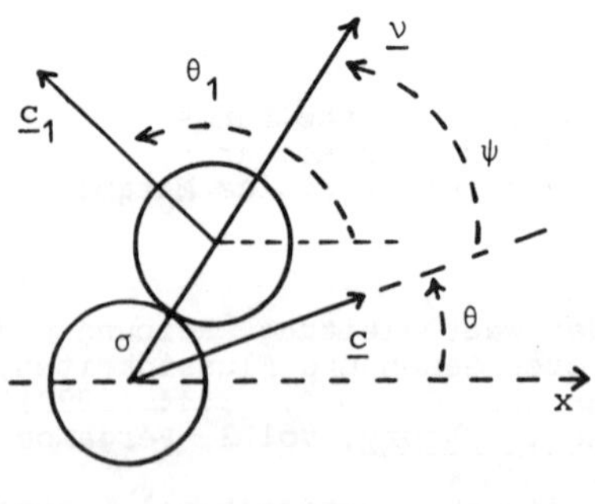

Fig.1 - "Exchange" collisions

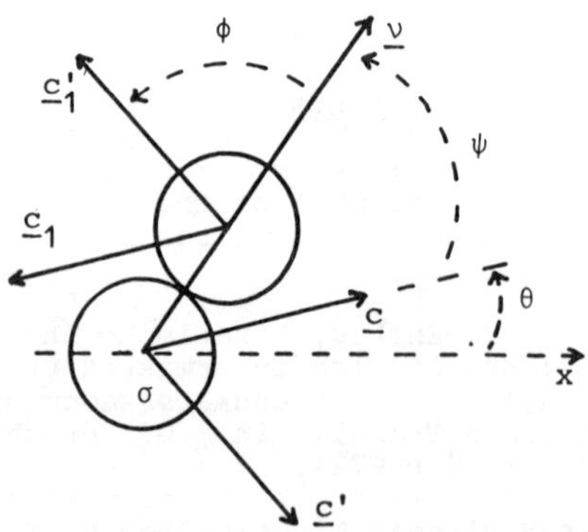

Fig.2 - "Head-on" collisions

INTERFACIAL STABILITY PROBLEMS IN LUBRICATED PIPELINING

L. Preziosi[1], *F. Rosso*[2], *K. P. Chen*[3] *and D. D. Joseph*[3]

[1]*Dipartimento di Matematica ed Applicazioni "R. Caccioppoli", Universitá degli Studi di Napoli, Italy.*
[2]*Dipartimento di Matematica ed Applicazioni "U. Dini", Universitá degli Studi di Firenze, Italy.*
[3]*Department of Aerospace Engineering and Mechanics, University of Minnesota, Minneapolis, U. S. A.*

ABSTRACT We analyze the stability of the flow of two immiscible fluids divided by a cylindrical interface in a pipe (core-annular flow, abbreviated CAF in the sequel) when the flow is driven by a pressure gradient or by a moving inner cylinder. The effect of surface tension and density is included, but gravity is excluded. We find upper and lower branches of the neutral curve in a Reynolds number ($\mathcal{R}$) vs. wave number (α) plane. A window of parameters is identified in which CAF is stable to small disturbances. When $\mathcal{R}$ is inferior to the lower critical value, CAF is destabilized by surface tension and long waves break up into slugs and bubbles. This long wave instability is a capillary instability, modified by shear, which reduces to Rayleigh's instability in the appropriate limit. At higher $\mathcal{R}$, the capillary instability is stabilized by shear. At yet higher $\mathcal{R}$, above the upper critical value, the flow is unstable to generally shorter waves leading to emulsification, water droplets in oil. The theory agrees with published experiments and data from Shell Oil Co. and seems applicable to the design of lubricated pipelines.

1. MATHEMATICAL FORMULATION

Two liquids are flowing in a region Ω between two coaxial pipes of radii R_1 and R_2 ($R_1 < R_2$), respectively. Formally $R_1 = 0$ means that there is no inner cylinder, in which case the flow is driven only by a pressure gradient. We use cylindrical coordinates (r, θ, x) and denote $r = R(\theta, x, t)$ the interface between the two fluids. Liquid 1 occupies the region $\Omega_1(t) = \{(r, \theta, x)/\theta \in [0, 2\pi),\ x \in \Re\ ,\ R_1 \leq r \leq R(\theta, x, t)\}$ and liquid 2 occupies the region $\Omega_2 = \Omega - \Omega_1$. Density and viscosity of the l–component are denoted by ρ_l and μ_l, respectively. The mean radius of $r = R(\theta, x, t)$ is a constant $\tilde{R}$ fixed by the presribed volume of each of the two liquids, independent of t.

The two newtonian incompressible fluids have to satisfy the Navier–Stokes equations and the usual interface conditions (continuity of velocity, kinematic constraint, stress conditions), the no–slip boundary condition at the pipe walls and, if $R_1 = 0$, the boundness of the velocity at the pipe axis.

It is convenient to rescale lengths with $\tilde{R}$, velocities with the velocity V of the inner cylinder (or the centerline velocity if there is no inner cylinder), time with $\tilde{R}/V$ and pressure with $\rho_1 V^2$.

We are going to study the stability of the following basic flows

$$(1) \qquad W(r) = \begin{cases} (\ln a_2 - m \ln r)/(\ln a_2 - m \ln a_1) & \text{if } a_1 \le r \le 1 \\ (\ln a_2 - \ln r)/(\ln a_2 - m \ln a_1) & \text{if } 1 \le r \le a_2 \end{cases} ,$$

$$(2) \qquad W(r) = \begin{cases} 1 - mr^2/(a_2^2 + m - 1) & \text{if } 0 \le r \le 1 \\ (a_2^2 - r^2)/(a_2^2 + m - 1) & \text{if } 1 \le r \le a_2 \end{cases}$$

where W is the axial velocity, $a_1 = R_1/\tilde{R} < 1$, $a_2 = R_2/\tilde{R} > 1$, $m = \mu_2/\mu_1$. The former represent core–annular flow of two fluids driven by an inner cylinder moving at a constant velocity with no pressure gradient, the latter represent the case in which the bicomponent flow is driven by a pressure gradient along the axis. In this case there is no inner cylinder.

We won't write here neither the long system of equations needed to study the linear stability of the flows nor the details of the long wave or the energy analysis since they are carefully described in [1] and [2]. We will therefore jump directly to the description of the mechanism of instability.

The stability properties depend on 8 parameters: the Reynolds number $\mathcal{R} = \rho_1 V \tilde{R}/\mu_1$, the surface tension parameter $J = \rho_1 T \tilde{R}/\mu_1^2$, the viscosity ratio $m = \mu_2/\mu_1$, the density ratio $\varsigma = \rho_2/\rho_1$, the dimensionless radii a_1 and a_2, the axial wave number α and the azimuthal mode n. We found,however, that the axisymmetric mode is always the most unstable one.

2. STABILITY OF THE FLOW DRIVEN BY AN INNER CYLINDER

We first consider the linear stability of a Couette flow of two immiscible liquids between sliding coaxial cylinders when they form one layer on each cylinder and the interface is perfectly cylindrical. In this case the basic flow is given by (1).

The long wave analysis of this flow tells us that, for a fixed a_1 (the inner radius), as a_2 (the outer radius) increases the region of stability increases when $m > 1$ (that is the more viscous fluid is outside) and decrease when $m < 1$. The same happens when a_2 is fixed and a_1 increases. This means that *the flow is stabilized when the flow rate of the less viscous fluid versus the more viscous one is reduced, independently of where they are located.*

A numerical study of the the whole range of α confirms the previous statement but reveals that the mechanism of stability is much more complex than what could be understood by a simple long–wave analysis.

In fig. 1a we show a fairly representative situation with two disjoint neutral curves: usually a lower and an upper branch. We may define the stability limits $\mathcal{R}_L = \max_\alpha \mathcal{R}_{CL}$ and $\mathcal{R}_U = \min_\alpha \mathcal{R}_{CU}$ where $\mathcal{R}_{CL}$ (resp. $\mathcal{R}_{CU}$) is the critical Reynolds number of the lower (resp. upper) branch. The lower branch is associated with long waves and low Reynolds numbers instability. It meets the α–axis ($\mathcal{R} = 0$) at $\alpha = 1$ and the $\mathcal{R}$–axis ($\alpha = 0$) at the theoretical value $\mathcal{R}(0)$ coming out from the long–wave analysis. If $\mathcal{R} < \mathcal{R}_L$, the flow is unstable to capillary instability caused by surface tension modified by shear giving rise to oil slugs and bubbles in water (see fig. 1 of [1]). The upper branch is associated with shorter wavelengths and higher Reynolds numbers. If $\mathcal{R} > \mathcal{R}_U$ the flow is unstable to an instability due either to the viscosity stratification or to the Reynolds stress in the less viscous fluid. This instability gives rise to emulsions and traveling waves which may be equilibrated by non linear terms.

Usually the maximum of $\mathcal{R}_{CL}$ occurs at $\alpha = 0$ at the value predicted by the long–wave analysis $\mathcal{R}_L = \mathcal{R}(0)$ (see fig. 1), but not always. This is evident in fig. 2a where we plot neutral curves at different viscosity ratios ranging between 15 and 30. When m increases, both the regions of instability (to short and to long waves) enlarge. At $m = 15$ we have $\mathcal{R}_L = \mathcal{R}(0) = 97.7$, but at $m = 20$, $\mathcal{R}_L$ is no longer obtained for $\alpha = 0$ but for $\alpha = .6$. If we further increase m, the critical wave number also increases and, for example, when $m = 25$, it turns out to be even greater than 1 (precisely 1.25). Things seems to go as if the region of instability to long waves were attracted by the short wave one as they get close to each other. When $m = 30$ a funnel connecting the to regions appears. Thus when $a_1 = .5$, $a_2 = 2$, $J = 1000$ there is no range of $\mathcal{R}$ for which CAF is stable if $m \geq 30$. This critical value of m should be compared with the value $m \approx 70$ obtained via the long wave analysis.

To understand the feedback mechanism of this instability we studied the evolution of the energy of the most unstable mode. We found that the stability or instability of the flow depends on the sign of the sum of 3 terms: the Reynolds stress term $I - 1$, the viscosity stratification term B_m and the surface tension term B_J. All of them may be either stabilizing or destabilizing according to the flow parameters.

The B_J–instability is always dominant at low $\mathcal{R}$ while the B_m–instability is less important at the lowest $\mathcal{R}$. The situation reverses for larger but still small $\mathcal{R}$. The Reynolds stress is stabilizing at small $\mathcal{R}$ and destabilizing at large $\mathcal{R}$. Moreover, at very large $\mathcal{R}$, instability is essentially due to $I - 1$ or B_m with negligible contribution due to B_J.

Fig. 1a shows that when ($a_1 = .1$, $a_2 = 1.1$, $m = .1$) CAF is stable only if $53.7 < \mathcal{R} < 730$.

Fig. 1b shows the behaviour of the energy terms at high $\mathcal{R}$ in this case. When $\mathcal{R} \approx 400$ surface tension has a strong destabilizing tendency, but it is stabilized by shear (namely $I < D = 1$). Increasing $\mathcal{R}$, B_J decreases while B_m, due to the viscosity stratification, is still destabilizing. This interfacial instability is soon overcome by the Reynolds stress $I - 1$ which increases very rapidly leading to emulsions and travelling waves. It can be proved moreover that the energy supplied by I in Ω_1 is lower than the one supplied in Ω_2, so the instability at high $\mathcal{R}$ is associated *with the water, not the oil.*

In fig 2b we consider a case with $m > 1$. The flow is unstable when $\mathcal{R} > 885$. B_m is stabilizing for $0 < \mathcal{R} < 100$ and B_J is destabilizing for $0 < \mathcal{R} < 900$. At small $\mathcal{R}$ surface tensin is the leading source of instability. The destabilizing effect of B_m is dominant for $\mathcal{R} > 400$. $I - 1$ is always negative. This stabilizes for a while the tendency to emulsify given by B_m, but when $\mathcal{R} > 885$ instability occurs and is not stabilized even when B_J becomes negative since the stabilizing effect of $I - 1$ decreases with $\mathcal{R}$.

In fig. 3 we plot $\mathcal{R}_L$, $\mathcal{R}_U$ and $\mathcal{R}(0)$ versus m when the less viscous fluid is inside ($a1 = .5$, $a2 = 2$, $J = 1000$). $\mathcal{R}_U$ is a decreasing funcion of m, that is by increasing the viscosity ratio short waves are destabilized, infact by the energy analysis we have just seen that this instability is due to the viscosity stratification (see fig. 2b). On the other hand $\mathcal{R}(0)$ has an asyntote when m=1, then decreases with m and increases again (when, say, $m > 10$) tending to infinity when $m \approx 70$. $\mathcal{R}_L$ coincide with $\mathcal{R}(0)$ when $m \leq 18$, then a protuberance forms in the capillarity instability region and $\mathcal{R}_L$ is no longer obtained at $\alpha = 0$. Therefore $\mathcal{R}_L$ detaches from the theoretical curve predicted by the long wave analysis and increases more rapidly connecting at last with the short wave region when m is about 29 (see fig. 2a). Thus a window forms in which CAF is stable.

In fig. 4 we plot $\mathcal{R}(0)$, $\mathcal{R}_L$ and $\mathcal{R}_U$ versus a_1 when the inner fluid is less viscous ($m = 100$) and $J = 1000$, $a_2 = 5$. Both $\mathcal{R}_U$ and $\mathcal{R}_L$ have an asyntote when $a_1 = 1$, but while $\mathcal{R}_U$ increases, $\mathcal{R}_L$ decreases with a_1 so that the smaller is the layer of less viscous lubricant the larger is the range of $\mathcal{R}$ for which the flow is stable. As before $\mathcal{R}_L = \mathcal{R}(0)$ when $a_1 > .66$, then $\mathcal{R}_L$ increases more rapidly because of the formation of the bump which connects the two instability regions when $a_1 \approx .635$.

In fig. 5 the reverse situation is represented: the less viscous fluid is outside ($m = .1$) with $a_1 = .1$ and $J = 500$. Also in this case by decreasing the amount of lubricant both short and long waves are stabilized.

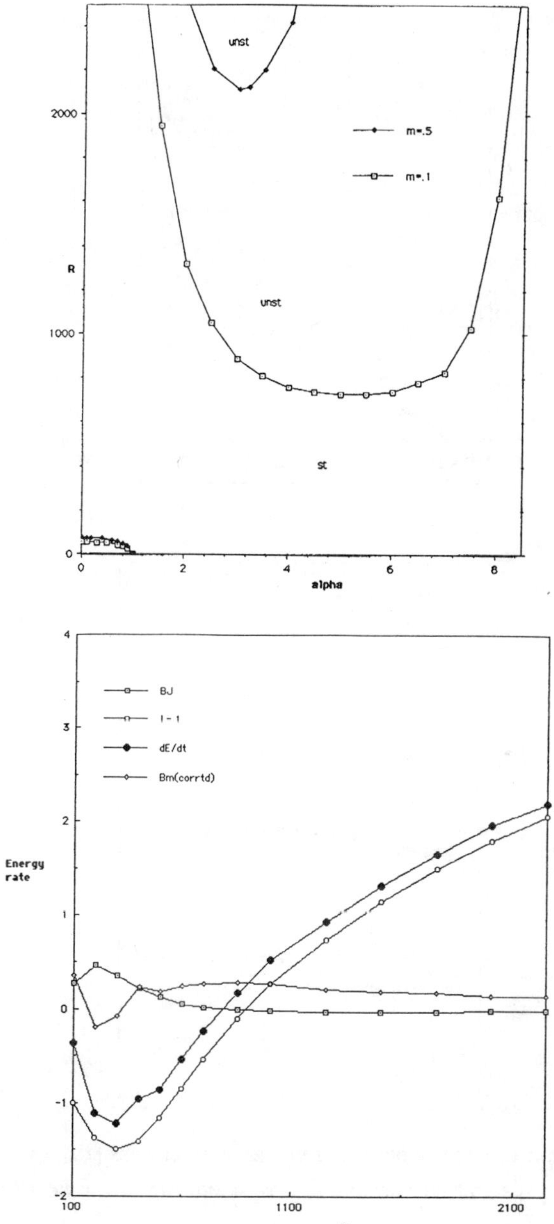

FIG. 1 A – NEUTRAL CURVES FOR $M = .1, .5$ WHEN $(j, A_1, A_2) = (500, .1, 1.1)$. FOR $54 < \mathcal{R} < 730$ CAF WITH $M = .1$ IS STABLE.

B – ENERGY TERMS WHEN $M = .1$ AT HIGH $\mathcal{R}$

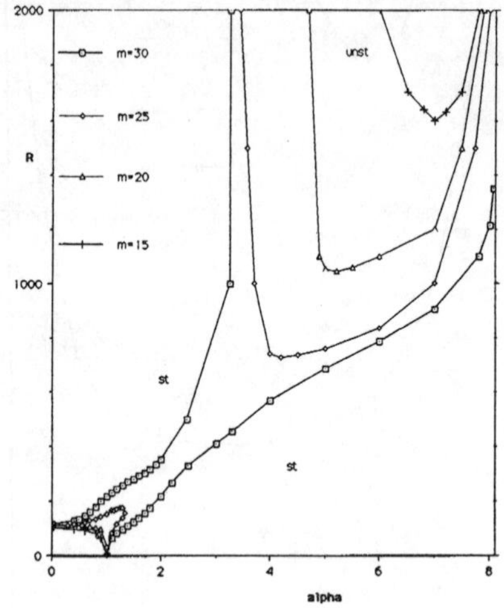

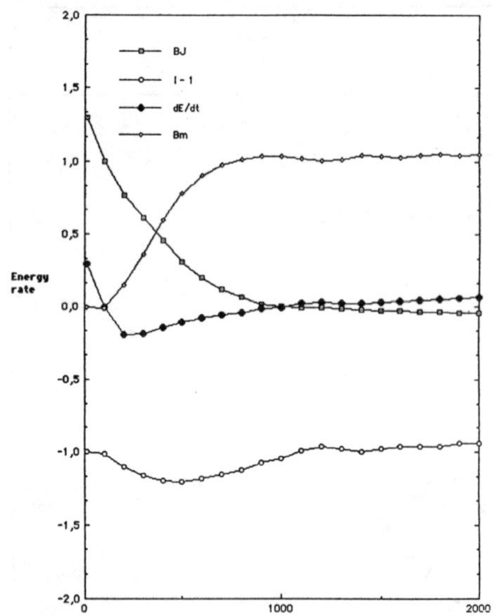

FIG. 2 A – NEUTRAL CURVES FOR DIFFERENT VALUES OF $M > 1$ WHEN $(J, A_1, A_2) = (1000, .5, 2)$. THE REGION OF STABILITY DECREASES WITH M AND THE FLOW IS NEVER STABLE IF $M > 29$.

B – ENERGY TERMS WHEN $M = 20$

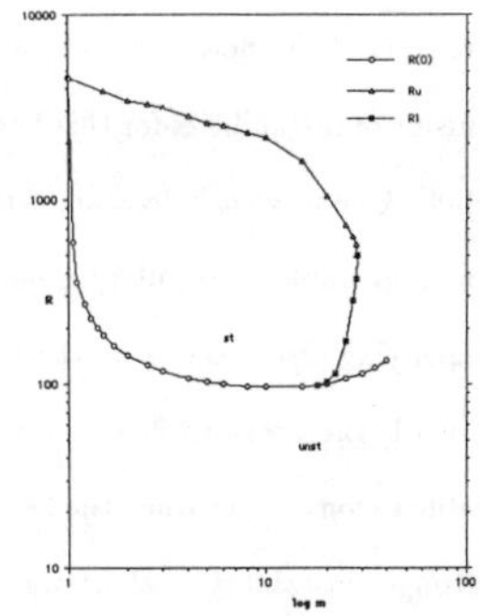

FIG. 3 $\mathcal{R}_L$, $\mathcal{R}_U$ AND $\mathcal{R}(0)$ ARE PLOTTED VERSUS M , $(J, A_1, A_2) = (1000, .5, 2)$. CAF IF STABLE IN THE ENCLOSED REGION. $\mathcal{R}_L = \mathcal{R}(0)$ FOR $1 < M < 18$.

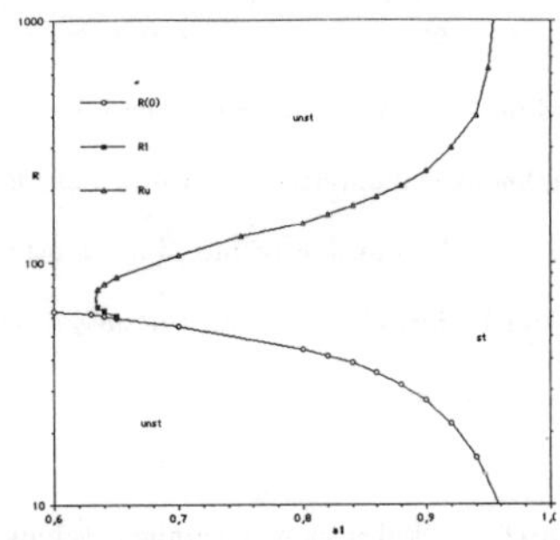

FIG. 4 $\mathcal{R}_L$, $\mathcal{R}_U$ AND $\mathcal{R}(0)$ ARE PLOTTED VERSUS A_1 , $(J, A_2, M) = (1000, 5, 100)$. THE RANGE OF STABILITY DECREASES WITH A_1, WHEN THE LESS VISCOUS FLUID IS CENTRALLY LOCATED. CAF IS NEVER STABLE WHEN $A_1 < .63$ AND $\mathcal{R}_L = \mathcal{R}(0)$ FOR $A_1 > .66$.

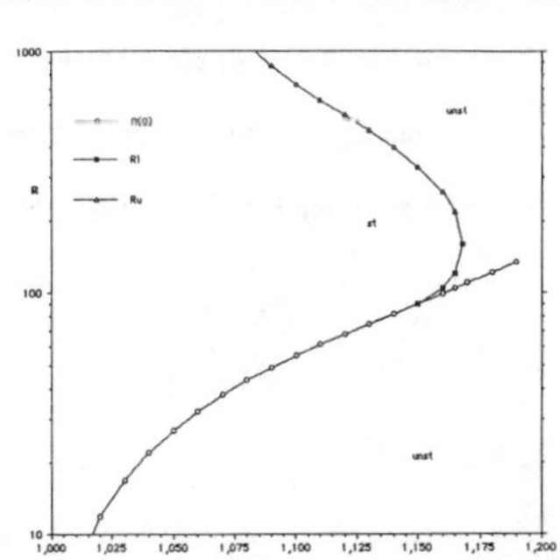

FIG. 5 $\mathcal{R}_L$, $\mathcal{R}_U$ AND $\mathcal{R}(0)$ ARE PLOTTED VERSUS A_2 , $(J, A_1, M) = (500, .1, .1)$. THE RANGE OF STABILITY DECREASES AS A_2 INCREASES, WHEN THE MORE VISCOUS FLUID IS CENTRALLY LOCATED. CAF IS NEVER STABLE WHEN $A_2 > 1.168$ AND $\mathcal{R}_L = \mathcal{R}(0)$ FOR $A_1 < 1.15$.

3. STABILITY OF CORE–ANNULAR FLOW DRIVEN BY A PRESSURE GRADIENT

When there is no inner cylinder and the flow is driven by a pressure gradient along the axis the basic flow is given by (2). The mechanisms of instabilities for this flow are essentially the same as before and are described in more detail in [1] and [4]. One new consideration is that we can achieve greater stability by using a heavy lubricant when the flow is unstable to capillary modes on the lower branch and a light lubricant when the flow is unstable to emulsifying disturbances on the upper branch.

The main difference is that while the previous flow could be stable in both cases $m < 1$ and $m > 1$, in this case the flow is always unstable to long waves when the less viscous fluid is in the core ($m > 1$). This instability is overcome by an even stronger instability to short waves when $\mathcal{R}$ increases.

This suggests that, apart from material considerations (hydrophobic pipes, etc.), the less viscous costituent tends to migrate to those regions of shear where there is a material discontinuity, in this case the pipe walls.

This gives rise to a kind of gift of nature in which the lubricated flows are stable and it opens up very interesting possibilities for technological applications in which one fluid is used to lubricate another.

We can imagine that it may be possible to introduce a beneficial effect in any flow of very viscous liquid important in applications by introducing small amounts of lubricating fluid. Nature's gift is evidently such that the lubricating fluid will migrate to the right places so as to do the desired job.

REFERENCES

[1] Preziosi L., Chen K. and Joseph D. D. "Lubricated pipelining: stability of core-annular flow" *J. Fluid Mech.* **201**, 323–356, (1989)

[2] Preziosi L., Rosso F. "Interfacial stability in a two layer shearing flow of immiscible liquids between sliding pipes" submitted to the European Journal of Applied Mathematics.

[3] Preziosi L., Rosso F. "Stability of a viscous liquid between sliding pipes" submitted to Phys. Fluids.

[4] Hu H. H. and Joseph D. D. "Lubricated pipelining: stability of core-annular flow. Part 2" *J. Fluid Mech.* **205**, 359–396, (1989)

STABILITY OF A SOLITARY WAVE SOLUTION TO A SET OF COUPLED BOUSSINESQ-LIKE EQUATIONS

Peter L. Christiansen*†, Salvatore Rionero*,
Peter S. Lomdahl**, Virginia Muto**.

* Università degli Studi di Napoli, Dipartimento di Matematica e Applicazioni "R. Caccioppoli", Via Mezzocannone 8, 80134 Napoli, Italy.

† Permanent address: Laboratory of Applied Mathematical Physics, The Technical University of Denmark, Building 303, DK-2800 Lyngby, Denmark.

** Theoretical Division and Center for Nonlinear Studies, Los Alamos National Laboratory, Los Alamos, NM 87545, U.S.A.

ABSTRACT

A set of coupled partial differential equations of Boussinesq type are derived in the continuum limit of a Toda lattice with a transverse degree of freedom. The stability of a new hybrid wave solution with longitudinal and transversal components is investigated by approximate calculation of the Hamiltonian for infinitesimal displacement of the longitudinal and transversal strains relatively and by numerical solution.

INTRODUCTION

In this paper we consider a particular solution to a set of Boussinesq-like partial differential equations. The equations were derived in the continuum limit for a Toda-lattice model of the dynamics of deoxyribonucleic acid (DNA) in which transversal motion of a single strand on the DNA-helix was included [1,2]. Actually, the longitudinal and transversal motions were assumed to be of the same order of magnitude. The Toda-lattice model of DNA is discussed elsewhere in these Proceedings.

THE COUPLED BOUSSINESQ-LIKE EQUATIONS

For the longitudinal and transversal strains, u and w, we derive the set of coupled nonlinear partial differential equations

$$u_{tt} = u_{xx} - \frac{\beta}{2}(u^2)_{xx} + \frac{1}{2}(w^2)_{xx} + \frac{1}{12}u_{xxtt} \tag{1.a}$$

$$w_{tt} = (uw)_{xx} + \frac{1}{12}w_{xxtt} \tag{1.b}$$

from the Hamiltonian

$$H = \int_{-\infty}^{\infty} dx \left[\left(\frac{1}{2} U_t^2 + \frac{1}{24} U_{xt}^2 + \frac{1}{2} W_t^2 + \frac{1}{24} W_{xt}^2\right) + \left(\frac{1}{2} U_x^2 - \frac{\beta}{6} U_x^3 + \frac{1}{2} U_x W_x^2\right)\right] \tag{2}$$

by means of the Euler-Lagrange equations. (Thus dH/dt = 0 when (u,w) is a solution to Eqs. (1)). In Eq. (2) U and W denote normalized longitudinal and transversal displacements, related to the strains by

$$U_x = u \quad \text{and} \quad W_x = w \tag{3.a,b}$$

where dimensionless space and time coordinates have been introduced via the transformations $x/\ell \rightarrow x$ and $t\sqrt{\frac{a}{\rho}}/\ell \rightarrow t$. In [1,2] we derive Eqs. (1) in the continuum limit of a Toda lattice with a transversal degree of freedom, ℓ being the lattice constant, a and $b = \beta/\ell$ being the parameters in the Toda potential, and ρ being the mass density in the lattice. Eqs. (1) are valid when the longitudinal and transversal strains are of the same order of magnitude. For small transversal field Eqs. (1) reduce to a single nonlinear partial differential equation, sometimes denoted the improved Boussinesq equation [3,4].

In Refs. [1,2] we show that the ansatz

$$w = Au + B \tag{4.a}$$

with

$$A = \pm\sqrt{2+\beta} \quad \text{and} \quad B = \mp\frac{\sqrt{2+\beta}}{1+\beta} \tag{4.b,c}$$

leads to the following improved Boussinesq equation

$$u_{tt} = \left(u^2 - \frac{1}{1+\beta} u\right)_{xx} + \frac{1}{12} u_{xxtt} = 0 \tag{5}$$

with the hybrid wave solution

$$u(x,t) = \frac{1}{1+\beta}\left[1 + \frac{3}{2}(s^2-1)\,\text{sech}^2 \frac{\sqrt{3(s^2-1)}}{s}\left(x - \frac{s}{\sqrt{1+\beta}} t - x_o\right)\right] \tag{6.a}$$

or

$$w(x,t) = \pm\frac{\sqrt{2+\beta}}{1+\beta}\,\frac{3}{2}(s^2-1)\,\text{sech}^2 \frac{\sqrt{3(s^2-1)}}{s}\left(x - \frac{s}{\sqrt{1+\beta}} t - x_o\right). \tag{6.b}$$

Here s is a constant ($|s| > 1$). In accordance with the ansatz (4.a) both waves travel with the same velocity, $s/\sqrt{1+\beta}$, which may be supersonic ($|s|/\sqrt{1+\beta} > 1$) or subsonic ($|s|/\sqrt{1+\beta} < 1$). The longitudinal wave, u(x,t), is an elongated solitary wave superposed on a constant elongation ($1/\sqrt{1+\beta}$), while the transversal wave is a positive or negative solitary wave. In cases where the two wave fields are not of the same order they will decouple and the longitudinal strain travels with velocity $s(|s|>1)$ and the transversal strain is a standing wave [1,2]. The hybrid wave, Eq. (6), can be viewed as a compromise between these extremes, travelling with the intermediate velocity, $s/\sqrt{1+\beta}$, where $s > s/\sqrt{1+\beta} > 0$. In the following sections we shall analyse the stability of the hybrid wave.

STABILITY ANALYSIS

As we shall see in the numerical work the longitudinal component tends to travel faster than the transversal component of the field. We therefore investigate what happens to the Hamiltonian (2) when the longitudinal component is a differential distance, dx_o, ahead of the transversal component. Insertion of

$$u = \frac{1}{1+\beta}\left[1 + \frac{3}{2}(s^2-1)\,\mathrm{sech}^2\,(X - X_o - dX_o)\right] \tag{7.a}$$

$$w = \frac{\sqrt{2+\beta}}{1+\beta}\,\frac{3}{2}\,(s^2-1)\,\mathrm{sech}^4\,(X - X_o) \tag{7.b}$$

with $X = \frac{\sqrt{3(s^2-1)}}{s}\left(x - \frac{st}{\sqrt{1+\beta}}\right)$ and $X_o = \frac{\sqrt{3(s^2-1)}}{s}\,x_o$ yields

$$dH = -\,\frac{18\sqrt{3}}{35}\,\frac{2+\beta}{(1+\beta)^4}\,(s^2-1)^{5/2}\,(dX_o)^2\ . \tag{8}$$

Here only the interaction term, $\frac{1}{2}uw^2$, contributes. Since the differential, dH, is negative for $dX_o \neq 0$, H has a <u>maximum</u> for $dX_o = 0$, indicating that separation of the longitudinal and transversal components leads to a decrease of the total energy of the solitary wave, which is thus unstable. However, it must be remembered that the analysis is only approximate because it has been assumed that the shape of the separated wave components is unchanged by the separation. Only if the Hamiltonian contains a negative interaction term, $-\frac{1}{2}uw^2$, a minimum corresponding to a stable wave can be obtained. However, such a Hamiltonian cannot be obtained in the continuum limit for the Toda lattice.

NUMERICAL RESULTS

The partial differential equations (1) with initial conditions (7) are integrated by means of a pseudo-spectral method, which employs a Fourier transformation of the spatial variable and a finite-difference approximation to the time variable. In a typical run we used space and time steps, $\Delta x = 0.1$ and $\Delta t = 0.001$, and 512 Fourier modes. For $dX_o = 0$ we get different behavior of the solutions in two distinct regions of (β, s) parameter space as shown in Fig. 1. In the region closest to the

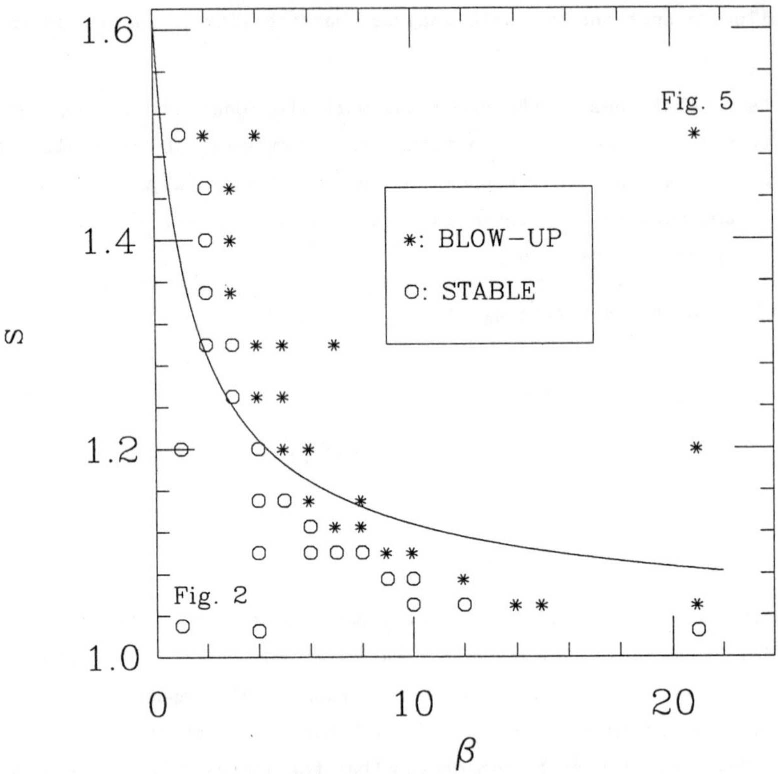

Figure 1. Stability of numerical solutions to Eqs. (1) and (7) with $dX_o = 0$.

axes stable behavior occurs, while the solution blows up in the region away from the axes. In both cases step lengths in space and time were

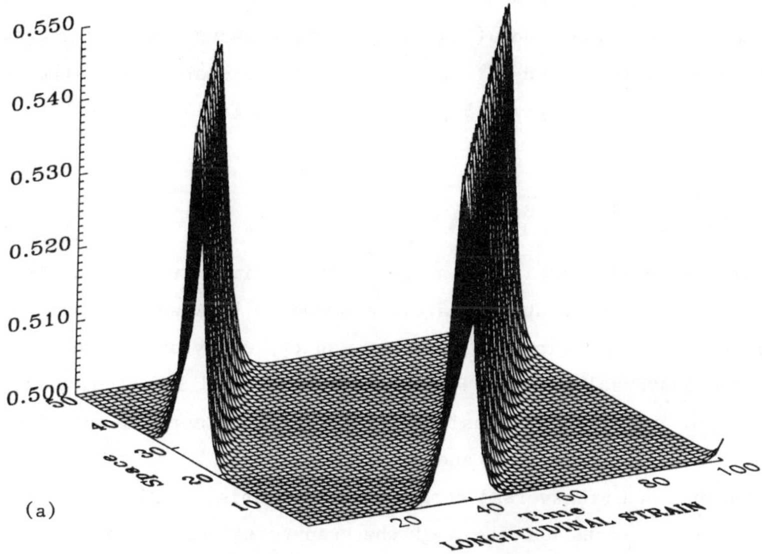

(a)

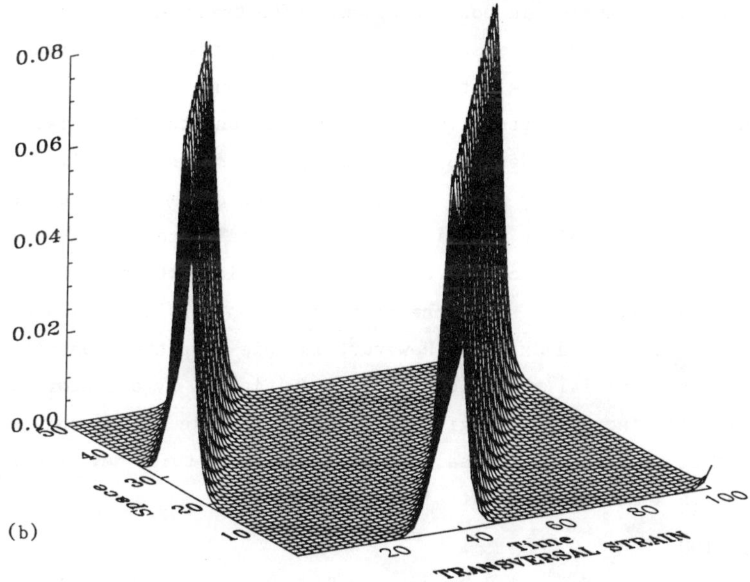

(b)

Figure 2. Longitudinal (a) and transversal (b) strains for $\beta = 1$, $s = 1.03$, $dx_o = 0$.

chosen sufficiently small to ensure numerical accuracy. Also the number of Fourier modes was chosen high enough to reproduce the abrupt rise in the longitudinal amplitude (Fig. 5a) in the blow-up case. Furthermore, addition of noise turned out not to disturb the stability of the solution in the stable region (Fig. 2). Along the full curve in Fig. 1, given by

$$\frac{2}{3(s^2-1)} + 1 = \sqrt{2+\beta} \ , \tag{9}$$

the maximal longitudinal strain equals the maximal transversal strain. This curve is seen to approximate the boundary between the regions of stable behavior and blow-up. Thus blow-up typically occurs when the maximal transversal field is larger than the maximal longitudinal field.

Figs. 2-4 show 3-dimensional plots of the numerical solutions to Eqs. (1) and (7) with $\beta = 1$ and $s = 1.03$. In Fig. 2 $dx_o = 0$ and the longitudinal and transversal strains remain stable. In Fig. 3 the longitudinal field starts out ahead of the transversal field ($dx_o = 1$). Dispersive tails develop in both components. The transversal strain, which decreases initially, is most damaged. In Fig. 4 the longitudinal strain is behind the transversal strain, ($dx_o = -1$), but travels faster and catches up with the latter. In this case the longitudinal strain decreases initially and develops the strongest dispersive tail. Fig. 5 illustrates the blow-up for $\beta = 21$, $s = 1$, and $dx_o = 0$. The event occurs at approximately $t = 10$ in the longitudinal strain. Changes in dx_o in (7a) give rise to much earlier blow-up in both strain components. Finally, Fig. 6 shows the time evolution for Eqs. (1) and (7), $\beta = 1$, $s = 1.03$, and $dx_o = 0$. However, in this case the transverse field in the initial condition (7.b) is multiplied by a factor 1.5. An initially increasing longitudinal strain is able to absorb the excess energy in the transversal strain. Dispersive tails develop in both strains in a manner somewhat similar to Fig. 3.

CONCLUSION

For a set of coupled Boussinesq-like equations, derived in the continuum limit for a Toda lattice with a transversal degree of freedom, the stability of a new hybrid wave solution is investigated. An approximate calculation of the Hamiltonian for the system is based on

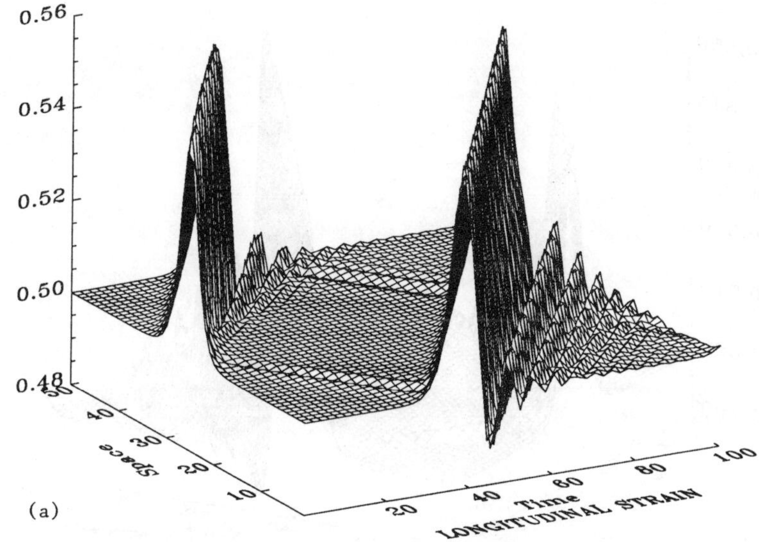

(a)

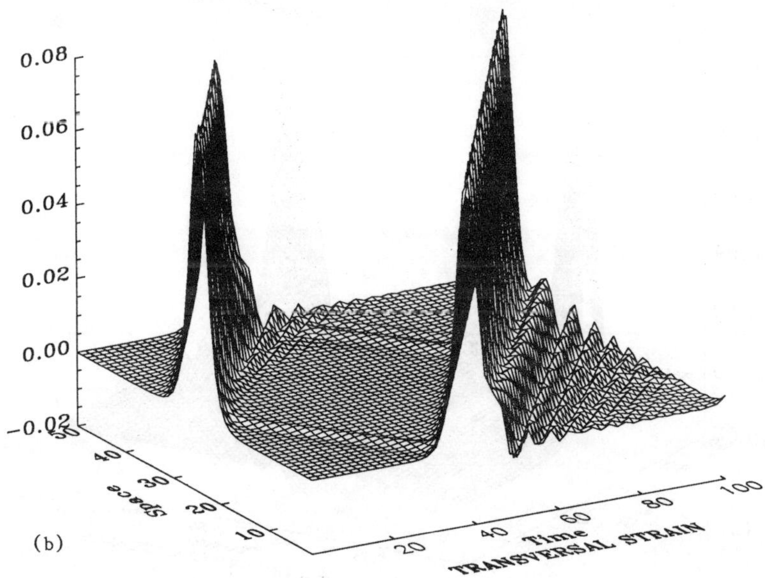

(b)

Figure 3. Longitudinal (a) and transversal (b) strains for $\beta = 1$, $s = 1.03$, $dx_o = 1$.

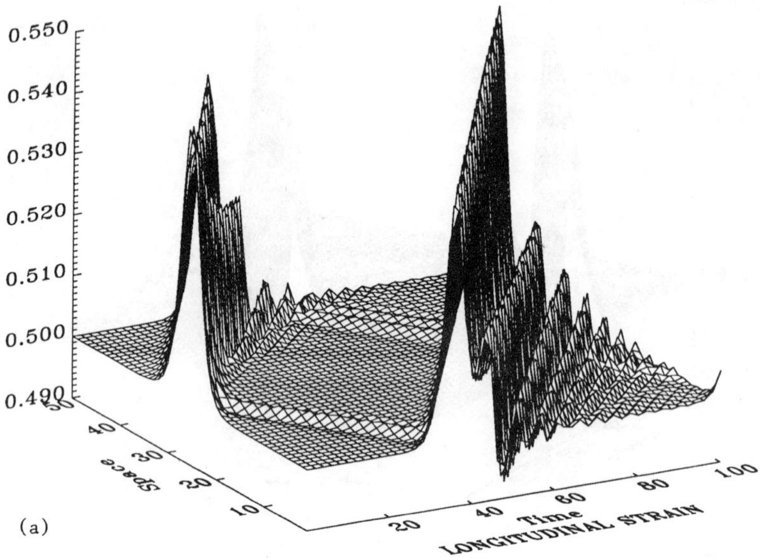

(a)

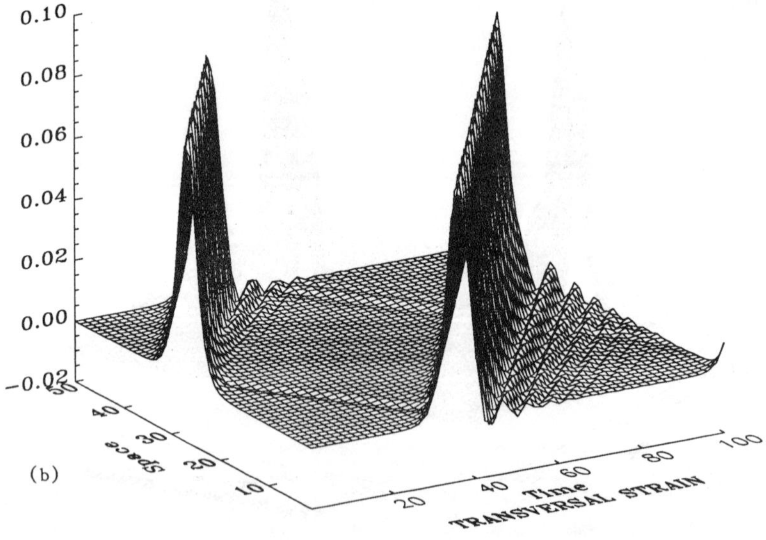

(b)

Figure 4. Longitudinal (a) and transversal (b) strains for $\beta = 1$, $s = 1.03$, $dx_o = -1$.

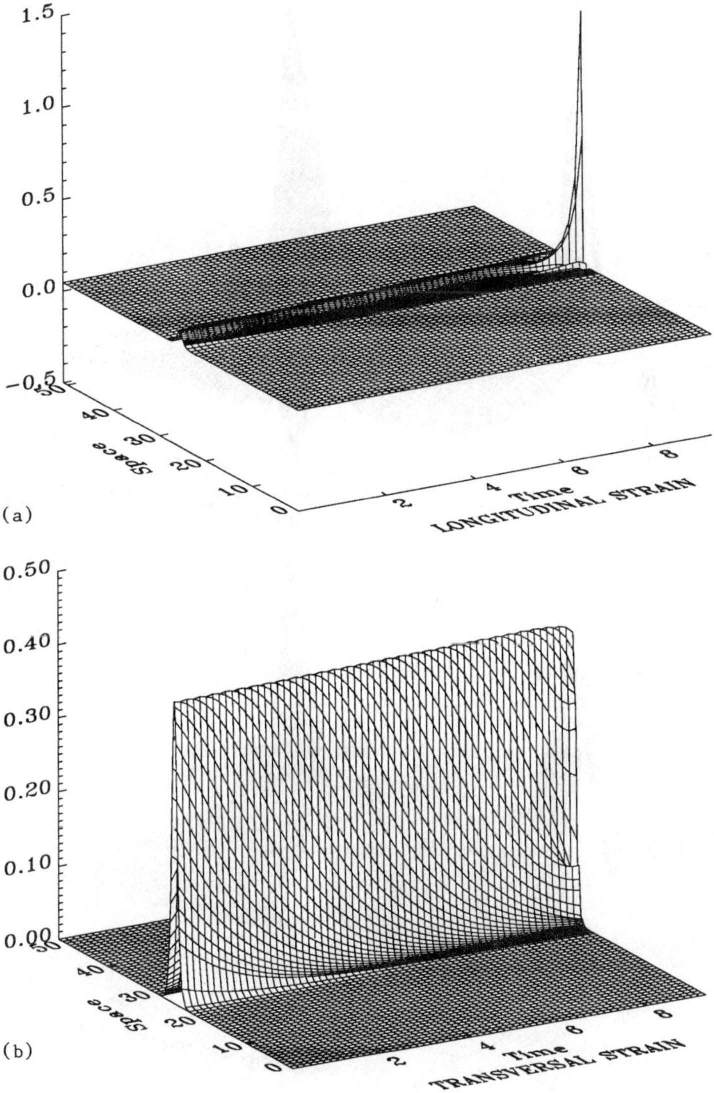

Figure 5. Longitudinal (a) and transversal (b) strains for β = 21, s = 1.5, dx_o = 0. Blow-up occurs at t ≃ 10.

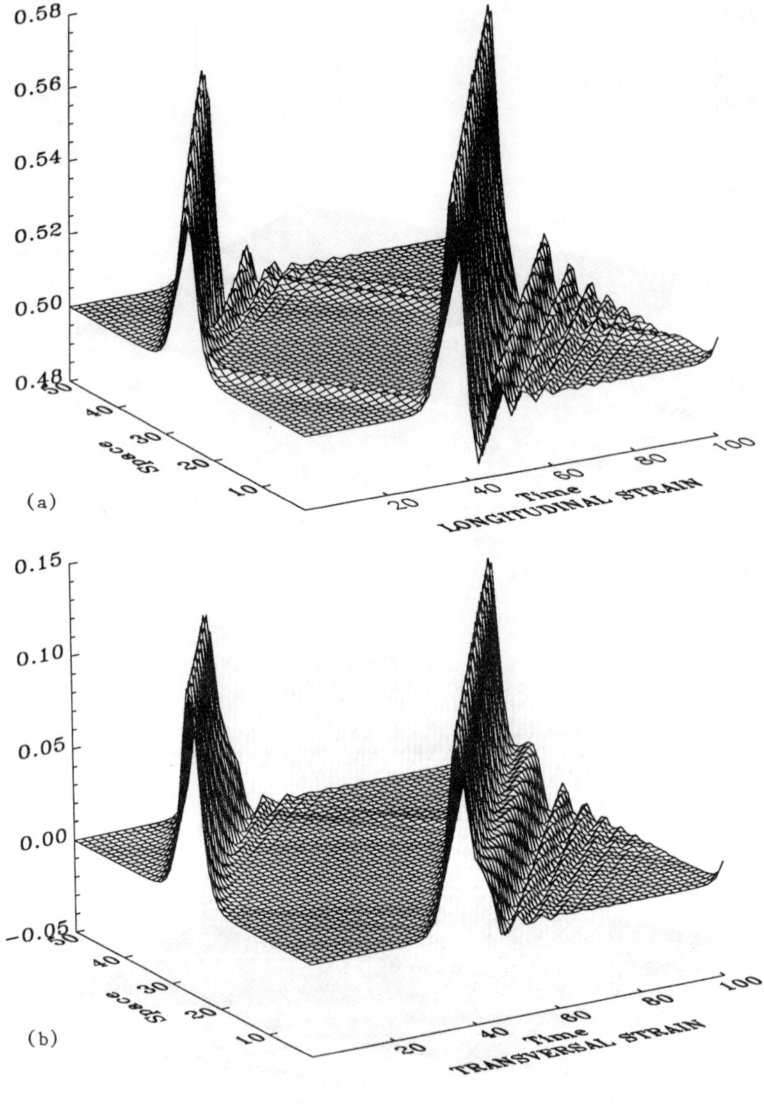

Figure 6. Longitudinal (a) and transversal (b) strains for β = 1, s = 1.03, dx_o = 0, and transversal strain in (7.b) multiplied by 1.5.

the assumption of unchanged longitudinal and transversal strain profiles when these two components are infinitesimally displaced relatively. As a result the Hamiltonian exhibits a maximum indicating instability of the strain. However, numerical solutions of the problem lead to stable solutions essentially when the longitudinal strain is dominant and blow-up in finite time when the transversal strain is dominant. The effect of perturbation of the relative position and amplitude of the longitudinal and transversal strains is investigated.

ACKNOWLEDGEMENTS

Parts of these studies were performed under the auspices of the US Department of Energy. One of us (PLC) expresses thanks for the warm hospitality of the Center for Nonlinear Studies, Los Alamos National Laboratory, and of the Dipartimento di Matematica e Applicazioni, Università degli Studi di Napoli, and acknowledges support from Julie Damms Studiefond, the Danish Technical Research Council, Denmark, and Consiglio Nazionale delle Riserche, Rome, Italy.

REFERENCES

[1] P.L. Christiansen, P.S. Lomdahl, and V. Muto, "Toda Lattice with Transverse Degree of Freedom" (to appear).

[2] P.L. Christiansen, P.S. Lomdahl, and V. Muto, "On a Toda Lattice Model with a Transversal Degree of Freedom" (to appear).

[3] M.P. Soerensen, P.L. Christiansen, and P.S. Lomdahl, "Solitary waves on nonlinear rods", J. Acoust. Soc. Am. 76, 871 (1984).

[4] M.P. Soerensen, P.L. Christiansen, P.S. Lomdahl, and O. Skovgaard, "Solitary waves on nonlinear rods II", J. Acoust. Soc. Am. 81, 1718 (1987).

NONLINEAR DISCRETE MODELS FOR DNA DYNAMICS

V. Muto*, A.C. Scott**, P.L. Christiansen** and P.S. Lomdahl*

* Center for Nonlinear Studies and Theoretical Division, Los Alamos National Laboratory, Los Alamos, 87545 New Mexico, USA

** Laboratory of Applied Mathematical Physics, The Technical University of Denmark, DK-2800 Lyngby, Denmark

ABSTRACT

In this paper we report the investigations on a problem related to *DNA* dynamics: thermal generation of localized pulses that are called *solitons*. The thermal generation of solitons in circular homogeneous *DNA* is investigated by calculating the number of solitons as a function of the absolute temperature. These calculations are effected by using two different models for the *DNA* molecule. In both models the parameters are chosen to match experimentally measured properties of the molecule. We find that a significant number of solitons is generated at physiological temperatures, and a $T^{1/3}$ law is followed at low temperatures.

THE ONE-DIMENSIONAL MODEL

Here the *DNA* is modeled as a Toda lattice [1-3]. Each particle in the lattice represents a single base pair, and the nonlinear spring represents the Van der Waals potential between adjacent base pairs. Therefore considering a homogeneous system of N masses (each with mass M), whose *longitudinal positions* are denoted by $y_n(t)$, $n = 1, 2, ..., N$, and connected to their nearest neighbours with massless nonlinear springs of potential $V(y_n - y_{n-1})$, Newton's second law for the longitudinal motion of the masses is the set of ordinary differential equations

$$M\ddot{y}_n = V'(y_{n+1} - y_n) - V'(y_n - y_{n-1}), \quad n = 1, 2, ..., N. \tag{1}$$

Here dot denotes time differentiation and prime indicates a derivative with respect to the argument. y_0 and y_{N+1} satisfy the periodicity condition. The exponential potential used by Toda [4] is

$$V(y_{n+1} - y_n) = \frac{a}{b}\exp[-b(y_{n+1} - y_n)] + a(y_{n+1} - y_n) \tag{2}$$

which has the nature of physical atomic forces for a and b positive parameters.

Moreover the Toda lattice is exactly integrable [4], and well developed analytical and numerical techniques can be used to efficiently count the number of solitons. Indeed in what follows, the method to count the solitons in a periodic homogeneous chain is the *spectral analyzer* (described in detail in [1,3]), that is based on the inverse spectral theory for the Toda lattice [4].

By fitting the Toda potential to a 6-12 Van der Waals potential and considering experimentally measured properties of *DNA* [1-3], the following values for the parameters are obtained: $M = 1.282 \times 10^{-24} kg$, $b = 6.176 \times 10^{10} m^{-1}$ and $a = 5.127 \times 10^{-10} N$.

Two different approaches to the thermalization of the *DNA* molecule have been considered. At first, by assuming the amount of energy per degree of freedom in the system equal to $k_B T$ (here k_B is the Boltzmann constant and T is the absolute temperature measured in K), the system is thermalized by choosing the initial conditions from Gaussian random distributions, whose variances are related to the absolute temperature T [1-3]. It is found that a significant number of solitons, N_S, is generated at physiological temperature. Moreover, at low temperatures, a $T^{1/3}$ law is obtained for the dependence of the number of solitons on the temperature [1-3]. A typical curve is shown in figure 1 where the ratio N_S/N is plotted versus the temperature T in a logaritmic plot. The dotted line is the curve computed for the unperturbed equations (1), for $N = 65000$ and with initial conditions such that the energy is only kinctic one. At physiological temperature, $310K$, the ratio N_S/N has a value approximately equal to 0.31.

In the second approach to thermalization, in order to describe the interaction of the system with a thermal reservoir at a finite temperature, a damping and a noise force which simulate a thermal bath are added to the equations of motion for the molecular displacement, thereby obtaining Langevin equations [5]. Then [3] the equation of motion of each particle is given by the perturbed Toda equation

$$M\ddot{y}_n = V'(y_{n+1} - y_n) - V'(y_n - y_{n-1}) - M\Gamma\dot{y}_n + \eta_n(t) \, , \tag{3}$$

where Γ is the damping coefficient and $\eta_n(t)$ is the random force, assumed to be a Gaussian process with zero mean and correlation function $< \eta_n(t) \; \eta_{n'}(t') > = 2M\Gamma k_B T \delta_{nn'} \delta(t - t')$. Here $\delta_{nn'}$ is the Kronecker delta symbol and $\delta(t - t')$ is the Dirac delta function.

Numerical integration of the perturbed system (3), for $N = 32$ masses, with different values for the normalized damping coefficient $\alpha = \Gamma\sqrt{\frac{M}{ab}}$, shows that, when thermal equilibrium is reached, the same significant number of solitons is generated in *DNA* at physiological temperature [3], namely $N_S/N \simeq 0.31$. Moreover from dynamical simulations for a system with $N = 32$, at different temperatures T, and at various α's it is found a good agreement with the $T^{1/3}$ law obtained as dependence of the number of solitons on the temperature [3]. Typical cases are shown in figure 1. The values denoted by the circle, the plus and the cross have been calculated for the perturbed system (3), with a number of masses $N = 32$ and damping coefficients $\alpha = 5.0$, $\alpha = 3.0$ and $\alpha = 0.85$, respectively.

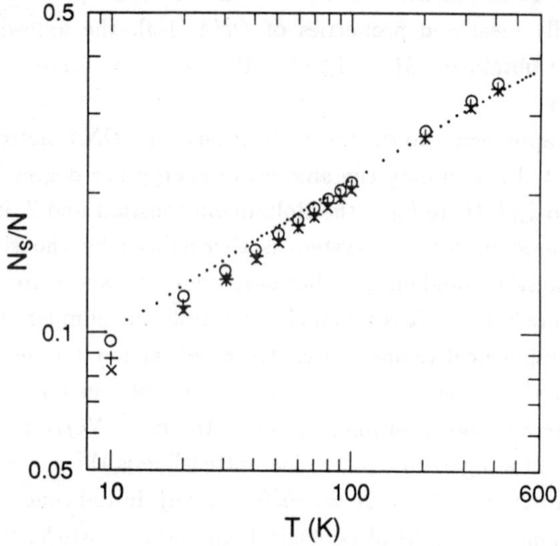

Figure 1. Logarithmic plot (to base 10) of the ratio N_S/N versus the temperature T measured in K. The dotted line is the curve calculated for the unperturbed equations (1), with a number of masses in the Toda chain $N = 65000$, and initial conditions such that the total energy is only kinetic energy. The values denoted by the circle, the plus and the cross are calculated for the perturbed equations (3) for the values of the damping coefficient $\alpha = 5.0$, $\alpha = 3.0$ and $\alpha = 0.85$, respectively, and with a number of masses $N = 32$.

THE TWO-DIMENSIONAL MODEL

Here the *DNA* is described by two chains transversally coupled, where each chain simulate one of the two polynucleotide strands of the *DNA* molecule [6]. Each of the two chains is a spring and mass system. Each mass of the model represents a single base of the base pair. As in the above mentioned one-dimensional model, the longitudinal springs, connecting masses of the same strand, represent the Van der Waals potential between adjacent base pairs. The transverse springs, connecting corresponding masses of the two strands, represent the hydrogen bonds that connect the two bases in a pair. Again we assume to deal with a homogeneous *DNA* molecule, therefore each particle in our mass and spring system has mass $\tilde{M}$, and the springs are assumed to be massless.

For each base pair, the model includes four degrees of freedom, u_n, x_n, and v_n, y_n, for the two strands, respectively. The $u_n = u_n(t)$ and $v_n = v_n(t)$, $n = 1, 2, ..., N$ denote the ***transverse diplacements***, (displacements along the direction of the hydrogen bonds). The $x_n = x_n(t)$ and $y_n = y_n(t)$, $n = 1, 2, ..., N$ denote the ***longitudinal diplacements***, (displacements along the direction of the backbone).

As in the previous model, the anharmonic Van der Waals potentials are described by the Toda potential. Denoting by I the first strand, the anharmonic potential is given by

$$V_{T,I}(l_n^{'} - l_L) = \frac{\tilde{a}}{\tilde{b}} \exp[-\tilde{b}(l_n^{'} - l_L)] + \tilde{a}(l_n^{'} - l_L) \,. \tag{4}$$

A similar expression holds for the potential of the second strand denoted by II.

In the above expression l_L is the equilibrium distance between two bases in the same strand, namely it is the distance along the helix axis between adjacent base pairs, ($l_L = 3.4$Å [7]). And $l_n^{'}$ denotes the distance between two bases in the strand denoted by I. Its expression is given by $l_n^{'} = \sqrt{(l_L + x_{n+1} - x_n)^2 + (u_{n+1} - u_n)^2}$. A similar expression holds for $l_n^{''}$ which denotes the distance between two bases in the strand denoted by II.

The two bases in a pair are connected through hydrogen bonds which we model by a Lennard-Jones potential [8,9]. Thus the anharmonic potential for the transverse springs is given by

$$V_{LJ}(t_n^{'} - l_T + l_H) = 4\,\varepsilon_{LJ} \left[\left(\frac{\sigma}{t_n^{'} - l_T + l_H} \right)^{12} - \left(\frac{\sigma}{t_n^{'} - l_T + l_H} \right)^{6} \right] . \tag{5}$$

The lenght of the hydrogen bond between the two bases in a pair is given by $t_n^{'} - l_T + l_H$. Here $t_n^{'}$ denotes the distance between two bases of the two strands and its expression is given by $t_n^{'} = \sqrt{(l_T + v_n - u_n)^2 + (y_n - x_n)^2}$. Moreover, l_T is the equilibrium distance between adjacent bases in a pair, namely the diameter of the helix, ($l_T = 20$Å [7]), and $l_H = 2^{1/6}\sigma$ [9] is the equilibrium length of the hydrogen bond.

As above, by fitting the Toda potential to a 6-12 Van der Waals potential and considering experimentally measured properties of *DNA*, the following values for the parameters are obtained: $\tilde{M} = 6.41 \times 10^{-25}\ kg$, $\tilde{b} = 6.176 \times 10^{10} m^{-1}$, $\tilde{a} = 2.5635 \times 10^{-10} N$, $\varepsilon_{LJ} = 0.22\ eV = 0.35244 \times 10^{-19}\ Nm$ and $\sigma = 4.01 \times 10^{-10}\ m$.

The equations of motion for the system are obtained from the Hamiltonian

$$H = \sum_{n=1}^{N} \left[T_n + V_{T,I}(l_n^{'} - l_L) + V_{T,II}(l_n^{''} - l_L) + V_{LJ}(t_n^{'} - l_T + l_H) \right] . \tag{6}$$

Here again N is the number of base pairs in the *DNA* molecule. The kinetic energy T_n is given by $T_n = \frac{1}{2}\tilde{M}(\dot{x}_n^2 + \dot{u}_n^2) + \frac{1}{2}\tilde{M}(\dot{y}_n^2 + \dot{v}_n^2)$, and the anharmonic potentials $V_{T,I}(l_n' - l_L)$, $V_{T,II}(l_n'' - l_L)$ and $V_{LJ}(t_n' - l_T + l_H)$ are given above.

If we denote by $d_n^{(i)}$, $i = 1,2,3,4$, the displacement variables, namely $d_n^{(1)} = x_n$, $d_n^{(2)} = u_n$, $d_n^{(3)} = y_n$, $d_n^{(4)} = v_n$, the equations of motion derived from the Hamiltonian (6) can be briefly written (for $i = 1,2,3,4$, and $n = 1,2,...,N$) as

$$\tilde{M}\ddot{d}_n^{(i)} = -\frac{\partial H}{\partial d_n^{(i)}} \,. \tag{7}$$

In order to describe the interaction of the system with a thermal reservoir at a finite temperature T, a damping force and a noise force are added to the equations of motion (7) for the molecular displacements [5], namely

$$F_n^{(i)} = -\tilde{M}\Gamma \, \dot{d}_n^{(i)} + \eta_n^{(i)}(t) \,. \tag{8}$$

Here Γ is the damping coefficient and $\eta_n^{(i)}(t)$, $i = 1,2,3,4$, are the random forces acting on the bases, with zero mean and correlation functions $\langle \eta_n^{(i)}(t) \; \eta_{n'}^{(j)}(t') \rangle = 2\tilde{M}\Gamma k_B T \delta_{nn'} \delta_{ij} \delta(t - t')$.

Then the equations of motion for the system of particles are given by the perturbed version of system (7), namely (for $i = 1,2,3,4$, and $n = 1,2,...,N$)

$$\tilde{M}\ddot{d}_n^{(i)} = -\frac{\partial H}{\partial d_n^{(i)}} - \tilde{M}\Gamma \, \dot{d}_n^{(i)} + \eta_n^{(i)}(t) \,. \tag{9}$$

In order to compute the number of solitons (N_S) in the double chain, we have averaged on the values for the number of solitons in each one of the two chains, computed using the spectral analyzer of the Toda lattice. Our results are shown in figure 2, where the ratio N_S/N is plotted versus the temperature T in a logaritmic plot. In particular the values denoted by the circle are computed for the perturbed system (9) chosen as model of the double chain immerged in a thermal bath. System (9) has been integrated for a number of masses $N = 32$ for each of the two identical chains, and with a damping coefficient $\alpha = 1.0$.

In figure 2 the present results are compared with the ones obtained with the one-dimensional model. In particular the dotted line describes the results obtained using the simple unperturbed Toda lattice. And the values denoted by the cross are the results obtained considering the perturbed model, with a damping coefficient $\alpha = 1.0$ and a number of masses $N = 32$.

These results show that, when thermal equilibrium is reached, in a chain of $N = 32$ masses, a significant number of solitons is generated in *DNA* at physiological temperature, namely $N_S/N \simeq 0.4$. Moreover, it is found a good agreement with the $T^{1/3}$ law obtained as dependence of the number of solitons on the temperature [1-3,6].

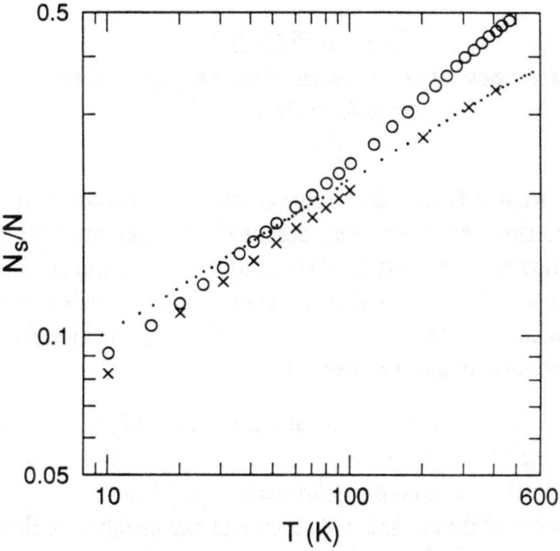

Figure 2. Logarithmic plot (to base 10) of the ratio N_S/N versus the temperature T measured in K. The dotted line is the curve calculated for the unperturbed (one-dimensional) equations (1), with a number of masses in the Toda chain $N = 65000$, and initial conditions such that the total energy is only kinetic energy. The values denoted by the crosses are the ones obtained from the perturbed (one-dimensional) system (3) for the damping coefficient $\alpha = 1.0$ and a number of masses $N = 32$. The values denoted by the circle are calculated for the perturbed (two-dimensional) system (9) with damping coefficient $\alpha = 1.0$, and $N = 32$ masses in each chain.

REFERENCES

[1] V. Muto, *Nonlinear Models for DNA Dynamics*, Ph.D. Thesis, The Danish Center for Applied Mathematics and Mechanics, report N. S 47, October 1988.

[2] V. Muto, A.C. Scott and P.L. Christiansen, Phys. Lett. A 136 (1,2), 33 (1989).

[3] V. Muto, A.C. Scott and P.L. Christiansen, *A Toda lattice model for DNA: thermally generated solitons*, submitted for publication on Physica D.

[4] M. Toda, *Theory for Nonlinear Lattices*, Springer Series in Solid State Sciences, 20, Springer-Verlag (1981).

[5] R. Kubo, Rep. Prog. Phys. 29, 255 (1966).

[6] V. Muto, P.S. Lomdahl and P.L. Christiansen, *A two-dimensional discrete model for DNA: Longitudinal wave propagation and denaturation*, in preparation.

[7] L. Stryer, *Biochemistry*, W.H. Freeman and Co., (1981).

[8] P. Perez and N. Theodorakopoulos, Phys. Lett. A 124, 267 (1987).

[9] J. Halding and P.S. Lomdahl, Phys. Lett. A 124, 37 (1987).

VECTORIAL THERMODYNAMIC VARIABLES AND HEAT PROPAGATION

Vincenzo CIANCIO
Department of Mathematics, University of Calabria,
Arcavacata di Rende (CS),
ITALY

Abstract. We study what influence the vectorial dynamic variables have on heat propagation. We point out that the phenomena described this way are heat propagation by electromagnetic radiation and/or acoustic waves and/or electron waves. Assuming the existence of a maximal velocity for the signal propagation, the heat flux is split into parts, each of which obeying a Cattaneo-Vernotte type equation, that may be identified with waves of different nature and different frequency intervals.

1. Introduction. The dynamic variables can be applied successfully for several phenomena, among others, for dielectric polarization and electric conduction [2], for heat conduction and diffusion [3], for viscoelastic and anelastic deformations [4],[5],etc..

The main convenience of the method is that account can be taken on the influence of the dynamic structures developing during dissipative processes without an *a priopri* and detailed knowledge on them, only the topology of the non-equilibrium state space has to be specified.

In [1] we study what influence the vectorial dynamic variables that are odd functions of microscopic particle velocities have on heat propagation. For the sake of simplicity we assume that the medium is at rest. We assume a medium the local state of which is given by the specific internal energy u, and n independent vector variables $\vec{\beta}^{(i)}, (i = 1, 2, \ldots, n)$.

According to the maximum propety of the entropy in an equilibrium state and to the Morse lemma [6], the entropy function can be casted into the form

$$s = s^{(e)}(u) - \frac{1}{2}\sum_{k=1}^{n}\left(\vec{\beta}^{(k)}\right)^2. \tag{1.1}$$

where $s^{(e)}(u)$ is the equilibrium entropy function depending on the internal energy, the only equilibrium state variable.

Making use of the relations

$$\frac{1}{T} = \frac{\partial s}{\partial u} = \frac{ds^{(e)}(u)}{du}; \qquad \frac{\partial s}{\partial \beta_\alpha^{(k)}} = -\beta_\alpha^{(k)} \qquad (k = 1, 2, \ldots, n), \tag{1.2}$$

we obtain

$$\varrho T \frac{ds}{dt} = \varrho \frac{du}{dt} - \varrho T \sum_{k=1}^{n} \vec{\beta}^{(k)} \cdot \frac{d\vec{\beta}^{(k)}}{dt}, \tag{1.3}$$

Equation (1.3) is usually called the Gibbs relation.

2. Entropy balance and phenomenological equations.

The first law of thermodynamics reads

$$\varrho \frac{du}{dt} + div\vec{\mathbf{J}}^{(q)} = 0, \tag{2.1}$$

where $\vec{\mathbf{J}}^{(q)}$ is the heat flow. By eliminating the internal energy u from the first law (2.1) and the Gibbs relation (1.3), one obtains for the entropy balance the form

$$\varrho \frac{ds}{dt} + div\left(\frac{\vec{\mathbf{J}}^{(q)}}{T}\right) = \sigma^{(s)}, \tag{2.2}$$

where

$$\sigma^{(s)} = T^{-1}\left[-T^{-1}\vec{\mathbf{J}}^{(q)} \cdot gradT - \varrho T \sum_{k=1}^{n} \vec{\beta}^{(k)} \cdot \frac{d\vec{\beta}^{(k)}}{dt}\right]. \tag{2.3}$$

The equation (2.2) is the balance equation for the entropy, $\vec{\mathbf{J}}^{(q)}/T$ is the entropy flow and $\sigma^{(s)}$ is the entropy production. According to the usual procedure of non-equilibrium thermodynamics, we have the following phenomenological equations for an anisotropic medium by virtue of the form (2.3) for the entropy production

$$J_{\alpha}^{(q)} = -T^{-1} L_{\alpha\gamma}^{(q)(q)} \frac{\partial T}{\partial x^{\gamma}} - \varrho T \sum_{k=1}^{n} L_{\alpha\gamma}^{(q)(k)} \frac{d\beta_{\gamma}^{(k)}}{dt}, \tag{2.4}$$

$$\beta_{\alpha}^{(i)} = -T^{-1} L_{\alpha\gamma}^{(i)(q)} \frac{\partial T}{\partial x^{\gamma}} - \varrho T \sum_{k=1}^{n} L_{\alpha\gamma}^{(i)(k)} \frac{d\beta_{\gamma}^{(k)}}{dt} \quad (i = 1, 2, \ldots, n). \tag{2.5}$$

The tensors L are called phenomenological tensors. The phenomenological tensors are material properties of the particular medium and so they are invariant against the isotropy transformations of the body, i.e. they are invariants of the isotropy group of the particular crystal lattice. If the material is isotropic, we obtain from (2.4), (2.5)

$$\vec{J}^{(q)} = -\lambda\, gradT - \varrho \sum_{k=1}^{n} \mu^{(q)(k)} \frac{d\vec{\beta}^{(k)}}{dt}, \tag{2.6}$$

$$\vec{\beta}^{(i)} = -\mu^{(i)(q)}\, gradT - \varrho \alpha^{(i)} \frac{d\vec{\beta}^{(i)}}{dt} \quad (i = 1, 2, \ldots, n). \tag{2.7}$$

where μ's, α's and λ are new phenomenological quantities replacing L's. According to the second law of thermodynamics the phenomenological coefficients make a positive definite form

$$T^2 \sigma^{(s)} = \lambda (gradT)^2 + 2\varrho T \sum_{i=1}^{n} \sqrt{\mu^{(q)(i)}\mu^{(i)(q)}} \cdot \frac{d\vec{\beta}^{(i)}}{dt} gradT$$

$$+ \varrho^2 T^2 \sum_{i=1}^{n} \alpha^{(i)} \left(\frac{d\vec{\beta}^{(i)}}{dt}\right)^2 \geq 0. \tag{2.8}$$

From here, in accordance with the Hurwitz criteria for the determinant of the matrix of the coefficients, we obtain

$$\lambda - \sum_{k=1}^{n} \frac{\mu^{(k)(q)}\mu^{(q)(k)}}{\alpha^{(k)}} = \lambda^{(0)} \geq 0; \quad \alpha^{(k)} > 0 \ (k = 1, 2, \ldots, n), \tag{2.9}$$

Introducing the notations

$$\mathbf{J}^{(0)} = -\lambda^{(0)} \nabla T \quad ; \quad \mathbf{J}^{(k)} = \frac{\mu^{(q)(k)}}{\alpha^{(k)}} \vec{\beta}^{(k)} \quad (k = 1, 2, \ldots, n); \tag{2.10}$$

$$\tau^{(k)} = \varrho \alpha^{(k)} \quad , \quad \lambda^{(k)} = T^2 \frac{(\mu^{(k)(q)})^2}{\alpha^{(k)}} \quad (k = 1, 2, \ldots, n), \quad \tau^{(0)} = 0, \tag{2.11}$$

and making use of (2.9) we have

$$\mathbf{J}^{(q)} = \sum_{k=0}^{n} \mathbf{J}^{(k)}, \tag{2.12}$$

with

$$\mathbf{J}^{(k)} + \tau^{(k)} \frac{d\mathbf{J}^{(k)}}{dt} = -\lambda^{(k)}\, gradT \quad (k = 0, 1, 2, \ldots, n). \tag{2.13}$$

This equation means, that the heat current is split into parts each of which is governed by Cattaneo-Vernotte type constitutive equation except for the part $\mathbf{J}^{(0)}$ as the relaxation time for it is zero.

3. Dispersion relations and signal propagation.

We assume that there exists a state of the medium, the reference state, in which the temperature is uniform and constant (T_0). In order to linearize the equations we introduce the specific heat c

$$\frac{du}{dt} = c\frac{dT}{dt}, \tag{3.1}$$

which is assumed, together with the heat conductivity coefficients λ_i and the relaxation times τ_i, constant. Solutions of form

$$T = T_0 + A exp[i(\vec{\mathbf{k}} \cdot \vec{\mathbf{r}} - \omega t)] \tag{3.2}$$

are investigated where $\vec{\mathbf{k}}$ is the complex wave number vector and ω is the real angular frequency.

It is seen from (3.2) that we consider the wave like solutions that propagate in the direction of the vector $\vec{\mathbf{k}}$. The solutions for the partial heat flows are obtained from (2.13)

$$\vec{\mathbf{J}}^{(j)} = \frac{-i\lambda^{(j)}\vec{\mathbf{k}}A}{1 - i\tau^{(j)}\omega} exp[i(\vec{\mathbf{k}} \cdot \vec{\mathbf{r}} - \omega t)] \quad (j = 0, 1, 2, \ldots, n). \tag{3.3}$$

It is our purpose to find the magnitude of $\vec{\mathbf{k}}$ as a function of ω, i.e. to find the dispersion law. To this end we combine (2.1),(2.12),(3.1),(3.2) and (3.3) and get

$$k^2 = \frac{i c \varrho \omega}{\sum_{j=o}^{n} \frac{\lambda^{(j)}}{1 - i\tau^{(j)}\omega}}. \tag{3.4}$$

In the range of low frequencies, the $\tau^{(j)}\omega$'s in the denominator of the right hand side are negligible and the dispersion relation reduces to that for Fourier's equation with

$$\lambda = \sum_{j=0}^{\infty} \lambda^{(j)} \tag{3.5}$$

while for high frequencies the sum in the denominator reduces to

$$\lambda^{(0)} + \frac{i}{\omega} \sum_{j=1}^{n} \frac{\lambda^{(j)}}{\tau^{(j)}}, \tag{3.6}$$

because the one can be neglected with respect to $\tau^{(j)}\omega$ except for the zero-th term as $\tau^{(0)} = 0$.

We have to distinct two possibilities. If $\lambda^{(0)}$ does not vanish, it turns dominating for sufficiently high frequencies and the dispersion law reduces to that of Fourier's equation with $\lambda = \lambda^{(0)}$, hence the velocity of signal propagation tends to infinity with increasing frequency but with vanishing $\lambda^{(0)}$ the dispersion relation for high frequencies reduces to

$$k^2 = v^2\,\omega^2\,, \tag{3.7}$$

with

$$v^2 = \frac{\varrho c}{\sum_{j=1}^{n} \frac{\lambda^{(j)}}{\tau^{(j)}}}\,, \tag{3.8}$$

which is the dispersion relation for the ordinary wave equation and v is both the phase and the signal velocity. So we incline to declare that $\lambda^{(0)}$ is always zero.

Our result is in perfect agreement with the statistical theories of heat propagation and the vectorial dynamic variables used here can be identified with the parts of the heat flow carried by electromagnetic, acoustic or electron waves in a frequency interval for which the phase velocity and the absorption coefficient are constant with appreciable accuracy.

Bibliography.

[1] Ciancio, V., Verhás, J., A thermodynamic theory for radiating heat transfer, J. Non-Equilib. Thermodyn., 15 (1989).
[2] Ciancio, V., On the generalized Debye equation for media with dielectric relaxation phenomena described by vectorial internal thermodynamic variable, J. Non-Equilib. Thermodyn., 14 (1989), 239.
[3] Gyarmati, I., On the wave approach of thermodynamics and some problems of non-linear theories, J. Non-Equilib. Thermodyn., 2 (1977), 233.
[4] Ciancio, V., Restuccia, L., The generalized Burgers equation in viscoanelastic media with memory, Physica, 142A (1987), 309.
[5] Kluitenberg, G.A., Plasticity and Non-equilibrium Thermodynamics, CISM , Lecture notes, Springer - Verlag - New York, 1984.
[6] Morse, M., Relation between the critical points of a real function of n independent variables, Trans. Am. Math. Soc., 27 (1925), 345.

A Survey of Results on the Loss of Regularity of Shearing Flows of Viscoelastic Fluids

Bernard D. Coleman
Department of Mechanics and Materials Science
Rutgers University
Piscataway, New Jersey

ABSTRACT

A review is given of those parts of the theory of rectilinear shearing flows of simple fluids with fading memory that appear to have relevance to the search for a theory of melt fracture of polymers at high rates of shear. The emphasis is laid on unbounded growth of shear-acceleration waves and breakdown of initially smooth solutions.

1. INTRODUCTION

Some years ago, Morton Gurtin and I [1] proposed that the phenomenon rheologists call "melt fracture" [2–10] or "elastic turbulence" may be an example of the formation of shear-shock waves (also called "vortex sheets") in viscoelastic fluids. Experimenters have not yet decided whether such is the case or whether the phenomenon is caused by failure of adherence to bounding surfaces. The matter has been complicated by the realization that there are several phenomena that are occasionally grouped under the appellation, "melt fracture"; among these is the "spurt" phenomenon studied by Vinogradov and co-workers [11][1] which appears to be due to an apparent loss of monotonicity of the shear-stress function. In this paper I do not discuss the experiments on melt fracture but do survey the mathematical theory of the formation and growth of singularities in viscoelastic fluids with monotone shear-stress functions.

[1] See also Bagley, Cabott, and West [12].

2. GENERAL THEORY OF SIMPLE FLUIDS IN SHEAR

Rectilinear Shearing Flows

A motion is called a *rectilinear shearing flow* if there is a Cartesian coordinate system in which the components of the velocity field have the form

$$v^x = 0, \qquad v^y = v(x,t), \qquad v^z = 0. \tag{2.1}$$

Such a flow is isochoric. Of frequent occurrence in the theory of rectilinear shearing flows in viscoelastic fluids is the real-valued function, λ^t, defined by

$$\lambda^t(s) = -\partial_x \int_{t-s}^{t} v(x,\xi)\, d\xi, \qquad (0 \le s < \infty); \tag{2.2}$$

λ^t is called the *history up to time t (at x) of the relative shear strain.*[2] Shortly after Noll [13] formulated his definition of a simple fluid, Coleman and Noll [14] showed that for a general incompressible simple fluid it follows from the principle of material frame indifference that the components of the stress tensor $\boldsymbol{T}$ in a rectilinear shearing flow obey the relations

$$\left.\begin{aligned} &T^{xy}(t) = \mathfrak{t}(\lambda^t),\\ &T^{yy}(t) - T^{xx}(t) = \mathfrak{n}_1(\lambda^t),\\ &T^{yy}(t) - T^{zz}(t) = \mathfrak{n}_2(\lambda^t),\\ &T^{xz} = T^{yz} = 0, \end{aligned}\right\} \tag{2.3}$$

in which $\mathfrak{t}$, $\mathfrak{n}_1$, and $\mathfrak{n}_2$ are real-valued functionals that depend on the material under consideration and satisfy the identities

$$\mathfrak{t}(-\lambda^t) = -\mathfrak{t}(\lambda^t), \tag{2.4}$$

$$\mathfrak{n}_i(-\lambda^t) = \mathfrak{n}_i(\lambda^t), \qquad (i = 1,2). \tag{2.5}$$

If one assumes that the body force density $\boldsymbol{b}$ has a potential ψ, *i.e.*, that

$$\boldsymbol{b} = -\mathrm{grad}\psi, \tag{2.6}$$

then an elementary analysis shows that it follows from (2.1)–(2.3) that the dynamical equations,

$$\mathrm{div}\boldsymbol{T} + \rho\boldsymbol{b} = \rho\dot{\boldsymbol{v}}, \tag{2.7}$$

[2] $\partial_x = \frac{\partial}{\partial x}$.

in which ρ is the mass density, are equivalent to the assertions that

$$\partial_x T^{xy} + \alpha(t) = \rho \partial_t v, \tag{2.8}$$

and

$$T^{xx} - \rho\psi = \alpha(t)y + \beta(t) \tag{2.9}$$

with α and β functions of t only. It is easily shown that α is the driving force (per unit volume) in the direction of flow. When the specific driving force α is specified, (2.8) is a functional-differential equation for the function v in (2.1) and (2.2).

Steady Rectilinear Shearing Flows

As we shall here be concerned with the dynamical stability of steady rectilinear shearing flows, a brief review of the properties of such solutions of (2.8) is needed.

If the flow is steady, then

$$\partial_t v = 0, \tag{2.10}$$

(2.2) reduces to

$$\lambda^t(s) = -\kappa s \tag{2.11}$$

with[3]

$$\kappa = \kappa(x) = v'(x), \tag{2.12}$$

and (2.8) becomes

$$\frac{d}{dx}\tau(\kappa) + \alpha = 0, \tag{2.13}$$

where τ is the *shear-stress function* which is familiar in the theory of steady viscometric flows[4] [15,16] and which is determined by the functional $\mathfrak{t}$ through the relation

$$\tau(\kappa) = \mathfrak{t}(\kappa\iota), \qquad \iota(s) \equiv s. \tag{2.14}$$

The number κ is called the *rate of shear.* It follows from (2.4) that the function τ, which takes real numbers into real numbers, is an *odd* function, *i.e.,*

$$\tau(-\kappa) = -\tau(\kappa), \tag{2.15}$$

[3] An apostrophe indicates a derivative.

[4] For a survey of that theory see [17].

and hence $\tau(0) = 0$. Thermodynamics requires that $\tau(\kappa)\kappa$ be never negative [16], and hence that $\tau'(0) \geq 0$. Let us here assume

$$\tau'(\kappa) > 0 \tag{2.16}$$

for all κ, which implies that τ is invertible. One writes τ^{-1} for the inverse of τ.

It follows from (2.10) and (2.13) that a steady rectilinear shearing flow can occur only when the specific driving force α is independent of t and

$$\tau(\kappa) = -\alpha x + \nu \tag{2.17}$$

with ν, like α, constant (in space and time). In view of (2.16), specification of α and ν determines the rate of shear as a function of x:

$$\kappa = \tau^{-1}(-\alpha x + \nu). \tag{2.18}$$

If the driving force α is zero, κ is independent of x, and, to within an added constant, v has the form

$$v = \kappa x. \tag{2.19}$$

A steady rectilinear flow obeying (2.19) is called a *simple shearing flow,* and is a motion that can take place in a fluid that lies between and adheres to two infinite plates parallel to the (y, z)-plane with separation d, one of which, at $x = 0$, is at rest, and the other, at $x = d$, is moving with velocity V, for clearly, (2.19) with

$$\kappa = V/d \tag{2.20}$$

is compatible with such boundary conditions.

Steady channel flow is a steady rectilinear shearing flow between two infinite plates of separation d which are again parallel to the (y, z)-plane, but are both at rest. For such a motion the driving force α is not zero. If we place the (y, z)-plane halfway between the plates, then the assumption that the fluid adheres to the plates gives the boundary conditions

$$v\,(d/2) = v\,(-d/2) = 0. \tag{2.21}$$

In view of (2.15) and (2.16), these boundary conditions are compatible with (2.18) only if $\nu = 0$, and hence[5] $v'(x) = \kappa(x) = -\tau^{-1}(\alpha x)$, *i.e.*,

$$v(x) = \int_x^{\frac{1}{2}d} \tau^{-1}(\alpha x)\, dx. \tag{2.22}$$

[5] *Cf.* [15,17].

3. FADING MEMORY AND INSTANTANEOUS ELASTICITY

The postulate of fading memory introduced by Coleman and Noll[6] [18,19] is an assumption of smoothness for constitutive functionals that relate certain variables, such as stress, to the history of others, such as strain. The postulate asserts that such functionals have continuous differentials with respect to a particular norm $\|\cdot\|$ on a space of histories. This norm has two important properties. (*i*) For a history such as λ^t, values $\lambda^t(s)$ at large s, *i.e.*, occurring in the "remote past", have less influence on $\|\lambda^t\|$ than values at small s. (*ii*) A material whose constitutive functionals are smooth with respect to this norm shows *instantaneous elasticity:* a jump in strain results in a jump in stress that depends smoothly on the jump in strain, albeit the response function governing such dependence is determined by the material's history prior to the jump.

As the present discussion is restricted to shearing motions governed by the functional-differential equation (2.8), *i.e.*,

$$\partial_x \mathfrak{t}(\lambda^t) + \alpha = \rho \partial_t v, \tag{3.1}$$

it is not necessary to present the theory of fading memory in full generality, and we may confine our attention to its implications for the functional $\mathfrak{t}$.

The elements λ^t of the domain of $\mathfrak{t}$ are histories of the relative shear strain. By (2.2), each such history is a function on $[0,\infty)$ obeying

$$\lambda^t(0) = 0, \tag{3.2}$$

and, therefore, the value $\mathfrak{t}(\lambda^t)$ of $\mathfrak{t}$ is determined by the restriction of λ^t to $(0,\infty)$. Thus, in a discussion of the functional $\mathfrak{t}$, we need not distinguish between a relative shearing history and its restriction to $(0,\infty)$.

For each (finite) $p \geq 1$ and each influence function k, *i.e.*, each positive, monotone-decreasing, measureable function k with $s^p k(s)$ integrable over $(0,\infty)$, let $\mathcal{L}_p(k)$, with norm $\|\cdot\|$, be the Banach space formed in the usual way from real-valued functions g on $(0,\infty)$ for which

$$\|g\|^p = \int_0^\infty g(s)^p k(s)\, ds \tag{3.3}$$

is finite.

---

[6] A more axiomatic treatment is given in the work Coleman and Mizel [20]. The presentation given below draws on that of Coleman and Gurtin [1].

In the present context, the postulate of fading memory asserts that there is a $p \geq 1$ and an influence function k such that $\mathfrak{t}$ has $\mathcal{L}_p(k)$ for its domain and is twice continuously differentiable in the following sense: For each g in $\mathcal{L}_p(k)$ there is, on $\mathcal{L}_p(k)$, a bounded linear form $\delta\mathfrak{t}(g|\cdot)$ and a bounded, symmetric, bilinear form $\delta^2\mathfrak{t}(g|\cdot,\cdot)$ such that

$$\mathfrak{t}(g+l) = \mathfrak{t}(g) + \delta\mathfrak{t}(g|l) + \frac{1}{2}\delta^2\mathfrak{t}(g|l,l) + o(\|l\|^2). \tag{3.4}$$

It is assumed that each of the functionals $\delta\mathfrak{t}(\cdot\,|\,\cdot)$ and $\delta^2\mathfrak{t}(\cdot\,|\,\cdot,\cdot)$ is jointly continuous in all its arguments.

Since $\delta\mathfrak{t}(g|\cdot)$ is both linear and continuous on $\mathcal{L}_p(k)$, there is a function K in $\mathcal{L}_q(k)$, with $q^{-1}+p^{-1}=1$, such that for each l in $\mathcal{L}_p(k)$

$$\delta\mathfrak{t}(g|l) = \int_0^\infty K(s)l(s)k(s)\,ds. \tag{3.5}$$

Of course, K depends on g. One puts

$$G'(s;g) = K(s)k(s), \tag{3.6}$$

so that

$$\delta\mathfrak{t}(g|l) = \int_0^\infty G'(s;g)l(s)\,ds. \tag{3.7}$$

It is assumed that the mapping $G'(s;\cdot)$ is, for each s, a continuous functional on $\mathcal{L}_p(k)$ and that, for each g, $G'(\cdot\,;g)$ has a bounded derivative $G''(\cdot\,;g)$. It is further assumed that the mapping $g \mapsto G''(\cdot\,;g)k(\cdot)^{-1}$ is continuous from $\mathcal{L}_p(k)$ into $\mathcal{L}_q(k)$.

If we put

$$G(s;g) = -\int_s^\infty G'(\sigma;g)\,d\sigma = -\int_s^\infty K(\sigma;g)k(\sigma)\,d\sigma, \quad (0 < s < \infty), \tag{3.8}$$

then for each g,

$$\lim_{s\to\infty} G(s;g) = 0. \tag{3.9}$$

The function $G(\cdot\,;g)$ is called the *relaxation modulus* (for perturbations about the history g). As we shall see below, $G(0;g)$ is an “instantaneous modulus”.

Let $\mathcal{R}$ be a configuration that occurs at some time, say $t=0$, during the rectilinear motion of the fluid, and let $\gamma(x,t)$ and $u(x,t)$ be the shear strain and displacement at x at time t computed using $\mathcal{R}$ as the reference configuration:

$$\gamma(x,t) = \partial_x u(x,t), \qquad u(x,t) = \int_0^t v(x,\xi)\,d\xi. \tag{3.10}$$

By (2.2), the relative shear strain at time $t-s$ (taken relative to the configuration at time t) is

$$\lambda^t(s) = \gamma(x, t-s) - \gamma(x,t). \tag{3.11}$$

Suppose now that at a particular value of x the shear strain γ suffers a jump θ at time t, so that

$$\gamma(x, t^+) = \gamma(x, t^-) + \theta. \tag{3.12}$$

Then, the histories of the relative shear strain immediately before and immediately after the jump are given, for each $s > 0$, by

$$\left.\begin{aligned} \lambda^{t^-}(s) &= \gamma(x, t-s) - \gamma(x, t^-), \\ \lambda^{t^+}(s) &= \gamma(x, t-s) - \gamma(x, t^+) = \lambda^{t^-}(s) - \theta; \end{aligned}\right\} \tag{3.13}$$

i.e.,

$$\lambda^{t^+} = \lambda^{t^-} - \theta 1^\dagger \tag{3.14}$$

with $1^\dagger$ the constant function on $(0,\infty)$ with value 1. For the shear-stresses immediately before and after the jump in strain, we have

$$\left.\begin{aligned} T^{xy}(t^-) &= \mathfrak{t}(\lambda^{t^-}), \\ T^{xy}(t^+) &= \mathfrak{t}(\lambda^{t^+}) = \mathfrak{t}(\lambda^{t^-} - \theta 1^\dagger). \end{aligned}\right\} \tag{3.15}$$

Thus the jump in stress can be expressed as a function of the jump θ in strain and this function depends on the history λ^{t^-} as a parameter:

$$[\![T^{xy}(t)]\!] := T^{xy}(t^+) - T^{xy}(t^-) = \mathfrak{t}(\lambda^{t^-} - \theta 1^\dagger) - \mathfrak{t}(\lambda^{t^-}) =: \mathfrak{f}(\theta; \lambda^{t^-}). \tag{3.16}$$

The functions $\theta \mapsto \mathfrak{f}(\theta; \lambda^{t^-})$ and $\theta \mapsto \mathfrak{t}(\lambda^{t^-} - \theta 1^\dagger)$ are called *instantaneous response functions* and characterize the instantaneous elasticity of the material. For each relative history λ^t, the *instantaneous modulus* E_1 and *instantaneous second-order modulus* E_2 are defined by

$$E_1 = E_1(\lambda^t) = \frac{\partial}{\partial\theta}\mathfrak{f}(\theta; \lambda^t)\bigg|_{\theta=0}, \qquad E_2 = E_2(\lambda^t) = \frac{\partial^2}{\partial\theta^2}\mathfrak{f}(\theta; \lambda^t)\bigg|_{\theta=0}. \tag{3.17}$$

In view of (3.4) and (3.16), we have

$$E_1(\lambda^t) = \delta\mathfrak{t}(\lambda^{t^-}|1^\dagger), \qquad E_2(\lambda^t) = \delta^2\mathfrak{t}(\lambda^{t^-}|1^\dagger, 1^\dagger), \tag{3.18}$$

and (3.7) and (3.8) yield

$$E_1(\lambda^t) = G(0; \lambda^t). \tag{3.19}$$

It follows from (2.4) that E_1 is an even function of λ^t and E_2 an odd function:

$$E_1(-\lambda^t) = E_1(\lambda^t), \qquad E_2(-\lambda^t) = -E_2(\lambda^t). \tag{3.20}$$

When g in (3.4) has the special form $g(s) = -\kappa s$, *i.e.*, when $g = -\kappa\iota = \lambda^t$, one writes simply $G(s;\kappa)$ for $G(s;g)$, $G'(s;\kappa)$ for $G'(s;g)$, and $E_i(\kappa)$ for $E_i(\lambda^t)$ in the above equations, *e.g.*,

$$E_1(\kappa) = \delta\mathfrak{t}(-\kappa\iota|\,1^\dagger) = E_1(-\kappa), \quad E_2(\kappa) = \delta^2\mathfrak{t}(-\kappa\iota|\,1^\dagger, 1^\dagger) = -E_2(-\kappa). \tag{3.21}$$

As E_2 is an odd function,

$$E_2(0) = 0; \tag{3.22}$$

i.e., the instantaneous second-order modulus E_2 is zero for a fluid that has never been sheared. Let us here assume

$$\left.\frac{d}{d\kappa}E_2(\kappa)\right|_{\kappa=0} \neq 0. \tag{3.23}$$

Thermodynamics requires that the instantaneous first-order modulus E_1 be not negative for a fluid that has not been sheared, *i.e.*, $E_1(0) \geq 0$. For simplicity, let us assume that

$$E_1(\lambda^t) > 0 \tag{3.24}$$

for all λ^t in $\mathcal{L}_p(k)$.

4. RECTILINEAR SHEAR-ACCELERATION WAVES

In our present subject the word *wave* refers to a propagating singular surface. A rectilinear shear wave is, at each instant, a planar surface perpendicular to the x-axis of the fixed Cartesian system in which (3.1) holds. If we write x_t for the value of x on this surface at time t, the velocity of the wave is

$$c = c(t) = \frac{d}{dt}x_t. \tag{4.1}$$

It is usual to assume that $v(x,t)$ and all its derivatives $\partial_t^i\partial_x^j v(x,t)$ are continuous functions of (x,t) for $(x,t) \neq (x_t,t)$ and that these quantities experience, at worst, jump discontinuities, $[\![\partial_t^i\partial_x^j v]\!]$, across the surface $x = x_t$.

The rectilinear shear waves for which $[\![v]\!] = 0$, but $[\![\partial_t v]\!] \neq 0$ and $[\![\partial_x v]\!] \neq 0$ are called *shear-acceleration waves*.

The velocity $c(t)$ of a shear-acceleration wave propagating in a fluid obeying the postulate of smoothness laid down in the previous section obeys the formula [1,21],

$$c(t)^2 = E_1(\lambda^t)/\rho, \tag{4.2}$$

in which $E_1(\cdot)$ is as in (3.17)–(3.19), and λ^t is the history up to t (at x_t) of the relative shear strain; *i.e.*, $E_1(\lambda^t)$ is the instantaneous modulus at the wave. (The $\mathcal{L}_p(k)$-valued function $(x,t) \mapsto \lambda^t$ is continuous across such a wave.)[7]

The goal of the research reported in [1] was to derive an exact formula for the amplitude

$$a = a(t) = [\![\partial_t v]\!] \tag{4.3}$$

of a shear-acceleration wave assuming that the wave is propagating into a region undergoing a steady rectilinear shearing flow. Thus, taking $c(t)$ to be positive, it was assumed that for $x \geq x_t$, $t \geq 0$, and $-\infty < \sigma \leq t$

$$v^x(x,\sigma) = 0, \qquad v^y(x,\sigma) = v(x), \qquad v(x,\sigma) = 0, \tag{4.4}$$

and hence ahead of the wave, for all $\sigma \leq t$, the history λ^σ is given by $\lambda^\sigma(s) = -\kappa s$ with $\kappa = v'(x)$, and at the wave,

$$\lambda^t(s) = -\kappa_t s \qquad (0 \leq s < \infty) \tag{4.5}$$

with

$$\kappa_t = v'(x_t). \tag{4.6}$$

In such a case, (4.2) becomes, in the notation of equation (3.21),

$$c(t)^2 = E_1(\kappa_t)/\rho. \tag{4.7}$$

It was shown that the quantity,

$$b = b(t) = a(t)c(t)^{1/2}, \tag{4.8}$$

obeys a differential equation of Bernoulli type,

$$\frac{db}{dt} + \mu(t)b + \varphi(t)b^2 = 0, \tag{4.9}$$

with

$$\mu(t) = -\frac{G'(0;\kappa_t)}{2E_1(\kappa_t)}, \qquad \varphi(t) = \frac{E_2(\kappa_t)}{2E_1(\kappa_t)c(t)^{3/2}}. \tag{4.10}$$

---

7 See [22], pp. 253, 254.

Solution of this equation gave the following theorem [1, p. 178]: *The amplitude $a(t)$ of a shear-acceleration wave which since time $t = 0$ has been advancing into a region undergoing a steady rectilinear shearing flow* (4.4) *is given by the explicit formula*

$$a(t)c(t)^{1/2} = \frac{a(0)c(0)^{1/2}\exp(-\Psi(t))}{1 + a(0)c(0)^{1/2}\int_0^t \varphi(\sigma)\exp(-\Psi(\sigma))d\sigma}, \qquad \Psi(t) = \int_0^t \mu(\sigma)\,d\sigma \quad (4.11)$$

with μ and φ as in (4.10).

In the same paper it was observed that methods developed by Coleman, Greenberg, and Gurtin [23, §3] can be employed to show that if the motion of the fluid ahead of the wave is a rectilinear shearing flow that is *not* steady, *i.e.*, if v^y in (4.4) depends on time, then (4.9) and (4.11) still hold, but μ and φ are not given by (4.10).

The implications of equation (4.11) are particularly transparent in the case in which the motion ahead of the wave is a simple shearing flow so that (4.4) becomes

$$v^x(x,\sigma) = 0, \qquad v^y(x,\sigma) = V + \kappa_0 x, \qquad v^z(x,\sigma) = 0, \qquad (4.12)$$

for all $x \geq x_t$, $t \geq 0$, and $-\infty < \sigma \leq t$. Here κ_0 is independent of x, and $\kappa_t = \kappa_0$ for $t \geq 0$. We then have

$$c = c(\kappa_0) = (E_1(\kappa_0)/\rho)^{1/2} = \text{const.}, \qquad (4.13)$$

and (4.11) reduces to

$$a(t) = \frac{\Lambda}{\left(\frac{\Lambda}{a(0)} - 1\right)e^{\mu t} + 1} \qquad (4.14a)$$

with

$$\mu = \mu(\kappa_0) = \frac{-G'(0;\kappa_0)}{2E_1(\kappa_0)} = \text{const.}, \qquad (4.14b)$$

$$\Lambda = \Lambda(\kappa_0) = \frac{c(\kappa_0)G'(0;\kappa_0)}{E_2(\kappa_0)} = \text{const.}; \qquad (4.14c)$$

$E_1(\kappa_0)$ is the initial value, and $G'(0;\kappa_0)$ the initial slope, of the stress relaxation function for shearing perturbations about κ_0. By (3.24), $E_1(\kappa_0;0) > 0$; let us also assume $G'(0;\kappa_0) < 0$, so that $\mu(\kappa_0) > 0$.

When $E_2(\kappa_0) = 0$, as is the case when $\kappa_0 = 0$, (4.14a) reduces to

$$a(t) = a(0)e^{-\mu t}. \qquad (4.15)$$

In particular, the amplitude of a shear-acceleration wave propagating into a region that has not been sheared previously decays to zero exponentially.

In view of (3.23), there will be a range of values of the rate of shearing for which

$$\kappa_0 \neq 0 \implies E_2(\kappa_0) \neq 0. \tag{4.16}$$

When $E_2(\kappa_0) \neq 0$, the number $|\Lambda(\kappa_0)|$ is finite and plays the role of a critical amplitude. Because we here take c to be positive and have $\mu(\kappa_0) > 0$, if $|a(0)| < |\Lambda(\kappa_0)|$ or if $a(0)E_2(\kappa_0) > 0$, then $a(t) \to 0$ monotonically as $t \to \infty$. If $a(0) = \Lambda(\kappa_0)$, then $a(t) = a(0)$. But, *if both* $|a(0)| > |\Lambda(\kappa_0)|$ *and* $a(0)E_2(\kappa_0) < 0$, *then* $|a(t)| \to \infty$ *monotonically and in a finite time* t_∞, *given by*

$$\begin{aligned} t_\infty &= \frac{-1}{\mu(\kappa_0)} \ln\left(1 - \frac{\Lambda(\kappa_0)}{a(0)}\right) \\ &= \frac{2G(0;\kappa_0)}{G'(0;\kappa_0)} \ln\left(1 - \frac{c(\kappa_0)G'(0;\kappa_0)}{E_2(\kappa_0)a(0)}\right). \end{aligned} \tag{4.17}$$

Thus, for a shear-acceleration wave propagating into a region undergoing the simple shearing motion (4.4), we have the following result: Although it is impossible for such a wave to grow in amplitude when $\kappa_0 = 0$, the wave can achieve infinite amplitude in finite time if $\kappa_0 \neq 0$, provided $[\![\partial_t v]\!]$ (or $-c[\![\partial_x v]\!]$) is of proper sign and exceeds in magnitude the critical amplitude $|\Lambda|$. One expects that for κ_0 near to zero, this critical amplitude, which is infinite for $\kappa_0 = 0$, will decrease as the magnitude of κ_0 is increased.

If the region ahead of the wave is undergoing a steady rectilinear shearing flow for which $v'(x)$ does not reduce to a constant κ_0 independent of x, the formula (4.11) does not reduce to (4.14) but can be analyzed[8] if $E_1(\kappa)$, $E_2(\kappa)$, and $G'(0;\kappa)$ are known as functions of κ and $\kappa = v'$ is specified as a function of x as it is in channel flow for which (2.21) and (2.22) hold and $\kappa = -\tau^{-1}(\alpha x)$. In channel flow, $|\kappa|$ is a maximum at the bounding surfaces $x = \pm\frac{1}{2}d$. From this and the observations made above for simple shearing flow, one expects that a shear-acceleration wave propagating into a fluid undergoing steady channel flow is more likely to attain infinite amplitude when it is near rather than far from the bounding surfaces. It is conjectured, but a proof is lacking, that the unbounded growth of the magnitude of $[\![\partial_t v]\!]$ as $t \to t_\infty$ signifies the appearance of a jump discontinuity in v, *i.e.*, the appearance of a shear-shock wave.

[8] As in [23, §3].

5. FORMATION OF SINGULARITIES IN INITIALLY SMOOTH MOTIONS

The results on the amplitude of acceleration waves suggest that when the functional $\mathfrak{t}$ obeys the postulate of fading memory adopted here, the solutions of equation (3.1) corresponding to smooth initial data with $|\partial_x v|$ or $|\partial_t v|$ large will be such that these quantities become infinite in finite time. Until recently, the only proven theorems of this type were for very special cases of fluids with fading memory.

Among the fluids obeying the postulate of fading memory as stated in Section 3 are those for which the functional $\mathfrak{t}$ has the form

$$\mathfrak{t}(\lambda^t) = \sum_{n \text{ odd}} \int_0^\infty \cdots \int_0^\infty K_n(s_1, \ldots, s_n)\lambda^t(s_1)\ldots\lambda^t(s_n)\,ds_1\ldots ds_n \tag{5.1}$$

with each function K_n decaying rapidly to zero for large values of its arguments s_i, $i = 1, \ldots, n$, and with these functions tending to zero in an appropriate sense as $n \to \infty$. The postulate is obeyed also by BKZ fluids [24], for which

$$\mathfrak{t}(\lambda^t) = \int_0^\infty H(\lambda^t(s), s)\,ds = \int_{-\infty}^t H\left(\gamma(\tau) - \gamma(t), t - \tau\right)\,d\tau, \tag{5.2}$$

provided H is sufficiently smooth and $|H(a,b)|$ decays sufficiently rapidly with increasing b at fixed a and does not grow too rapidly with increasing $|a|$ at fixed b.[9] Also obeying the postulate are the extensions of the BKZ theory proposed recently by Coleman and Zapas [25].

Slemrod [26] was the first to obtain a theorem showing the non-existence of global smooth solutions to the dynamical equation (2.8) for rectilinear shearing flows of simple fluids under the postulate of fading memory. He considered the (albeit unlikely) special case of (5.1) in which

$$K_n(s_1, \ldots, s_n) = -\beta^n h_n e^{-\beta(s_1 + \cdots + s_n)}, \qquad (n = 1, 3, 5, \ldots) \tag{5.3}$$

which yields

$$T^{xy}(x,t) = h\left(\int_0^\infty e^{-\beta s}\partial_x v(x, t-s)\,ds\right) \tag{5.4}$$

with

$$h(\xi) = \sum_{n \text{ odd}} h_n \xi^n, \qquad h(-\xi) = -h(\xi), \tag{5.5}$$

[9] For BKZ fluids, the existence of $\delta^2\mathfrak{t}$ in (3.4) requires that p in (3.3) be ≥ 2.

and he studied finite perturbations $\hat{v}$ of steady simple shearing flow. Thus he considered velocity fields v of the form,

$$v(x,t) = Vx/d + \hat{v}(x,t), \tag{5.6}$$

for which (5.4) becomes

$$T^{xy}(x,t) = h\left(\int_0^\infty e^{-\beta s}\partial_x \hat{v}(x,t-s)\,ds + \frac{V}{\beta d}\right) \tag{5.7}$$

It was assumed that $\hat{v}(x,t)$ is given as a smooth function for $0 \le x < d$ and $-\infty < t \le 0$. Slemrod's analysis was based on the observation that after the change of variables

$$\left.\begin{aligned} u(x,t) &= \hat{v}(x,t) - \beta\int_0^\infty e^{-\beta s}\hat{v}(x,t-s)\,ds, \\ w(x,t) &= \partial_x v(x,t), \end{aligned}\right\} \tag{5.8}$$

the functional-differential equation (3.1) with $\alpha = 0$ becomes

$$\partial_t w = \partial_x u, \qquad \partial_t u = \partial_x \hat{h}(w) - \beta u \tag{5.9}$$

with

$$\hat{h}(\zeta) = \frac{1}{\rho}\left[h\left(\zeta + \frac{V}{\beta d}\right) - h\left(\frac{V}{\beta d}\right)\right]. \tag{5.10}$$

After subjecting (5.9) to the prescribed data,

$$u(d,t) = u(0,t) = 0, \qquad u(x,0) = u_0(x), \qquad w(x,0) = w_0(x), \tag{5.11}$$

and studying the evolution of the Riemann invariants along the characteristic curves of (5.9), he obtained results which may be summarized as follows: *If* $h'(0) > 0$, $h''(0) \neq 0$, *if* $\max_x |u_0(x)|$ and $\max_x |w_0(x)|$ are sufficiently small, *if* $|\partial_x u_0(x)|$ or $|\partial_x w_0(x)|$ are sufficiently large, and *if* appropriate sign conditions, depending on the sign of $\sigma''(0)$, hold for $\partial_x u_0$ and $\partial_x w_0$ where $|\partial_x u_0|$ and $|\partial_x w_0|$ attain their maxima, *then* (5.9), (5.11) has a classical solution with u and w in $C^1([0,d])$ for only a finite time; *i.e.*, there is a number t_∞ for which

$$\lim_{t\to t_\infty} \max_x \Big[|\partial_x \hat{v}(x,t)| + |\partial_t \hat{v}(x,t)|\Big] = \infty. \tag{5.12}$$

This result was extended to channel flows by Hattori [27] and further extended by Gripenberg [28]. (Related results for a fluid of the rate type are given in the recent monograph[10] of Renardy, Hrusa, and Nohel.)

[10] [29, pp. 67–69].

Many results on "blow up" of smooth solutions were subsequently obtained using constitutive equations appropriate for solids. In particular there is the numerical study of Markowich and Renardy [30] and the recent analyses of Dafermos [31], Ramaha [32], and Nohel and Renardy [33], all of which are described in Chapter II, Section 4 of the monograph of Renardy, Hrusa, and Nohel [29].[11]

The method published by Dafermos [31] in 1986 is applicable to Cauchy problems for many types of viscoelastic fluids, including BKZ fluids, and, more generally, fluids obeying the postulate of fading memory. The limitation of available space does not permit me to give Dafermos' treatment the full description it deserves. My discussion must conclude with the observation that, when applied to rectilinear shearing flows of fluids with fading memory, the method developed by Dafermos permits one to show that *if*, for a perturbation $\hat{v}$ of a flow v in which E_1 and E_2 are both positive, one has (i) an appropriate norm of a relative-strain history (constructed from $\hat{v}$) everywhere small initially, (ii) $\sup_x(-\partial_x\hat{v})$ not too large initially, and (iii) $\sup_x(\partial_x\hat{v})$ large enough initially, *then* a classical solution does not exist for all $t > 0$, and hence $\sup_x |\partial_x\hat{v}| \to \infty$ in finite time.

Here, as in the theory of acceleration waves, one expects, but cannot as yet prove,[12] that the blow up of $|\partial_x\hat{v}|$ or $|\partial_t\hat{v}|$ in finite time implies the formation of a shear-shock wave.

Acknowledgments. I am grateful to William J. Hrusa and Daniel C. Newman for help and suggestions. While this article was in preparation my research was supported by the National Science Foundation and by the Donors of the Petroleum Research Fund, administered by the American Chemical Society.

REFERENCES

[1] B.D. Coleman and M.E. Gurtin, *J. Fluid. Mech.* **33**, 165–181 (1968).
[2] H.K. Nason, *J. Appl. Phys.* **16**, 338–343 (1945).
[3] R.S. Spencer and R.E. Dillon, *J. Colloid Sci.* **4**, 241–255 (1949).
[4] J.P. Tordella, *J. Appl. Phys.* **27**, 454–458 (1956).
[5] J.P. Tordella, *Trans. Soc. Rheology* **1**, 203–212 (1957).
[6] J.P. Tordella, *J. Appl. Polymer Sci.* **7**, 215–229 (1963).
[7] J.P. Tordella, in *Rheology, Vol. 5*, F.R. Eirich, ed., pp. 57–92, Academic Press, New York, 1969.
[8] E.B. Bagley, *J. Appl. Phys.* **28**, 624–627 (1957).

[11] See also the short survey [34].

[12] The formation of shocks is strongly suggested by numerical studies. See, in particular, the study of a viscoelastic solid performed by Markowich and Renardy [30].

[9] E.B. Bagley, *J. Appl. Polymer Sci.* **7**, S7–S8 (1963).

[10] J.J. Benbow, R.V. Charley, and P. Lamb, *Nature, London* **192**, 223–224 (1961).

[11] G.V. Vinogradov, A. Ya. Malkin, Yu. G. Yanooskii, E.K. Borisenkova, B.V. Yarlykov, and G.V. Berezhnaya, *J. Poly. Sci. A2* **10**, 1061–1084 (1972).

[12] E.B. Bagley, I.M. Cabott, and D.C. West, *J. Appl. Phys.* **29**, 109–110 (1958).

[13] W. Noll, *Arch. Rational Mech. Anal.* **2**, 197–226 (1958).

[14] B.D. Coleman and W. Noll, *Ann. N.Y. Acad. Sci.* **89**, 672–714 (1961).

[15] B.D. Coleman and W. Noll, *Arch. Rational Mech. Anal.* **3**, 289–303 (1959).

[16] B.D. Coleman, *Arch. Rational Mech. Anal.* **9**, 273–300 (1962).

[17] B.D. Coleman, H. Markowitz, and W. Noll, *Viscometric Flows of Non-Newtonian Fluids*, Springer, Berlin, Heidelberg, New York, 1966.

[18] B.D. Coleman and W. Noll, *Arch. Rational Mech. Anal.* **6**, 355–370 (1960).

[19] B.D. Coleman and W. Noll, *Rev. Mod. Phys.* **33**, 239–249 (1961); Errata, *ibid.*, **36**, 239 (1964).

[20] B.D. Coleman and V.J. Mizel, *Arch. Rational Mech. Anal.* **23**, 87–123 (1966).

[21] B.D. Coleman, M.E. Gurtin, and I. Herrera R., *Arch. Rational Mech. Anal.* **19**, 1–19 (1965).

[22] B.D. Coleman and M.E. Gurtin, *Arch. Rational Mech. Anal.* **19**, 239–265 (1965).

[23] B.D. Coleman, J.M. Greenberg, and M.E. Gurtin, *Arch. Rational Mech. Anal.* **22**, 333-354 (1966).

[24] B. Bernstein, E.A. Kearseley, and L.J. Zapas, *Trans. Soc. Rheology* **7**, 391–410 (1963).

[25] B.D. Coleman and L.J. Zapas, *J. Rheology* **33**, 501–516 (1989).

[26] M. Slemrod, *Arch. Rational Mech. Anal.* **68**, 211-225 (1978).

[27] H. Hattori, *Quart. Appl. Math* **40**, 112–127 (1982).

[28] G. Gripenberg, *SIAM J. Math.* **13**, 954–961 (1982).

[29] M. Renardy, W.J. Hrusa, and J.A. Nohel, *Mathematical Problems in Viscoelasticity*, Longman, Harlow, Essex, and Wiley, New York, 1987.

[30] P. Markowich and M. Renardy, *SIAM J. Numer. Anal.* **21**, 24–51 (1984); Corrigenda, *ibid.*, **22**, 204 (1985).

[31] C.M. Dafermos, *Arch. Rational Mech. Anal.* **91**, 365–377 (1986).

[32] M. Rammaha, *Commun. in Partial Differential Equations* **12**, 243–262 (1987).

[33] J.A. Nohel and M. Renardy, in *Amorphous Polymers and non-Newtonian Fluids*, C. Dafermos, J.L. Ericksen, and D. Kinderlehrer, eds., *IMA Volumes in Mathematics and its Applications* Vol. 6, Springer, New York, *etc.*, pp. 139–152, 1987.

[34] W.J. Hrusa, J.A. Nohel, and M. Renardy, *Appl. Mech. Rev.* **41**, 371–378 (1988).

A HYPERBOLIC-PARABOLIC SINGULAR PERTUBATION PROBLEM WITH MOVING BOUNDARY

Berardino D'Acunto

Dipartimento di Matematica e Applicazioni, Facoltà di Ingegneria,
via Claudio, 21 - 80125 Napoli, Italy.

ABSTRACT

We discuss a hyperbolic-parabolic singular pertubation problem with moving boundary. We show rigorous approximation formulas and prove the convergence of the hyperbolic solution to the corresponding parabolic one.

0. INTRODUCTION

It is well known that classical heat conduction equation carries the paradox of instantaneous propagation of thermal waves, though the disturbance attenuates greatly with the distance. First, Catteneo [1]) and then Vernotte [10]) proposed to modify the heat flow equation using arguments of statistical mechanics. The new equation is hyperbolic and includes an inertial term with a small coefficient ε.

The analysis concerning the approximation of the solutions of these two equations falls in the category of singular pertubation problems, since, for example, the initial value problem requires two data in the hyperbolic case and only one in the parabolic one. In a series of papers Zlamal [11]) - [13]) and Fulks and Guenther [6]) have studied both the Cauchy problem and the initial-boundary value problem. However, a situation with moving boundary has never been treated.

Recently, it has been emphasized the importance of analysing the phenomenon of phase change basing on non-Fourier models and some papers have been published [2) - 5), 7) - 9)]. I proposed to study the singular pertubation problem related to the hyperbolic-parabolic free boundary problem. Preliminarily, the case with a moving (known) boundary must be analysed. This question is developed in the present work.

1. STATEMENT OF THE PROBLEM

Consider the following problem

$$\varepsilon u_{tt} - \alpha u_{xx} + u_t = 0 \ , \ \varepsilon > 0 \ , \ \alpha > 0 \ , \tag{1.1}$$

$$u(x,0) = u_o(x) \ , \ u_t(x,0) = u_1(x) \ , \ x > 0 \ , \tag{1.2}$$

$$u(at,t) = g(t) \ , \ 0 < t < T \ , \ |a| < (\alpha/\varepsilon)^{\frac{1}{2}} \ . \tag{1.3}$$

Under the hypotheses

$$u_o \in C^1(]0,+\infty[), \ u_1 \in C^\circ(]0,+\infty[) \ , \ g \in C^1(]0,T[) \ , T > 0, \tag{1.4}$$

equation (1.1) has a unique weak solution $u \in C^1(\{(x,t): 0<t<T, \ x>at\})$ satisfying initial and boundary conditions (1.2), (1.3).

For our purposes we are interested only to the solution on

$$\Omega_T^\varepsilon = \{(x,t) : 0 < t < T \ , \ at < x < t(\alpha/\varepsilon)^{\frac{1}{2}}\}. \tag{1.5}$$

This solution is given by

$$u(x_o,t_o) = u_1(x_o,t_o) + u_2(x_o,t_o) \ , \ (x_o,t_o) \in \Omega_T^\varepsilon, \tag{1.6}$$

where,

$$u_1(x_o,t_o)=(4\alpha\varepsilon)^{\frac{1}{2}}V(x_o,t_o-t^*,\varepsilon)g(t^*)+\int_0^{t^*} g(t)(2a\varepsilon\partial_t+2\alpha\partial_x-a)V(x_o-at,t_o-t,\varepsilon)dt, \tag{1.7}$$

$$u_2(x_o,t_o)= \tfrac{1}{2}e^{-t_o/2\varepsilon}[u_o(x_2)-u_o(x_1)]+\int_0^{x_2} \{u_o(x)(1-\varepsilon\partial_t)V(x_o-x,t_o,\varepsilon) +$$

$$+ \varepsilon u_1(x)V(x_o-x,t_o,\varepsilon)\}dx -\int_0^{x_1} e^{ax/\alpha a_2}\{u_1(x)V(x_o+a_3x/a_2,t_o+2ax\varepsilon/a_2,\varepsilon) +$$

$$+ u_o(x)a_2^{-1}(1-\varepsilon a_3\partial_t-2a\varepsilon\partial_x)V(x_o+a_3x/a_2,t_o+2ax\varepsilon/\alpha a_2,\varepsilon)\}dx \ . \tag{1.8}$$

In (1.7), (1.8), denoting by I_n the modified Bessel function of order n, it is

$$V(x_o,t_o,\varepsilon) = (4\alpha\varepsilon)^{-\frac{1}{2}}\ e^{-t_o/2\varepsilon} I_o((t_o^2/4\varepsilon^2 - x_o^2/4\alpha\varepsilon)^{\frac{1}{2}}), \tag{1.9}$$

$$t^* = (t_o - x_o(\varepsilon/\alpha)^{\frac{1}{2}})/(1-a(\varepsilon/\alpha)^{\frac{1}{2}}),\ a_1 = (1+a(\varepsilon/\alpha)^{\frac{1}{2}})/(1-a(\varepsilon/\alpha)^{\frac{1}{2}}), \tag{1.10}$$

$$a_2 = 1-a^2\varepsilon/\alpha,\ a_3 = 1+a^2\varepsilon/\alpha,\ x_2 = x_o + t_o(\alpha/\varepsilon)^{\frac{1}{2}},\ x_1 = a_1(t_o(\alpha/\varepsilon)^{\frac{1}{2}} - x_o). \tag{1.11}$$

In the next sections we will compare the solution of (1.1)-(1.3) on Ω_T^ε with the one of the following problem

$$-\alpha U_{xx} + U_t = 0\ ,\ \alpha > 0\ , \tag{1.12}$$

$$U(x,0) = u_o(x)\ ,\ x > 0\ , \tag{1.13}$$

$$U(at,t) = g(t)\ , |a| < (\alpha/\varepsilon)^{\frac{1}{2}},\ 0 < t < T\ . \tag{1.14}$$

Assume hypotheses (1.4) hold and, in addition, that

$$|u_i| \leq C_i\ ,\ i = 1,2\ ,\ |g| \leq C_g\ ,\ C_i, C_g = \text{const.} \tag{1.15}$$

Then the unique solution U of (1.12)-(1.14) on

$$\Omega_T = \{(x,t) : 0 < t < T\ ,\ x > at\} \tag{1.16}$$

is given by

$$U(x_o,t_o) = U_1(x_o,t_o) + U_2(x_o,t_o), \tag{1.17}$$

where

$$U_1(x_o,t_o) = \int_0^{t_o} g(t)(x_o - at_o)(t_o - t)^{-1} E(x_o - at, \alpha(t_o - t))dt, \tag{1.18}$$

$$U_2(x_o,t_o) = \int_0^{+\infty} u_o(x)[E(x_o - x, \alpha t_o) - E(x_o + x, \alpha t_o)e^{ax/\alpha}]dx. \tag{1.19}$$

Here,

$$E(x_o,\alpha t_o) = (4\pi\alpha t_o)^{\frac{1}{2}}\ e^{-x_o^2/4\alpha t_o}\ . \tag{1.20}$$

2. PRELIMINARY ESTIMATES

In this section we discuss some basic inequalities involving the fundamental solutions of the equations (1.1) and (1.12).

Th. 2.1 *There exist two positive constants* χ *and* χ' *such that*
$(4\alpha\varepsilon)^{-\frac{1}{2}} e^{-(x_o - at_o)/2\varepsilon(y-a)}(1-\varepsilon y^2/\alpha)^{-\frac{1}{2}} I_1[(1-\varepsilon y^2/\alpha)^{\frac{1}{2}}(x_o - at_o)/2\varepsilon(y-a)] \leq$

$$\leq \chi\ [(y-a)/4\pi\alpha(x_o-at_o)]^{\frac{1}{2}}\ e^{-y^2(x_o-at_o)/4\alpha(y-a)}, \tag{2.1}$$

with $|y| < (\alpha/\varepsilon)^{\frac{1}{2}}$, $a < x_o/t_o \leq y$,

and, moreover,

$$(4\alpha\varepsilon)^{-\frac{1}{2}}\ e^{-t_o/2\varepsilon} I_o[(1-\varepsilon y^2/\alpha t_o^2)^{\frac{1}{2}} t_o/2\varepsilon] \leq \chi'(4\pi\alpha t_o)^{-\frac{1}{2}}\ e^{-y^2/4\alpha t_o}, \tag{2.2}$$

with $t_o > 0$, $|y| < t_o(\alpha/\varepsilon)^{\frac{1}{2}}$.

Here, $\chi = (2\pi)^{\frac{1}{2}}(2 + 3/e)^{3/2}$, $\chi' = [2\pi(2 + 1/e)]^{\frac{1}{2}}$.

Proof. The result is achieved by employing known inequalities on the Bessel functions, following the methodology introduced in [6]).

3. THE SINGULAR PERTUBATION PROBLEM

Now, we want to compare the solutions $u(x,t)$ and $U(x,t)$ for small ε. In this §, the symbols K_{ij}, H_{ij} $(i,j = 1,2...)$ denote constants which do not depend on ε, x_o, t_o.

Th. 3.1 *Under the assumptions (1.4),(1.15) the hyperbolic solution* u_1 *converges to the parabolic solution* U_1 *as* $\varepsilon \to 0$. *Precisely, there exist three constants* K_i *independent of* ε, x_o, t_o *such that*

$$|u_1-U_1| \leq K_1\varepsilon + K_2\varepsilon^{\frac{1}{2}}(x_o-at_o)^{-1} + K_3[\varepsilon/(x_o-at_o)]^{\frac{1}{2}}.$$

Proof. From (1.7),(1.9),(1.10) and (1.18), (1.20) we have

$$\Big|\int_0^{t^*} g(t)(2a\varepsilon\partial_t - 2\alpha\partial_x - a)V(x_o - at, t_o-t,\varepsilon)dt -$$

$$-\int_0^{t_o} g(t)(x_o - at_o)(t_o - t)^{-1}E(x_o - at, \alpha(t_o - t))dt\Big| = D. \tag{3.1}$$

Assume, first, $x_o/t_o < -\frac{1}{2}(\alpha/\varepsilon)^{\frac{1}{2}}$ and introduce the variable $y=(x_o-at_o)\times$ $\times(t_o-t)^{-1}$. Considering the inequality (2.1) we obtain

$$D \leq D_1 + D_2 + D_3 \tag{3.2}$$

where

$$D_1 = K_{11}\int_{\frac{1}{2}(\alpha/\varepsilon)^{\frac{1}{2}}}^{+\infty} [(x_o-at_o)/(4\pi\alpha(y-a))]^{\frac{1}{2}} e^{-y^2(x_o-at_o)/4\alpha(y-a)}dy, \tag{3.3}$$

$$D_2+D_3 = (C_g\int_{-\frac{1}{2}(\alpha/\varepsilon)^{\frac{1}{2}}}^{\frac{1}{2}(\alpha/\varepsilon)^{\frac{1}{2}}} + C_g\int_{x_o/t_o}^{-\frac{1}{2}(\alpha/\varepsilon)^{\frac{1}{2}}})(x_o-at_o)(y-a)^{-1}|(4\alpha\varepsilon R)^{-\frac{1}{2}} \times$$

$$\times\ e^{-(x_o-at_o)/2\varepsilon(y-a)} I_1[R^{\frac{1}{2}}(at_o-x_o)/2\varepsilon(a-y)] - [(y-a)/4\pi\alpha(x_o-at_o)]^{\frac{1}{2}} \times$$

$$\times\ e^{-y^2(x_o-at_o)/4\alpha(y-a)}|dy\ , \tag{3.4}$$

with $R = 1 - \varepsilon y^2/\alpha$.

From (3.3), with $\rho = y[(x_o-at_o)/4\alpha(y-a)]^{\frac{1}{2}}$, $\rho_1 = \frac{1}{2}(\alpha/\varepsilon)^{\frac{1}{2}}[(x_o-at_o) \times$

$\times\ (4\alpha(\frac{1}{2}(\alpha/\varepsilon)^{\frac{1}{2}}-a)^{-1}]^{\frac{1}{2}}$, we deduce

$$D_1 \leqq K_{12}\int_{\rho_1}^{+\infty} e^{-\rho^2}d\rho + K_{13}\int_{\frac{1}{2}(\alpha/\varepsilon)^{\frac{1}{2}}}^{+\infty} y^{-3}dy \leqq K_{14}\varepsilon^{\frac{1}{2}}/(x_o-at_o) + K_{15}\varepsilon. \tag{3.5}$$

Consider, now, D_2 and note that

$$D_2 \leqq D_{21} + D_{22}\ , \tag{3.6}$$

where

$$D_{21} = C_g\int_{-\frac{1}{2}(\alpha/\varepsilon)^{\frac{1}{2}}}^{\frac{1}{2}(\alpha/\varepsilon)^{\frac{1}{2}}} (x_o-at_o)(y-a)^{-1}(4\alpha\varepsilon R)^{-\frac{1}{2}} e^{-(x_o-at_o)/2\varepsilon(y-a)}|I_1[R^{\frac{1}{2}}(at_o-x_o) \times$$

$$\times\ (2\varepsilon(y-a))^{-1}] - e^{R^{\frac{1}{2}}(x_o-at_o)/2\varepsilon(y-a)}R^{-1/4}[\varepsilon(y-a)/\pi(x_o-at_o)]^{\frac{1}{2}}|dy, \tag{3.7}$$

$$D_{22} = C_g\int_{-\frac{1}{2}(\alpha/\varepsilon)^{\frac{1}{2}}}^{\frac{1}{2}(\alpha/\varepsilon)^{\frac{1}{2}}} [(x_o-at_o)/4\pi\alpha(y-a)]^{\frac{1}{2}}|R^{-3/4}\ e^{[(x_o-at_o)/2\varepsilon(y-a)](R^{\frac{1}{2}}-1)} -$$

$$-\ e^{-y^2(x_o-at_o)/4\alpha(y-a)}|dy. \tag{3.8}$$

From (3.7), using $|I_1(w)-e^w/(2\pi w)^{\frac{1}{2}}| \leqq Le^w/w^{3/2}$, $w > 0$, L=constant, we have

$$D_{21} \leqq K_{21}\varepsilon\int_{-\frac{1}{2}(\alpha/\varepsilon)^{\frac{1}{2}}}^{\frac{1}{2}(\alpha/\varepsilon)^{\frac{1}{2}}} [(y-a)/(x_o-at_o)]^{\frac{1}{2}}\ e^{-y^2(x_o-at_o)/4\alpha(y-a)}dy \leqq$$

$$\leqq K_{22}\varepsilon^{\frac{1}{2}}/(x_o-at_o)^{\frac{1}{2}}. \tag{3.9}$$

From (3.8) we obtain

$$D_{22} \leq D_{221} + D_{222} \tag{3.10}$$

where

$$D_{221} = C_g \int_{-\frac{1}{2}(\alpha/\varepsilon)^{\frac{1}{2}}}^{\frac{1}{2}(\alpha/\varepsilon)^{\frac{1}{2}}} [(x_o - at_o)/4\pi\alpha(y-a)]^{\frac{1}{2}} |e^{[(x_o-at_o)/2\varepsilon(y-a)]^{\frac{1}{2}}(R^{\frac{1}{2}}-1)} \times$$

$$\times (R^{-3/4}-1)|dy, \tag{3.11}$$

$$D_{222} = C_g \int_{-\frac{1}{2}(\alpha/\varepsilon)^{\frac{1}{2}}}^{\frac{1}{2}(\alpha/\varepsilon)^{\frac{1}{2}}} [(x_o - at_o)/4\pi\alpha(y-a)]^{\frac{1}{2}} |e^{[(x_o-at_o)\ /2\varepsilon(y-a)](R^{\frac{1}{2}}-1)} -$$

$$- e^{-y^2(x_o-at_o)/4\alpha(y-a)}|dy. \tag{3.12}$$

Moreover,

$$D_{221} \leq K_{23} \int_{-\frac{1}{2}(\alpha/\varepsilon)^{\frac{1}{2}}}^{\frac{1}{2}(\alpha/\varepsilon)^{\frac{1}{2}}} [(x_o - at_o)/(y-a)]^{\frac{1}{2}}\ e^{-y^2(x_o-at_o)/4\alpha(y-a)}dy \leq$$

$$\leq K_{24}[\varepsilon/(x_o-at_o)]^{\frac{1}{2}}, \tag{3.13}$$

$$D_{222} \leq \pi^{-1}C_g \int_{-\frac{1}{2}(\alpha/\varepsilon)^{\frac{1}{2}}}^{\frac{1}{2}(\alpha/\varepsilon)^{\frac{1}{2}}} [(x_o - at_o)/4\alpha(y-a)]^{3/2}\ e^{-y^2(x_o-at_o)/4\alpha(y-a)}(4/3)^{3/2} \times$$

$$\times (\varepsilon y^4/4\alpha)dy,$$

and, therefore,

$$D_{222} \leq K_{25}[\varepsilon/(x_o-at_o)]^{\frac{1}{2}}. \tag{3.14}$$

Then, we note that

$$D_3 \leq K_{31} \int_{x_o/t_o}^{-\frac{1}{2}(\alpha/\varepsilon)^{\frac{1}{2}}} [(x_o - at_o)/4\pi\alpha(y-a)]^{\frac{1}{2}}\ e^{-y^2(x_o-at_o)/4\alpha(y-a)}dy \leq$$

$$\leq \varepsilon^{\frac{1}{2}}K_{32}(x_o-at_o)^{-1} + K_{33}\ \varepsilon. \tag{3.15}$$

Now, assume $x_o/t_o \geq -\frac{1}{2}(\alpha/\varepsilon)^{\frac{1}{2}}$. By using a similar proceeding it is not difficult to achieve exatly the same result.

Finally, we consider the finite term of (1.7). We easily have

$$|(4\alpha\varepsilon)^{\frac{1}{2}}\ V(x_o,t_o-t^*,\varepsilon)g(t^*)| \leq C_g\ e^{-(x_o-at_o)/4(\alpha\varepsilon)^{\frac{1}{2}}} \leq$$

$$\leq 4C_g(\alpha\varepsilon)^{\frac{1}{2}}/(x_o-at_o). \tag{3.16}$$

In conclusion, we can state that Th. 3.1 follows from (3.1)-(3.16)

An analogous result holds for u_2. Indeed, we can prove the following

Th. 3.2 *Under the same assumptions as in Th. 3.1 also the solution* u_2 *converges to* U_2 *as* $\varepsilon \to 0$. *Indeed, there exist two constants* H_i *independent of* ε, x_o, t_o *such that*

$$|u_2 - U_2| \leqq H_1\varepsilon + H_2\varepsilon/t_o.$$

REFERENCES

1) Cattaneo, C., Atti Sem. Mat. Fis. Univ. Modena, 3 83(1948).

2) D'Acunto, B., Ricerche Mat., 37(1), 122(1988).

3) D'Acunto, B., Meeting "Waves and Stability in Continuous Media", Taormina, 5-10 oct. 1987.

4) De Socio, L.M., and Gualtieri, G., Quart. Appl. Math., 41(2), 253 (1983).

5) Friedman, A., and Hu, B., J. Math. Anal. Appl., 138(1), 249(1989).

6) Fulks, W., and Guenther, R.B., Czechoslovak Math. J., 21(96) 683 (1971).

7) Greenberg, J.M., IMA J. Appl. Math., 38, 1(1987).

8) Sadd, M.H., and Didlake, J.E., J. Heat Transfer, Trans. ASME, 99(2), 25(1977).

9) Salomon, A.D., Alexiades, V., Wilson, B. G., and Drake, J., Quart. Appl. Math., 43, 295(1985).

10) Vernotte, P., C.R. Acad. Sc. Paris, 246(1958).

11) Zlamal, M., Atti Accad. Naz. Lincei Rend. Cl. Sci. Fis. Mat., 27(8), 324(1959).

12) Zlamal, M., Czechoslovak Math. J., 10(85), 83(1960).

13) Zlamal, M. , Differential Eq.s and their Applications. Proc. Conf., Prague, 1962, Academic Press, N.Y., 1963.

GENESIS OF A DRIVEN-DAMPED NON-LINEAR SCHROEDINGER EQUATION

E.D'Ambrogio

Facolta' di Scienze dell'Universita' degli Studi di Trieste
Trieste,Italy

ABSTRACT
For a collisionless 1-D plasma model,a formal derivation of a generalized,cubically non-linear equation of Schrödinger type is discussed to describe deterministic chaos in the modulational instability of coupled waves.The energy influx arising from the additional terms is depicted as a memory effect associated with the beating wave-particle resonances.

1.INTRODUCTION

The slow temporal modulation of electrostatic waves in a warm plasma may be modelled according to the cubically non-linear Schrödinger (NLS)equation.It appears as the adiabatic approximation of the set of equations describing the soliton-caviton structure of the Langmuir condensate [1].The range of validity of the NLS- model,in describing electrostatic turbulence is matter of current research.The integrability of the equation [2],[3],suggests that the model is nonstochastic.On the other hand,dissipative systems,involving cooperative phenomena in active media are known to be chaotically driven unstable by energy influx.

Being interested in the competition between coherence and chaos in collisionless plasmas,the following topics require our attention:

- the influence of the transients and the memory phenomena,
- the role of the interactions between non-linear waves and particles.

A possible approach to these (related)questions will be presented here,within the context of the fixed-phase description [4] of the non-linear wave-interactions in a Vlasov plasma.By means of a reductive perturbation procedure,a two-mode set of improved amplitude equations has been derived in the self-consistent field approximation,in order to describe decay-diffusion problems.As a result ofthe adiabatic elimination procedure,a driven-damped cubically non-linear equation of Schrödinger type emerges,in which additional terms are described as memory effects,arising from resonant wave-particle processes.

2.THE FORMAL APPROACH

We are considering the propagation of a set of finite-amplitude wave-packets $a(x,t) = a_k(x,t)\exp(ikx-iwt)$ + c.c. in a 1-D plasma model,in which ions provide static neutralizing background and electrons are described by Vlasov cquation with selfconsistent electrostatic field.Circular frequency w is expressed as a simple root of dispersion relation for plasma waves.The amplitudes are (complex) slowly-varying functions of x(space) and t(time). For each mode a_k ,the envelope equation for slow processes,in compact form ,may be written as

$$D_k^{(n)}(k-i\partial_x,w+i\partial_t)a_k = B_k + N_k \quad . \tag{1}$$

Here n is the order of the truncated differential operator D in the (x,t)- space,whereas B and N refer to the initial-external sources and,respectively,to the non-linear contribution to the field strength.

Our approach relies on an abridged version of the multiple-time-spacc scale method and involves truncation to a finite number of (Fourier)modes in the treatment of the fully non-linear Vlasov-Poisson problem.An important point here is the consideration of the du al decay of the electric field in collisionless plasmas.This is an old question [5],involving aspects of intrinsic interest as well as observations in laboratory plasmas [6],[7].We shall remove the assumption that free-streaming terms phase-mix to zero.The corresponding contribution to the asymptotic (linearized) Vlasov disturbance,allows for an approximate estimate of the non-linear resonant wave-particle process within the context of the coherent (fixed-phase-)description of the non-linear wave-coupling theory.Assuming convergence,separate terms of the formal expansion of the electronic distribution perturbation are written as ascending powers of the amplitudes

$$f_k^{(n)} = \delta_{n1}(b_k + m_k a_k) + (1-\delta_{n1})g_k^{(n)} \quad , (n=1,2,..) \tag{2}$$

where, δ is the Kronecker delta,

$$b_k = f_k(0,v)\exp(-it/r), \quad f(0)=f(x(t=0),t=0), r=1/(kv-w) \tag{3}$$

and

$$m_k - m_k^{(o)} = \sum_{n=1} m_k^{(n)} = \sum_{n=1} (1/n!)(i\partial_w m_k^{(o)}\partial_t - i\partial_k m_k^{(o)}\partial_x)^{(n)}. \quad (4)$$

According to our assumption, free-streaming term in quantity

$$m_k^{(o)} = -irF'(v)(1-\exp(t/ir)) \quad (4')$$

has been retained. The fully non-linear Vlasov-Poisson response, in presence of n-order dispersive effects, may be cast into form (1) by putting

$$D_k^{(n)} = ik + \partial_x + M_k^{(n)}, \quad M_k^{(n)} = \Omega^2 \sum_{n=0}^{n} \int dv \, m_k^{(n)}, \quad (5)$$

$$B_k = \Omega^2 \int dv \, f_k(0,v)\exp(t/ir), \quad (\Omega\text{=Langmuir frequency}). \quad (6)$$

For a three-mode system, h,q,k(=h+q), the n-order non-linear contribution reads

$$N_k^{(n)} = [h,q]\, \Omega^2 \int dv \, S_k \partial_v (m_h a_h + (1-\delta_{n1}) \sum_{n'=2}^{n} g_h^{(n')}) a_q, \quad (7)$$

(n=1,2..)

where [-,-] denotes symmetrization w.r.to the arguments,

$$S_k = -\int_0^t dt' \exp(-ist'+i(t'-t)/r), \quad (8)$$

$s=w_h+w_q-w$ is the detuning parameter and

$$g_h^{(n)} = [-q,k]\, S_{-q} \partial_v (\delta_{n2}(b_k + m_k a_k) + (1-\delta_{n2}) g_k^{(n-1)}) a_k. \quad (9)$$

(n=2,3..)

Linear coefficients like $\int dv \, S_p \, \partial_v \, b_j$ have been dropped for secular regularization[8].
Eq.(1), together with the corresponding ones to h and q modes provides a set of p.d.e., whose order and nonlinearities will depend on the truncation criteria, as dictated by dispersion properties and existing inhomogeneities. The subsequent discussion will be confined to the application which motivated this investigation, namely the onset of NLS-systems within the context of the decay instability [9].
Accordingly, we will retain on the L.H.S. of eq.(1), first-order in time and second-order in space derivatives, whereas, in the non-linear terms on the R.H.S. we will limit ourselves to the consideration of the lowest

significant contribution to dispersive effects as given by quantity (4').The resulting two-mode model equations describing decay-in-time and diffusion-in-space processes,are given by the following set of coupled equations of parabolic type

$$
\begin{aligned}
(D_1^{(2)} - C_1)a_1 &= 2\, M_{13}\, a_3^{\,2} + 2\, B_{13-3}\, a_1 a_3 \\
(D_3^{(2)} - C_3)a_3 &= (M_{31} + M_{3-3})a_1 a_{-3} + B_{3-33}\, a_1 a_{-1} + B_{3-3-1}\, a_1 a_3 \, .
\end{aligned}
\qquad (10)
$$

The contraction of the original three-body problem has been obtained by taking q+k=0 ,with the notation $h=k_1$ and $k=k_3$. Transients B as given by (6), have been disregarded for simplicity of the presentation.With regard to coefficients : the occurrence of

$$
M_{pj} = \Omega^2 \int dv\; S_p\, \partial_v m_j^{(\circ)} \qquad (11)
$$

is well known in the applications of the quadratically non-linear coupled-mode theory.As a result of third-order calculations , we get

$$
\begin{aligned}
C_1 &= (M_{13-3} + 2\, M_{131} + M_{13-1})a_3 a_{-3} \\
C_3 &= (M_{3-3-1} + M_{3-33})a_1 a_{-1} + 2\, M_{313}\, a_3 a_{-3}
\end{aligned}
\qquad (12)
$$

where

$$
\begin{aligned}
(M,B)_{pjh} &= \Omega^2 \int dv\; S_p \partial_v S_j \partial_v (m^{(\circ)}, b)_{-h} \;, \quad (h \neq k) \\
(M,B)_{pjp} &= \Omega^2 \int dv\; S_p \partial_v S_j \partial_v (m^{(\circ)}, b)_p
\end{aligned}
\qquad (13)
$$

are,in general,time-periodic quantities.Calculations show that factors M_{pjl} depend on both,the wave-wave coupling mechanism as well as resonant wave-particle effects,arising from the beating term in (4').Coefficients B_{pjl} are determined solely from the interactions of the latter type,involving the initial excitation rather than the equilibrium distribution function ,as shown by the coefficients M_{pjl} .In absence of the wave-particle processes,terms B_{pjl} vanish and M_{pjl} provide the non-linear frequency-shift as given by the conventional (coupled- mode) approach.

3.THE ADIABATIC APPROXIMATION

The history of the non-linear decay of a linearly growing amplitude a_1 to a linearly damped mode a_3 may be followed by taking the damping rate larger and larger again [10].To that extent, it is rather interesting to consider the adiabatic limit,within the context of the improved equations (10),with mode a_1 locked , in order to study the long-time behavior of mode a_3 .This signifies

$$D_1^{(2)} = 0 \tag{14}$$

in the first equation of set (10).If we confine ourselves , for simplicity's sake,to the consideration of the primary resonances in the wave-particle effects, as defined by $v = w_j/k_j$,j=1,3 , and neglect contributions arising from higher-order satellites, as given, e.g.,by the resonance $v = (w_1 + w_3)/3k_3$, the corresponding term vanishes,i.e.

$$B_{13-3} = 0 \quad , \tag{15}$$

and we are left with the following equation

$$(\partial_t + iP\,\partial^2_{xx} + \Gamma(t))a_3 + iQa_3^2\, a_{-3} = B\exp(-ist/2) \quad , \tag{16}$$

where P,Q and B are constants, whereas Γ(t) is a (complex) quantity having periodically modulated components,beating at the frequencies s/2 and s/3.In recent years much attention is being paid to computational simulations of NLS models, in investigations of non-linear dynamics and chaos, [11],[12].According to the structure of the transport coefficients B and Γ,and the physical meaning of coupling factors (12),(13),we infer that memory effects play a role in the competition between the coherent and chaotic response of a fully non-linear Vlasov-Poisson plasma.Within the context of the present approach , memory effects are mirrored by the recurrent-stochastic properties of a single -mode equation,modelled according to a generalized,cubically NLS equation,in which driving frequency is fixed by the detuning parameter s ,and forcing amplitude by the (saturated) energy level and the initial conditions.

ACKNOWLEDGMENT.
This work was performed as a part of the research program MPI,60% and 40% ,"Problemi di Evoluzione nei Fluidi e nei Solidi", under the auspices and partial support of the Gruppo Nazionale Fisica Matematica del C.N.R.

REFERENCES

1-Zakharov,V.E.,Zh.Eksp.Teor.Fiz.,62,1745 (1972); Sov.Phys. JETP 35,908 (1972)
2-Zakharov,V.E. and Shabat,A.B., Zh.Eksp.Teor.Fiz. 61,118 (1971); Sov.Phys.JETP 34,62 (1972)
3-Tracy,E.R.,Chen,H.H.and Lee,Y.C., Phys.Rev.Lett.53, 218 (1984)
4-Weiland J.and Wilhelmsson H.,"Coherent non-linear interaction of waves in plasmas",Pergamon Press 1977
5-Landau L.,J.Phys.USSR 10,25 (1946)
6-Krall,N.A.and Trivelpiece ,A.W.,"Principles of plasma physics",Mc-Graw Hill 1973,Chapters 8,10
7-Tidman ,D.A. and Krall,N.A.,"Shock waves in collisionless plasmas",J.Wiley Interscience,1971,Chapter 8
8-D'Ambrogio ,E.,Meccanica (to be published)
9-Davidson,R.C.,"Methods in nonlinear plasma theory ", Academic Press,1972,Chapter 6
10-Wersinger,J.M.,Finn,J.M. and Ott.,E.,Phys.Fluids 23,1142 (1980)
11-Bekki,N.and Nozaki,K.,in "Dynamical Problems in Soliton Systems",268-271,Ed.Takeno,Springer V.1985
12-Guo-Zheng Sun,Nicholson,D.R.and Harvey A.Rose,Phys.Fluids 29,1011 (1986)

STABILITY PROPERTIES FOR AN ALMOST PERIODIC EQUATION UNDER PERSISTENT PERTURBATIONS

ARMANDO D'ANNA(*)   ANNA MARIA MONTE (**)

1-.Introduction.

It is known that it is possible to obtain stability properties for an ordinary differential equation in $\mathbb{R}^n$

(1.1) $$x'=f(t,x)$$

by using techniques of topological dynamics. So, by introducing an appropriate topology [1,8], it is possible to deduce local stability properties and the asymptotic behavior of the solutions of (1.1) from analogous qualitative properties of the solutions of all the limiting equations

(1.2) $$x'=f^*(t,x) \quad , \quad f^*\in\Omega_f^* \ .$$

of (1.1) [1-14].

When f is Bohr almost periodic (B.a.p.) in t and the limit set Ω_f^* is obtained using the Bohr topology (B.t.), then it is proven [12] that,to get total stability properties for (1.1), it is sufficient to consider only an equation (1.2).

In this paper we analyse some stability properties in the large for (1.1) when f is B.a.p. and there are constantly acting perturbations.

(*) Dipartimento di Matematica e Applicazioni "R.Caccioppoli"- via Claudio 21, 80125 Napoli.

(**) Istituto di Matematica- Universita' della Basilicata.

2- Notations, definitions, preliminary results.

2.1 We suppose $f\in C(I\times\mathbb{R}^n;\mathbb{R}^n)$ where $I=[0,\omega[$ and $C(I\times\mathbb{R}^n;\mathbb{R}^n)$ denotes the collection of all continuous functions defined on $I\times\mathbb{R}^n$ with values in $\mathbb{R}^n$; f is *admissible* [1] if for every $(s,y)\in I\times\mathbb{R}^n$ there exists a unique non continuable solution $x(t;s,y;f)$ of (1.1) satisfying the initial condition $x(s;s,y;f)=y$.

For every $\tau\in I$, we indicate by $f_\tau(t,x)=f(\tau+t,x)$ the *translate* of the function $f(t,x)$ and put $\mathfrak{F} = \{f_\tau:\tau\in I\}$.

When f is Bohr almost periodic in t, we denote with $\mathfrak{H}(f)$ the closure of the set $\mathfrak{F}$ in the Bohr topology (B.t.) and by $\Lambda^*(f)$ the set

$\Lambda^*(f)=\{f^*|$there is a sequence $\{\tau_n\}\to\infty:\{f_{\tau_n}\}\to f^*$ in the B.t.$\}$.

If every f^* is admissible, f is said to be *regular*.

For the set $A\subset\mathbb{R}^n$ and the positive number ε, we define the ε-neighborhood of A, $A(\varepsilon)=\{y\in\mathbb{R}^n|d(x,A)<\varepsilon\}$.

By $|.|$ we indicate any norm in $\mathbb{R}^n$.

Let now $\phi:I\to\mathbb{R}^n$ and $\psi:I\to\mathbb{R}^n$ be respectively two solutions of (1.1) and (1.2) contained in the compact set $K\subset\mathbb{R}^n$.

With respect to a notation of [6], we put $(\psi,f^*)\in\Lambda^*(\phi,f)$ if there is a sequence $\{\tau_n\}\to\infty$ such that $\{f_{\tau_n}\}\to f^*\in\Lambda^*(f)$ in the B.t. and $\{\phi_{\tau_n}\}\to\psi$ uniformly on any compact set of I.

Finally, we consider perturbed systems of (1.1) and (1.2)

(2.1) $\quad x'=f(t,x)+p(t)+q(t,x) \qquad p(t)\in C(I),$

(2.2) $\quad x'=f^*(t,x)+p(t)+q(t,x) \qquad q(t,x)\in C(I\times\mathbb{R}^n).$ ∎

2.2- Definition. *The solution* $\phi(t)\in K$ *of* (1.1) *is said to*

be integrally totally stable **(I.T.S.)** *if for any* $\varepsilon > 0$ *there exists a* $\delta(\varepsilon) \in]0,\varepsilon[$ *such that, if* $p(t) \in C(I;\mathbb{R}^n)$ *and* $q(t,x) \in C(I\times\mathbb{R}^n;\mathbb{R}^n)$ *and verify*

$$(2.3) \qquad \int_t^{\infty}|p(\tau)|d\tau + |q(x,t)| < \delta \quad \text{for } t\geq s \text{ , } x \in K(\varepsilon),$$

then for $s\in I$, $|y-\phi(s)| < \delta$, *every solution* $x(t;s,y;f+p+q)$ *of* **(2.1)** *satisfies* $|x(t;s,y;f+p+q) - \phi(t)| < \varepsilon$ *for all* $t \geq s$.■

2.3- Definition. *The solution* $\phi(t)\in K$ *of* **(1.1)** *is said to be eventually integrally totally attractive in the large* **(E.I.T.A.L.)** *if for any* $\alpha>0$ *there exists* $\beta(\alpha)>\alpha$ *and besides for any* $\nu>0$ *there are* $\sigma(\alpha,\nu)>0$, $T(\alpha,\nu)>0$, $\eta(\alpha,\nu)>0$ *such that for* $s\geq\sigma$, $|y-\phi(s)|\leq\alpha$, *and if* $p(t)\in C(I,\mathbb{R}^n)$, $q(t,x)\in C(I\times\mathbb{R}^n,\mathbb{R}^n)$ *and verify*

$$\int_t^{\infty}|p(\tau)|d\tau + |q(t,x)| < \eta \quad \text{for } t\geq s \text{ and } x\in K(\beta),$$

then any solution of **(2.1)** *satisfies*

$$|x(t;s,y;f+p+q) - \phi(t)| < \nu \quad \text{for } t\geq s+T.$$■

With reference to the equation (1.1) and to the perturbed equation (2.1), we have the following lemma.

2.4- Lemma. *Suppose that the hull* $\mathfrak{H}$ **(f)** *of* **f** *is compact and* **f** *regular. Let* **T** *be a positive number,* **C** *and* **K** *two compact sets of* $\mathbb{R}^n$ *such that* $C\subset\overset{\circ}{K}$. *Moreover, admit that* $x(t;s,z;f)\in K$ *when* $z\in C$. *Then, for each* $\varepsilon>0$ *there exists a* $\delta(\varepsilon)\in]0,\varepsilon[$ *such that, if* $p(t)\in C(I)$ *and* $q(t,x)\in C(I\times\mathbb{R}^n)$ *and verify*

$$\int_t^{\infty}|p(\tau)d\tau + |q(t,x)|<\delta \quad \text{for } t\in[s,s+T],\ x\in K(\varepsilon),$$

we have $|x(t;s,y;f+p+q) - x(t;s,z;f)|< \varepsilon$ *for* $s\in I$, $t\in[s,s+T]$, $|y-z| < \delta$.

The proof of this lemma is obtained generalizing some results contained in [6].■

2.5- Lemma. *Let us suppose that the hull of* **f** *is compact and* **f** *regular. If* $\mathbb{R}^n$ *is region of attraction for a bounded solution* $\phi(t)$ *of* **(1.1)**, *then for any* $\alpha>0$ *there exists a number* $\beta(\alpha)>\alpha$ *such that*

$$|x(t;s,y;f)-\phi(t)|\leq\beta \quad \textit{for} \quad s\in I,\ |y-\phi(s)|\leq\alpha. \tag{2.4}$$

Proof. Let us assume that there is no β that satisfies (2.4). Then, if $\{\beta_m\}$ is a divergent sequence of positive real numbers with $\beta_m\geq\beta>0$, there exist $\alpha>0$ and the sequences $\{s_m\}\subset I$ and $\{y_m\}\subset K$ (compact set containing $\phi(t)$), with

$$|y_m-\phi(s_m)|\leq\alpha \quad \text{and} \quad \{T_m\}\subset I ,$$

such that

$$|x(s_m+T_m;s_m,y_m;f) - \phi(s_m+T_m)|\geq\beta_m . \tag{2.5}$$

Because $\phi(t)$ is attractive, there exists $T(\alpha,\beta)>0$ such that

$$s\in I,\ |y-\phi(s)|\leq\alpha,\ t\geq T \ \Rightarrow\ |x(t+s;s,y;f)-\phi(t+s)|<\beta.$$

Hence $T_m<T$ for every m. Without loss of generality, we can suppose $y_m\to\hat{y}\in K$.

Now we consider the following alternative: $\{s_m\}$ bounded, $\{s_m\}$ unbounded.

Admit $\{s_m\}\to\hat{s}$. By the continuous dependence on initial values, there exists a sequence of solutions $\{x(t+s_m;s_m,y_m;f)\}$ converging to the solution $x(t+\hat{s};\hat{s},\hat{y};f) = x(t)$ on compact intervals of I. Let be now $[t_1,t_2]$ a compact interval such that $[\hat{s},\hat{s}+T]\subset(t_1,t_2)$ and $M>0$ be a number for which $|x(t)-\phi(t)|\leq M$ for every $t\in[t_1,t_2]$. The assumed convergence implies that given $\varepsilon>0$ there is a positive integer m_ε such that

$$|x(t+s_m;s_m,y_m;f) - x(t)| < \varepsilon \quad \forall m\geq m_\varepsilon ,\ \forall t\in[t_1,t_2]. \tag{2.6}$$

By choosing m_ε in such a way that $s_m+T_m\in[t_1,t_2]$ and $\beta_m>M+\varepsilon$ for every $m\geq m_\varepsilon$, employing (2.6), we deduce

$$|x(s_m+T_m;s_m,y_m;f) - \phi(s_m+T_m)| \le M+\varepsilon<\beta_m$$

that contradicts (2.5).

On the other hand, if the sequence $\{s_m\}$ is unbounded, by considering $\{f_{s_m}\}$, we get the uniform convergence on each compact interval contained in I of the sequences of solutions $\{x(s_m+t;s_m,y_m;f) = x(t;0,y_m;f_{s_m})\}$ and $\{\phi(s_m+t)\}$ of equation (1.1) respectively to the unique solution $x(t;0,\hat{y};f^*)$ and to $\psi(t)$ of equation (1.2). Then, given $\varepsilon>0$, there exists an integer $m_\varepsilon>0$ such that

$$|x(s_m+t;s_m,y_m;f) - x(t;0,\hat{y},f^*)| < \varepsilon \text{ for } m\ge m_\varepsilon \text{ and } t\in[0,T].$$

Let M>0 be chosen such that

$$|x(t;0,\hat{y};f^*) - \phi(s_m+t)| \le M \quad \text{for } m\ge m_\varepsilon \quad \text{and } t\in[0,T].$$

Then

$$|x(s_m+t;s_m,y_m;f) - \phi(s_m+t)| \le \varepsilon+M$$

and for $\beta_m>\varepsilon+M$ we obtain the absurd again.∎

3-. The main results

First of all we demonstrate the following

3.1- Theorem. *Suppose that the solution* $\phi(t)$ *of* (1.1) *is* I.T.S.. *Then the solution* $\psi(t)$ *of* (1.2), *where* $f^*\in\Lambda^*(f)$, *is* I.T.S..

Proof. Let be $\{t_k\}\to\infty$ as $k\to\infty$ a sequence such that $\{f_{t_k}\}\to f^*$ in the B.t.. Because $\phi(t)$ is I.T.S., for every $\varepsilon>0$ there exists a $\delta(\varepsilon)\in]0,\varepsilon[$ such that, if $p(t)\in C(I,\mathbb{R}^n)$ and $q(x,t)\in C(\mathbb{R}^n\times I,\mathbb{R}^n)$

verify (2.3) written with $\varepsilon/2$ instead of ε, then for $s\in I$ and $|y-\phi(s)|<\delta(\varepsilon/2)$ every solution of (2.1) satisfies

$$(3.1)\qquad |x(t;s,y;f+p+q) - \phi(t)| < \frac{\varepsilon}{2} \quad \text{for } t\geq s.$$

Now we consider the following perturbed problem

$$(3.2)\qquad x'=f_{t_k}(t,x) + p(t) + q(x,t) \quad ,t\geq s; \quad x(s)=y.$$

We have

$$(3.3)\qquad |x(t;s,y;f_{t_k}+p(t)+q(x,t) - \phi_{t_k}(t)| =$$

$$|x(\vartheta;s+t_k,y;f(\vartheta)+p(\vartheta-t_k)+q(x,\vartheta-t_k)) - \phi(\vartheta)|, \quad t+t_k=\vartheta.$$

By setting $z-t_k=\tau$ and using (2.3), we get

$$\int_{\vartheta}^{\infty}|p(z-t_k)|dz + |q(x,\vartheta-t_k)| =$$

$$\int_{t}^{\infty}|p(\tau)|d\tau + |q(x,t)| < \delta(\varepsilon/2).$$

Because $\phi(t)$ is I.T.S., by (3.3), we deduce

$$|y-\phi_{t_k}(s)|<\delta(\varepsilon/2)/2,\ t\geq s \Rightarrow |x(t;s,y;f_{t_k}+p+q)-\phi_{t_k}(t)|<\varepsilon/2$$

which shows the total integral stability of $\phi_{t_k}(t)$.

Let k be a positive integer such that for $t\in I$ and $x\in K(\varepsilon)$ it results

$$(3.4)\qquad |\phi_{t_k}(s)-\psi(s)|<\delta(\varepsilon/2)/2, \quad |f_{t_k}(t,x)-f^*(t,x)|<\delta(\varepsilon/2)/2.$$

If we choose the functions p(t) and q(t) verifying

$$(3.5)\qquad \int_t^{\infty} |p(\tau)|d\tau + |q(x,t)| < \delta(\varepsilon/2)/2 \quad \text{for } t\geq s,\ x\in K(\varepsilon),$$

then, employing $(3.4)_2$, (3.5) and the integral total stabili ty of $\phi_{t_k}(t)$, we have

$$(3.6)\qquad |y-\phi_{t_k}(s)|<\delta(\varepsilon/2),\ |f^*(t,x)-f_{t_k}(t,x)| +$$

$$\int_t^{\infty}|p(\tau)|d\tau +|q(t,x)|<\delta(\varepsilon/2)\ ,\ x\in K(\varepsilon),\ t\geq s \Rightarrow$$

$$|x(t;s,y;f^*+p+q)-\phi_{t_k}(t)|<\varepsilon/2.$$

In particular, by exploiting (3.4) and the total stability

of $\phi_{t_k}(t)$, it follows

$$(3.7) \qquad |\psi(t) - \phi_{t_k}(t)| < \varepsilon/2 \quad \text{for} \quad t \geq s.$$

Futhermore

$$(3.8) \qquad |y-\psi(s)| < \delta(\varepsilon/2)/2 \quad \text{and} \quad (3.4)_1 \quad \Rightarrow \quad |y-\phi_{t_k}(s)| \leq$$

$$|y-\psi(s)| + |\psi(s)-\phi_{t_k}(s)| < \frac{\delta}{2} + \frac{\delta}{2} = \delta.$$

At last, if $|y-\psi(s)|<\delta(\varepsilon/2)/2$ and (3.5) holds, we can employ $(3.4)_2$, (3.8), (3.6) and (3.7) obtaining

$$|x(t;s,y;f^*+p+q)-\psi(t)| \leq |x(t;s,y;f^*+p+q)-\phi_{t_k}(t)| +$$

$$|\phi_{t_k}(t)-\psi(t)| < \varepsilon/2 + \varepsilon/2 = \varepsilon \quad \text{for} \quad t \geq s. \quad \text{q.d.} \blacksquare$$

By using Lemmas 2.4, 2.5 and Theorem 3.1 we can deduce

3.2-.Theorem. *Suppose equation* **(1.1) B.a.p.** *in* **t** *and regular. Let* ϕ**(t)**$\in$**K** *be an* **I.T.S.** *solution of* **(1.1)**. *If* **(1.1)** *admits a limiting equation* **(1.2)** *whose solution* ψ**(t)** *such that* (ψ,f^*) $\in\Lambda^*(\phi,f)$ *is uniformly attractive in the large, then* ϕ**(t)** *is* **E.I.T.A.L.**$\blacksquare$

For the pertinent proof and some other results see [14].

4-. References

1 -Sell G.R. *Nonautonomous differential equations and topological dynamics* I-II. Trans. Amer.Math.Soc.,127 (1967)

2 -Yoshizawa T. *Asymptotically almost periodic solutions of an almost periodic system*, Funkcial.Ekvac., 12 (1969)

3 -**Kato J.-Yoshizawa T.***A relationship between uniformly asym ptotically stability and total stability*,**Funk.Ekvac.,12(1969)**

4 -**Fink A.M.-Seifert G.** *Liapunov functions and almost perio dic solutions for almost periodic systems*,**J.Diff.Eq.,5 (1969)**

5 -**Fink A.M.** *Semi-Separated conditions for almost periodic solutions*, **J. Diff. Eq.,11 (1972)**

6 -**Yoshizawa T.***Stability theory and the existence of periodic solutions and almost periodic solutions*,**Ap.Math Sc.,14 (1975)**

7 -**Bondi P.-Moauro V.-Visentin F.** *Limiting equations in the stability problem*, **Nonlinear Anal. TMA, 1-2 (1977)**

8 -**Artstein Z.** *Uniform asymptotic stability via the limiting equations*, **J. Diff. Eq. , 27 (1978)**

9 -**D'Anna A.-Maio A.-Monte A.M.** *Limiting equations in asymp totic stability problem for nonautonomous differential equa tions*, **Ann. Univ.Ferrara, Sez VII, XXV (1979)**

10-**D'Anna A.-Maio A.-Moauro V.***Global stability properties by means of limiting equations*, **Nonlinear Anal. TMA, 4-2 (1980)**

11-**D'Anna A.** *Limiting equations in the first approximation stability problem for nonautonomous differential equations*, **B.U.M.I., 6-1-A, (1982)**

12-**D'Anna A.***Total stability properties for an almost periodic equations by means of limiting equations* **Funk.Ek.,127 (1984)**

13-**Martynyuk A.-Karimzhanov A.** *Limiting equations and stabi lity of non-stationary motions*,**J. Math. An. and Appl.,(1986)**

14-**D'Anna A.-Monte A.M.** *Global stability properties for a perturbed almost periodic equation*, **to appear**

Acknowledgment. **This research was supported by Italian Ministri of Education (M.P.I.).**

THE INTERFACE BETWEEN PHASES AS A LAYER, PART II
A H-ORDER MODEL FOR TWO DIMENSIONAL NONMATERIAL CONTINUA

Francesco dell' Isola
DIPARTIMENTO DI MATEMATICA E SUE APPLICAZIONI "RENATO CACCIOPPOLI"
UNIVERSITA' DI NAPOLI, NAPOLI, ITALIA
and
Witold Kosiński
POLSKA AKADEMIA NAUK
INSTYTUT PODSTAWOWYCH PROBLEMÓW TECHNIKI
IPPT PAN, WARSZAWA, POLSKA

ABSTRACT

We model the interlayer between phases as a 2D continuum accounting for its actual "3D" nature. This is done by defining on a reference surface interfacial quantities in the form of integrals along the thickness of the layer. Therefore it is possible to deduce the general surface balance law from the 3D one. The balance laws are derived for of all the first H moments of mass, linear momentum and angular momentum. This procedure yields surface angular momentum balances Galilean invariant. Finally a Kirchhoff-Love type interface is defined and shortly discussed by means of a first-order model.

1. INTRODUCTION

Capillarity or *Surface Phenomena* occur when significant amounts of physically relevant quantities are *localized* in a region, henceforth called *a layer* $\mathfrak{L}$, whose thickness is *small.*

1.1 Assumptions

We start considering a 3D continuum $\mathfrak{C}$ occupying at instant t the region C(t) and moving with the velocity $\underline{v}$. We assume that the layer $\mathfrak{L}$ is located in $\mathfrak{C}$. We shall model $\mathfrak{L}$ as a *2D nonmaterial continuum* $\mathcal{S}$ whose balance equations for surface densities of mass, linear and angular momenta, are formulated *taking into account the 3D nature* of $\mathfrak{L}$ itself. Here by the term *nonmaterial continuum* we mean that the motion of the layer can be different from that determined by the field $\underline{v}$.

Following [1)] we assume that:

There exists a regular surface $\Sigma \subset \mathfrak{L}$, which models the spatial evolution and geometrical properties of $\mathfrak{L}$ itself in such a way that

$$\mathfrak{L} = \bigcup_{\ell\in I} \Sigma^{\ell} \tag{1.1}$$

where $I \equiv [-\mathfrak{z}/2, +\mathfrak{z}/2]$ (we denote $\pm\mathfrak{z}/2 \equiv \mathfrak{z}^{\pm}$), $\mathfrak{z}$ being the thickness of the layer which in a first approach we assume to be constant, and

$$\Sigma^{\ell} \equiv \left\{ \underline{x} \in \mathfrak{E} \mid \exists!\ \underline{r} \in \Sigma : \underline{x} = \underline{r} + \ell\,\underline{n}\ ,\ \underline{n} \equiv \text{unit normal to } \Sigma \text{ at } \underline{r} \right\}$$

Let us notice that the regular surface Σ is chosen with no particular reference to thermomechanical quantities defined in $\mathfrak{E}$ as it is used to be done when introducing Gibbs' dividing surfaces.

The remaining material structure of $\mathfrak{L}$ is preserved by equipping the surface Σ with a structure of 2D continuum. It is done by defining surface densities and flux of physical quantities as suitable integrals of the corresponding volume ones along the thickness of $\mathfrak{L}$. The procedure makes possible an identification of surface quantities which appear in surface balance laws even when dealing with nonmaterial continua. Moreover, this allows for a more careful discussion of the Galilean invariance of the balance laws.

1.2 H-th Order Models For $\mathfrak{L}$.

Following [1)] for every volume density ψ whose flux is $\underline{w}$ and production is $\underline{p}$, we can define the corresponding surface density and production functions as

$$f^{S} \equiv \int_{z^-}^{z^+} j(\ell,\underline{r}) f(\underline{r}+\ell\underline{n}(\underline{r}),t)\,d\ell \equiv \langle j\ f\rangle \ \text{(if } f \text{ stands for } \underline{p} \text{ or } \psi) \tag{1.2}$$

and the surface flux as

$$\underline{W}^{S'}\{\ \psi\ \} \equiv \langle\psi \otimes (\underline{v}+\ell \operatorname{grad}_S c_n)\ \underline{A}_S(\ell)\rangle + \langle\underline{w}\ \underline{A}_S(\ell)\rangle, \tag{1.3}$$

where $j(\ell,\underline{r}) \equiv 1 - 2H(\underline{r})\ell + K(\underline{r})\ell$. Here $H(\underline{r})$ and $K(\underline{r})$ are, respectively, the mean and Gauss curvatures of Σ at $\underline{r}$, the tensors $\underline{b}_0$ and $\mathbb{I}_S$ are the surface curvature and metric tensors of the moving surface Σ, respectively, c_n its geometrical normal speed and $\underline{A}_S(\ell) \equiv (1-2H\ell)\mathbb{I}_S + \ell\underline{b}_0$ is a tensor field which depends of course on $\underline{r}$ and t.

The above definitions give at the same time the only possible relationships between the surface quantities and their bulk counterparts (better to say - their primitives), in order the interfacial balance law localized on the surface Σ were compatible and derivable from the 3D law. The latter is postulated for ψ in the integral form as well as in the following local one:

$$\frac{\partial}{\partial t}\psi + \mathrm{div}\,(\psi\otimes\underline{v} + \underline{w}) = \underline{p}\,. \tag{1.4}$$

On the other hand using the model developed in [1,2] one could not completely take into account [*1*] the influence of the thickness of the layer on the thermomechanical behaviour of phase interfaces. Quoted model we shall call a *0-th order model*.

In order to resolve the aforementioned difficulty an H-th order model is proposed, introducing the k-th moments ($k \le H$) of a typical quantity f by

$$^{k}f \equiv \ell^{k}(\underline{x},t)\, f \qquad \text{when } \underline{x} \in \mathfrak{L}. \tag{1.5}$$

here ℓ^k means the k-th power of ℓ. One should mention at this point the approach used by Dumais [3], who already introduced the concept of the higher order moments in modelling interface phenomena. The following local balance equation for ψ can be easily derived from (1.4) and properties of the function $\ell(\underline{x},t)$ (see Lemma 2.9 in [1]) :

$$\frac{\partial}{\partial t}{}^{k}\psi + \mathrm{div}\,({}^{k}\psi\otimes\underline{v} + {}^{k}\underline{w}) = {}^{k}\underline{p} + k(\,{}^{k-1}\psi\otimes(\underline{v}\cdot\underline{n}-c_n) + {}^{k-1}\underline{w}\cdot\underline{n}) \tag{1.6}$$

Regarding (1.6) as a particular case of (1.4) and recalling the results in Sec. 2 of [1], we get the following surface balance equation:

$$\begin{aligned}\frac{\delta n}{\delta t}\,{}^{k}\psi^{s} - 2Hc_n\,{}^{k}\psi^{s} + \mathrm{div}_s(\langle{}^{k}\psi\otimes(\underline{v}+\ell\,\mathrm{grad}_s c_n)\,\underline{A}_s(\ell)\rangle + \langle{}^{k}\underline{w}\,\underline{A}_s(\ell)\rangle) =\\ = [\![j({}^{k}\psi(\underline{v}-\underline{c}) - {}^{k}\underline{w}\,]\!]\cdot\underline{n} + {}^{k}p^{s} + \langle jk(\,{}^{k-1}\psi\otimes(\underline{v}\cdot\underline{n}-c_n) + {}^{k-1}\underline{w}\cdot\underline{n})\rangle,\end{aligned} \tag{1.7}$$

where $\frac{\delta n}{\delta t}$, grad_s and div_s denote respectively the surface Thomas derivative, gradient and divergence operators. This equation is valid for any $k \ge 1$. If $k = 0$ the primitive interfacial balance law found in [1,2] is the counterpart of (1.4). In this particular case the last term of its L.H.S. is given by (1.3) and its last term of the R.H.S. vanishes.

---

1 The authors gratefully acknowledge having this drawn to their attention by Professor S. RIONERO.

It can however arise a question concerning the mathematical completeness of this kind of approach. To answer this question one should first notice that the k-th moment of a typical function f in (1.5) (regarded as a function of ℓ only) defines the projection of f on the polynomial ℓ^k belonging to the basis formed by all polynomialsof the function space $L^2([\mathfrak{z}^-,\mathfrak{z}^+],\ d\mu = j d\ell)$. The measure μ is positive and absolutely continuous with respect to the Lebesgue measure as long as j is positive and H and K are finite; this corresponds to the hypothesis H3.1 in [1] on the thickness of the layer $\mathfrak{L}$. Therefore the H-order theory deals with truncated expansions along the thickness of the layer of physical quantities to be balanced.

2. MECHANICAL INTERFACIAL BALANCE LAWS

Let ρ be the mass density function for the continuum $\mathfrak{C}$. To make the presentation more compact and to make evident the Galilean invariance of surface balance laws we introduce the following denotation (compare 1.3)

$$\underline{W}^{s'}\{{}^k\rho\} \equiv {}^k\underline{M}^s := \langle {}^k\underline{m}(\ell)\rangle = \langle j\ell^k \rho \underline{v}\,\mathbb{I}_s\rangle + {}^k\underline{W}_\rho,$$

$${}^k\underline{S}_1\{\psi\} := \left\langle \left(\frac{{}^k\psi}{{}^k\rho} - \frac{{}^k\psi^s}{{}^k\rho^s} \right) \otimes {}^k\underline{m}(\ell)\right\rangle, \quad {}^k\underline{S}_2\{\psi\} := \langle {}^k\underline{w}\ \underline{A}_s(\ell)\rangle, \tag{2.1}$$

where ${}^k\underline{W}_\rho := \langle \ell^k \rho(\underline{A}_s(\ell) - j\mathbb{I}_s)\underline{v} + \ell^{k+1}\rho\ \mathrm{grad}_s c_n\ \underline{A}_s(\ell)\rangle$ can be proved to be Galilean invariant paralleling the proof of Lemma 4.2 in [1] . Moreover it is evident that the ${}^k\underline{S}_m$, m = 1,2, are Galilean invariant, of course, if $\underline{w}$ fulfills the same requirement.

2.1. Balance of Moments of Mass and Linear Momentum

With our denotation the balance equation for the moments of mass will take the following form, where $H \geq k \geq 1$,

$$\frac{\delta n}{\delta t}\,{}^k\rho^s - 2Hc_n\,{}^k\rho^s + \mathrm{div}_s\,{}^k\underline{M}^s = [\![j({}^k\rho(\underline{v}\cdot\underline{n} - c_n)]\!] + k\langle {}^{k-1}\rho\,(\underline{v}\cdot\underline{n} - c_n)\rangle. \tag{2.2}$$

To balance the first H moments of linear momentum with $\underline{w}$ equal to the minus Cauchy stress $\underline{T}$ the following set of equations must be satisfied

$$\frac{\delta n}{\delta t}\left({}^k\rho^s\,{}^k\underline{V}^s\right) - 2Hc_n\,{}^k\rho^s\,{}^k\underline{V}^s + \mathrm{div}_s\left({}^k\underline{V}\otimes{}^k\underline{M}^s + {}^k\underline{S}_1\{\rho\underline{v}\} + {}^k\underline{S}_2\{\rho\underline{v}\}\right) = {}^k(\rho\underline{b})^s$$

$$+ [\![({}^k\rho\underline{v}\otimes(\underline{v}\cdot\underline{n}-c_n) - {}^k\underline{T}\,\underline{n})j]\!] + \langle j\,k({}^{k-1}\rho\underline{v}\otimes(\underline{v}\cdot\underline{n}-c_n) - {}^{k-1}\underline{T}\cdot\underline{n})\rangle \qquad (2.3)$$

where ${}^k\rho^s\,{}^k\underline{V}^s := {}^k(\rho\underline{v})^s$, ${}^k\underline{S}_1\{\rho\underline{v}\} = \langle(\underline{v}-{}^k\underline{V}^s)\otimes{}^k\underline{m}(\ell)\rangle$, ${}^k\underline{S}_2\{\rho\underline{v}\} = -\langle{}^k\underline{T}\,\underline{A}_s(\ell)\rangle$.

Let us notice that for every $k \le H-2$ the term ${}^k\underline{W}_\rho$ is known in terms of moments of mass and linear momentum whose evolution is governed by the balance equations (2.2 -2.3) introduced in H-order models. Therefore in such models constitutive equations for ${}^{H-n}\underline{W}_\rho$ are required, together with constitutive equations for ${}^k\underline{S}_m\{\rho\underline{v}\}$ where $k=1,..,H$ and $n=0,1$.

2.2 Balance of Moments of Angular Momentum

We restrict now ourself only to nonpolar continua. The master angular momentum balance law is well known in the 3D theory; its interfacial counterpart requires to define two quantities for each k-th moment. Namely

$${}^k\Psi^s := \langle j(\underline{r} + \ell\,\underline{n}) \times \ell^k\rho\underline{v}\rangle\ , \quad {}^k\Psi_2^s := \langle j\ell\,\underline{n}\times\ell^k\rho\underline{v}\rangle \qquad (2.4)$$

and the corresponding flux and production terms are

$${}^k\underline{w}(\ell) = -\,(\underline{r} + \ell\,\underline{n})\times\ell^k\underline{T}, \qquad {}^k\underline{w}_2(\ell) = -\,\ell\,\underline{n}\times\ell^k\underline{T}$$

$${}^k\underline{p}^s = \langle j(\underline{r} + \ell\,\underline{n})\times\ell^k\rho\underline{b}\rangle, \qquad {}^k\underline{p}_2^s = \langle j\ell\,\underline{n}\times\ell^k\rho\underline{b}\rangle - \langle j\ell^k\underline{F}\rangle \qquad (2.5)$$

where (cf. [1]) $j\underline{F} := (j\underline{1}+\ell(\underline{b}_0 - K\ell\mathbb{I}_s)\times\underline{T}\rangle - j\underline{n}\times\underline{T}\cdot\underline{n} - \underline{m}(\ell)\times(\underline{v}\cdot\underline{n})\underline{n} + j\rho\underline{v}\times c_n\underline{n}$.
Making use of the balance law (1.7) substituting ${}^k\psi^s$ with those defined in (2.4) together with the corresponding terms from (2.5) we arrive at a pair of equations (for any $H \ge k \ge 1$) which generalize (4.17) and (4.22) in [1]. Subtracting the second from the first and using the cross product of $\underline{r}$ with the law (2.3) we end up with

$$\langle\, j(\underline{1}\times\ell^k\underline{T})\,\rangle = \underline{0} \qquad (2.6)$$

which is automatically satisfied if $\underline{T}$ is symmetric. On the other hand the $H + 1$ equations (2.6) with $k = 0,1,\ldots H$ represent the H-th order condition of non-polarity of the layer.

3. SHELL-LIKE INTERFACE OF KIRCHHOFF-LOVE TYPE

Let us consider the particular case of interfaces in which the layer $\mathfrak{L}$ is defined by the kinematics of the continuum $\mathfrak{C}$. We assume that a surface Σ can be found such that the velocity field has the particular form, namely

$$\underline{v}(\ell;\underline{r},t) = \underline{v}_0(\underline{r},t) - \ell(\mathrm{grad}_s w_n + \underline{b}_0\underline{v}_\tau), \quad \text{where } \underline{v}_\tau\cdot\underline{n} = 0 \tag{3.1}$$

This velocity field describes the motion in which particles appearing in the layer on a line tangent to a normal $\underline{n}$ move with equal normal speeds. However if w_n is equal to c_n and $\underline{v}_\tau$ equal to $\underline{v}_0\mathbb{I}_s$ and the both are independent of ℓ then the aforementioned particles remain on a common line. In this case it turns out that the following relations

$$\underline{v}_0(\underline{r},t)\mathbb{I}_s = j^{-1}(d_z)\underline{A}_s(d_z)(\,\underline{V}^s + d_z \,\mathrm{grad}\, c_n), \quad (\underline{V}^s - \underline{v}_0)\cdot\underline{n} = 0, \tag{3.2}$$

hold, where $d_{\mathfrak{z}}$ defined by ${}^1\rho^s = d_{\mathfrak{z}}\,\rho^s$ belongs to $[\mathfrak{z}^-,\ \mathfrak{z}^+]$.

Formula (3.2) shows that the first order model is required for describing the Kirchhoff-Love type interface. More precisely if we consider all together

1) the interfacial linear momentum and mass balance laws

2) the boundary values for ρ^- and ρ^+ (they play the role of constitutive quantities) consistent with the sign of $\underline{v}_0\cdot\underline{n}$,

we obtain a well posed problem for bulk mechanical balance laws. Kirchhoff-Love type interface is a first instance in which the higher order models show their applicability in modelling surface phenomena.

REFERENCES

1. dell' Isola, F. and Kosiński, W., "The Interface Between Phases as a Layer, Part I", submitted for publication.
2. Kosiński, W. and Romano, A. Evolution Balance Laws in Two-Phase Problems, IUTAM Congress in Grenoble, August 1988, also Arch. Mech., 41, (1989).
3. Alts, T. and Hutter, K., "Continuum Description of the Dynamics and Thermodynamics of Phase Boundaries Between Ice and Water, Parts I and II", I.Non-Equilib.Thermodyn., 13, 221-280 (1988).
4. Dumais, J-F., " Two and Three-Dimensional Interface Dynamics, Physica, 104 A, 143-180 (1980).

STABILITY PROBLEMS FOR NON-LINEAR DISSIPATIVE SYSTEMS

MAURO FABRIZIO (Dipartimento di Matematica Piazza di Porta San Donato 5 40127 Bologna, Italy)

BARBARA LAZZARI (Dipartimento di Automazione Industriale Via Valotti 9 25060 Brescia, Italy)

ABSTRACT: A new approach is showed that permits to study the stability of a non-linear problems for Maxwell's equations by means of results on the associated linearized differential system.

1. INTRODUCTION

The purpose of this paper is to study asymptotic behaviour for solutions of non-linear Maxwell's equations, when the boundary condition is given by Schelkunoff - Graffi dissipative relation (see [1], [2], [3])[1]

$$\mathbf{E}^t = \lambda\, \mathbf{H}^t \times \mathbf{n} + \mathbf{g}. \tag{1.1}$$

Local uniqueness and existence theorem was proved by Cesari [4] of the same initial-boundary problem, while an uniqueness theorem for a finite temporal interval was find by Fabrizio [5]. Here we use a new approach that permits to study the stability of the non-linear problem by means of results on the linearized differential system. In particular we prove existence, uniqueness and stability theorems for linear problem with not smooth coefficients.

2. STATEMENT OF THE PROBLEM

The non linear differential system for Maxwell's equations is given by:

$$\frac{\partial \mathbf{D}}{\partial t} = \nabla \times \mathbf{H} - \mathbf{J} \tag{2.1}$$

$$\frac{\partial \mathbf{B}}{\partial t} = -\nabla \times \mathbf{E} \tag{2.2}$$

where the functions **E, H, D, B** are defined on a smooth domain $Q = \Omega \times (t_1,t_2)$ and are respectively the *electric intensity*, the *magnetic intensity*, the *electric induction*, the

---

[1] $\mathbf{E}^t$, $\mathbf{H}^t$ represent the boundary tangential components of the vectors **E** and **H**.

magnetic induction, while the *electric current* **J** is a given function of $(x,t) \in Q$. The material behaviour is specified by the constitutive equations:

$$\mathbf{D}(x,t) = \hat{\mathbf{D}}(\mathbf{E}(x,t),\mathbf{H}(x,t)); \qquad \mathbf{B}(x,t) = \hat{\mathbf{B}}(\mathbf{E}(x,t),\mathbf{H}(x,t)) \tag{2.3}$$

We suppose the functions (2.3) smooth of C^1 and satisfying the following thermodynamics conditions [6]:

$$\frac{\partial D_i}{\partial E_j} = \frac{\partial D_j}{\partial E_i}; \qquad \frac{\partial D_i}{\partial H_j} = \frac{\partial B_j}{\partial E_i}; \qquad \frac{\partial B_i}{\partial H_j} = \frac{\partial B_j}{\partial H_i} \tag{2.4}$$

$$\sum_{i,j=1}^{3} \left(\frac{\partial D_i}{\partial E_j} a_i a_j + 2\frac{\partial D_i}{\partial H_j} a_i b_j + \frac{\partial B_i}{\partial H_j} b_i b_j\right) > \gamma^2 (a^2 + b^2) \tag{2.5}$$

Finally we join to the system (2.1), (2.2) the non-homogeneous boundary condition

$$\mathbf{E}^t(x,t) = \lambda(x)\, \mathbf{H}^t(x,t) \times \mathbf{n}(x) + \mathbf{g}(x,t) \tag{2.6}$$

where $\mathbf{g}$ is a given vector of $(x,t) \in \partial\Omega \times (t_1,t_2)$ that provides the electric flux through the boundary.

It is useful to consider the vectors of $\mathbb{R}^6$:

$$u = (\mathbf{E},\mathbf{H}), \quad G = (\mathbf{D},\mathbf{B}), \quad F = (\mathbf{J},\mathbf{0}) \tag{2.7}$$

that are joined, by (2.3), from the relation:

$$G = \hat{G}(u) \tag{2.8}$$

This function must be satisfying the conditions (2.4) and (2.5); namely we have:

$$\partial_u \hat{G} = (\partial_u \hat{G})^T; \qquad (\partial_u \hat{G}\, \xi\, , \xi) > \gamma^2 \xi^2 \tag{2.9}$$

where $\xi = (\mathbf{a},\mathbf{b})$ and $\mathbf{a},\mathbf{b}$ are general vectors of $\mathbb{R}^3$, while $(\cdot,\cdot)$ labels the scalar product of $\mathbb{R}^6$.

Let us write the Maxwell's equations in the form

$$\frac{\partial \hat{G}(u)}{\partial t} = \Lambda^i \frac{\partial u}{\partial x^i} + F \tag{2.10}$$

where Λ^i are 6×6 symmetric and constant matrices (see [7]).

We relate (2.10) with initial conditions

$$u(x,0) = u_0(x) = (\mathbf{E}_0(x),\mathbf{H}_0(x)) \tag{2.11}$$

and the boundary condition (2.6).

Finally, let us introduce the function spaces [2]:

$$\mathsf{H}(t_1,t_2;\Omega) = \{\ \mathbf{e} \in H^1(t_1,t_2; L^2(\Omega)) \cap L^2(t_1,t_2; H^1(\Omega));\ \mathbf{e}(x,t_2) = \mathbf{0}\} \tag{2.12}$$

---

[2] When the temporal interval (t_1,t_2) is given by $(0,\infty)$ then the space $\mathsf{H}(0,\infty;\Omega)$ is now defined $\mathsf{H}(0,\infty;\Omega) = H^1(0,\infty; L^2(\Omega)) \cap L^2(0,\infty; H^1(\Omega))$

$$H(t_1,t_2;\Omega) = \mathsf{H}(t_1,t_2;\Omega) \times \mathsf{H}(t_1,t_2;\Omega) \tag{2.13}$$

$$H_\Delta(t_1,t_2;\Omega) =$$
$$= \{(\mathbf{e},\mathbf{h}) \in H(t_1,t_2;\Omega);\ \mathbf{e}^t(x,t) = \lambda(x)\ \mathbf{h}^t(x,t) \times \mathbf{n}(x) \text{ a.e. on } \partial\Omega\times(t_1,t_2)\} \tag{2.14}$$

Now it is possible to formulate the following definition of solution:

Definition 2.1 - *A function* $u = (\mathbf{E},\mathbf{H}) \in H(t_1,t_2;\Omega)$ *is called a (weak) solution of the problem* (2.10), (2.11) *with the boundary condition* (2.6) *and supplies* $F \in (L^2(Q))^2$, $\mathbf{g} \in L^2(\partial\Omega \times (t_1,t_2))$ *if satisfies*

$$\int_Q [\mathbf{H}(x,t)\cdot\nabla \times \mathbf{e}(x,t) + \mathbf{E}(x,t)\cdot\nabla \times \mathbf{h}(x,t)]\,dx\,dt +$$

$$+ \int_Q [\mathbf{D}(x,t)\cdot\frac{\partial \mathbf{e}(x,t)}{\partial t} - \mathbf{B}(x,t)\cdot\frac{\partial \mathbf{h}(x,t)}{\partial t} - \mathbf{J}(x,t)\cdot\mathbf{e}(x,t)]\,dx\,dt - \tag{2.15}$$

$$- \int_\Omega [\mathbf{D}(x,0)\cdot\mathbf{e}(x,0) - \mathbf{B}(x,0)\cdot\mathbf{h}(x,0)]\,dx + \frac{1}{\lambda}\int_{t_1}^{t_2}\int_{\partial\Omega} \mathbf{h}^t(x,t)\cdot\mathbf{g}(x,t)\,d\sigma\,dt = 0$$

for all $(\mathbf{e},\mathbf{h}) \in H_\Delta(t_1,t_2;\Omega)$.

Remark - *The integral equality* (2.20) *can be represented using the system* (2.10), *also in the form:*

$$\int_Q \{(u,\Lambda^i \frac{\partial v}{\partial x^i}) + (\mathcal{G}(u), \frac{\partial v}{\partial t}) - (F,v)\}dx\,dt +$$

$$+ \int_\Omega [\ \mathcal{G}(u(x,0))\cdot v(x,0))\,dx + \frac{1}{\lambda}\int_{t_1}^{t_2}\int_{\partial\Omega} \mathbf{h}^t(x,t)\cdot\mathbf{g}(x,t)\,d\sigma\,dt = 0 \tag{2.16}$$

for all $v \in H_\Delta(t_1,t_2;\Omega)$.

3. LINEAR PROBLEM

Let us consider the linear problem joined to non-linear system (2.1), (2.2) on the domain $Q = \Omega \times (t_1,t_2)$:

$$\tilde{\varepsilon}(x,t)\ \frac{\partial \mathbf{e}(x,t)}{\partial t} = \nabla \times \mathbf{h}(x,t) - \mathbf{j}_1(x,t) \tag{3.1}$$

$$\tilde{\mu}(x,t)\,\frac{\partial \mathbf{h}(x,t)}{\partial t} = -\nabla \times \mathbf{e}(x,t) - \mathbf{j}_2(x,t) \tag{3.2}$$

$$\mathbf{e}(x,0) = \mathbf{e}_0(x); \qquad \mathbf{h}(x,0) = \mathbf{h}_0(x) \tag{3.3}$$

where $\tilde{\varepsilon}$, $\tilde{\mu}$ are two symmetric and positive tensors of of second order, namely there are two constants $\gamma_1, \gamma_2 \in \mathbb{R}$ for which for all $(x,t) \in Q$ we have:

$$\tilde{\varepsilon}(x,t) = (\tilde{\varepsilon}(x,t))^{\dagger}; \qquad \tilde{\mu}(x,t) = (\mu(x,t))^{\dagger} \tag{3.4}$$

$$\tilde{\varepsilon}(x,t)\, \mathbf{a}\cdot\mathbf{a} \geq \gamma_1^2\, \mathbf{a}^2; \quad \tilde{\mu}(x,t)\, \mathbf{a}\cdot\mathbf{a} \geq \gamma_2^2\, \mathbf{a}^2; \quad \forall\, \mathbf{a} \in \mathbb{R}^3 \tag{3.5}$$

moreover we assume the coefficients such that:

$$\tilde{\varepsilon}(x,t),\ \tilde{\mu}(x,t) \in BV(0,\infty;L^{\infty}(\Omega)) \tag{3.6}$$

and the homogeneous boundary condition holds on $\partial\Omega$, i.e.

$$\mathbf{e}^{t}(x,t) = \lambda(x)\, \mathbf{h}^{t}(x,t) \times \mathbf{n}(x) \qquad x \in \partial\Omega \tag{3.7}$$

Using the vectors of $\mathbb{R}^6$: $v = (\mathbf{e},\mathbf{h})$, $f = (\mathbf{j}_1,\mathbf{j}_2)$ and the tensor

$$A = \begin{bmatrix} \tilde{\varepsilon} & \mathbf{0} \\ \mathbf{0} & \tilde{\mu} \end{bmatrix}$$

we are able to write the system (3.1), (3.2) in compact form:

$$A(x,t)\frac{\partial v(x,t)}{\partial t} = \Lambda^{i}\frac{\partial v(x,t)}{\partial x^{i}} + f(x,t) \tag{3.8}$$

One of us, Lazzari [8] studied the problem (3.1) - (3.7), when the coefficients $\tilde{\varepsilon}$, $\tilde{\mu}$ are temporally homogeneous. Since she obtain the following:

Theorem 3.1. *When the coefficients* $\tilde{\varepsilon}(x)$, $\tilde{\mu}(x)$ *are such that:*

$$\tilde{\varepsilon} \in L^{\infty}(\Omega), \quad \tilde{\mu} \in L^{\infty}(\Omega) \tag{3.9}$$

and $\mathbf{j}_i, \mathbf{j}_2 \in L^2(0,\infty;L^2(\Omega))$, $\mathbf{e}_0, \mathbf{h}_0 \in L^2(\Omega)$, *then there exists one and only one solution for the problem* (3.1) - (3.7) *such that:*

$$\mathbf{e}, \mathbf{h} \in H^1(0,\infty; L^2(\Omega)) \cap L^2(0,\infty; H^1(\Omega)) \tag{3.10}$$

In the general case, namely when the coefficients are possibly temporally inhomogeneous, Arosio [9], [10] studied the asymptotic behaviour of equations (3.1), (3.2). As a result of conditions (3.6) he showed that the following limits exist:

$$\lim_{t\to\infty} |\tilde{\varepsilon}(x,t) - \tilde{\varepsilon}^{\infty}(x)|_{L^{\infty}(\Omega)}, \quad \lim_{t\to\infty} |\tilde{\mu}(x,t) - \tilde{\mu}^{\infty}(x)|_{L^{\infty}(\Omega)} \tag{3.11}$$

and he considered the new system:

$$\tilde{\varepsilon}^{\infty}(x)\frac{\partial \mathbf{e}(x,t)}{\partial t} = \nabla \times \mathbf{h}(x,t) \tag{3.12}$$

$$\tilde{\mu}^{\infty}(x)\frac{\partial \mathbf{h}(x,t)}{\partial t} = -\nabla \times \mathbf{e}(x,t) \tag{3.13}$$

then he claimed:

Theorem 3.2. *If the coefficients* $\tilde{\varepsilon}(x,t)$, $\tilde{\mu}(x,t)$ *satisfy the hypotheses* (3.4), (3.5), (3.6), (3.11) *and the new condition:*

$$\int_0^{\infty} \operatorname*{ess\,sup}_{x\in\Omega} \{ |\tilde{\varepsilon}(x,t) - \tilde{\varepsilon}^{\infty}(x)| + |\tilde{\mu}(x,t) - \tilde{\mu}^{\infty}(x)| \}\, dt < \infty$$

then, for every solution $(\mathbf{e},\mathbf{h}) \in H^1_{loc}(0,\infty; L^2(\Omega)) \cap L^1_{loc}(0,\infty; H^1(\Omega))$ *of the problem* (3.1) - (3.7), *there exists a unique solution* $(\mathbf{e}^\infty, \mathbf{h}^\infty)$ *of the system* (3.12), (3.13) *such that:*

$$\lim_{t\to\infty} \| \mathbf{e}^\infty(t) - \mathbf{e}(t)\|_{L^2(\Omega)} = 0\,, \quad \lim_{t\to\infty} \| \mathbf{h}^\infty(t) - \mathbf{h}(t)\|_{L^2(\Omega)} = 0 \tag{3.14}$$

Hence it is immediate to prove:

Theorem 3.3. *If* $\mathbf{j}_1, \mathbf{j}_2 \in L^2(0,\infty; L^2(\Omega))$ *and* $\mathbf{e}_0, \mathbf{h}_0 \in L^2(\Omega)$, *then the problem* (3.1) - (3.2) *has at most one solution:*

$$\mathbf{e}, \mathbf{h} \in H^1(0,\infty; L^2(\Omega)) \cap L^2(0,\infty; H^1(\Omega)) \tag{3.15}$$

4. STABILITY

We now consider the time interval $(t_1,t_2) = (0,T)$, then the Definition 1 can be written in the following form:

$$\int_0^T \int_\Omega [\, \mathbf{H}(x,t) \cdot \nabla \times \mathbf{e}^T(x,t) + \mathbf{E}(x,t) \cdot \nabla \times \mathbf{h}^T(x,t)]\, dx\, dt\, +$$

$$+ \int_0^T \int_\Omega [\mathbf{D}(x,t) \cdot \frac{\partial \mathbf{e}^T(x,t)}{\partial t} - \mathbf{B}(x,t) \cdot \frac{\partial \mathbf{h}^T(x,t)}{\partial t} - \mathbf{J}(x,t) \cdot \mathbf{e}^T(x,t)]\, dx\, dt\, - \tag{4.1}$$

$$- \int_\Omega [\mathbf{D}(x,0) \cdot \mathbf{e}^T(x,0) - \mathbf{B}(x,0) \cdot \mathbf{h}^T(x,0)]\, dx + \frac{1}{\lambda} \int_0^T \int_{\partial\Omega} (\mathbf{h}^t)^T(x,t) \cdot \mathbf{g}(x,t)\, d\sigma\, dt = 0$$

where $f^T(t) = f(T - t)$. We now use the properties of convolution product to conclude

$$\int_0^T \int_\Omega [\mathbf{H}^T(x,t) \cdot \nabla \times \mathbf{e}(x,t) + \mathbf{E}^T(x,t) \cdot \nabla \times \mathbf{h}(x,t) + \mathbf{D}^T(x,t) \cdot \frac{\partial \mathbf{e}(x,t)}{\partial t}]dx\, dt\, -$$

$$- \int_0^T \int_\Omega [\, \mathbf{B}^T(x,t) \cdot \frac{\partial \mathbf{h}(x,t)}{\partial t} + \mathbf{J}^T(x,t) \cdot \mathbf{e}(x,t))\, dx\, dt\, - \tag{4.2}$$

$$- \int_\Omega (\mathbf{D}(x,0) \cdot \mathbf{e}(x,T) - \mathbf{B}(x,0) \cdot \mathbf{h}(x,T)\,)\, dx + \frac{1}{\lambda} \int_0^T \int_{\partial\Omega} \mathbf{h}^t(x,t) \cdot \mathbf{g}^T(x,t)\, d\sigma\, dt = 0$$

Now let us consider two bounded solutions $(\mathbf{E}_1(x,t), \mathbf{H}_1(x,t))$, $(\mathbf{E}_2(x,t), \mathbf{H}_2(x,t))$ belonging to the spaces $BV(0,\infty; L^\infty(\Omega)) \cap L^2_{loc}(0,\infty; L^2(0,\infty; L^2(\Omega))$ corresponding to the following data:

a - *electric density currents*

$$\mathbf{J}_1, \mathbf{J}_2 \in L^2_{loc}(0,\infty; L^2(\Omega)), \text{ such that } \mathbf{J}_1 - \mathbf{J}_2 \in L^2(0,\infty; L^2(\Omega)) \tag{4.3}$$

b - *initial conditions*:

$$\mathbf{E}_1(x,0) = \mathbf{E}^0_1(x),\ \mathbf{H}_1(x,0) = \mathbf{H}^0_1(x),$$

$$\mathbf{E}_2(x,0) = \mathbf{E}^0_2(x),\ \mathbf{H}_2(x,0) = \mathbf{H}^0_2(x,0) \tag{4.4}$$

such that:

$$\hat{\mathbf{D}}(\mathbf{E}^0_1, \mathbf{H}^0_1), \hat{\mathbf{D}}(\mathbf{E}^0_2, \mathbf{H}^0_2), \hat{\mathbf{B}}(\mathbf{E}^0_1, \mathbf{H}^0_1), \hat{\mathbf{B}}(\mathbf{E}^0_2, \mathbf{H}^0_2) \in L^2(\Omega)$$

c - *electric fluxes*

$$\mathbf{g}_1, \mathbf{g}_2 \in H^1_{loc}(0,\infty; L^2(\partial\Omega)), \text{ such that:}$$

$$\mathbf{g}_1 - \mathbf{g}_2 \in H^1(0,\infty; L^2(\partial\Omega)). \tag{4.5}$$

Next, from (4.2), we see that

$$\int_0^T \int_\Omega [(\mathbf{H}^T_1(x,t) - \mathbf{H}^T_2(x,t)) \cdot \nabla \times \mathbf{e}(x,t) + (\mathbf{E}^T_1(x,t) - \mathbf{E}^T_2(x,t)) \cdot \nabla \times \mathbf{h}(x,t)] dx\, dt +$$

$$+ \int_0^T \int_\Omega [(\mathbf{D}^T_1(x,t) - \mathbf{D}^T_2(x,t)) \cdot \frac{\partial \mathbf{e}(x,t)}{\partial t} - (\mathbf{B}^T_1(x,t) - \mathbf{B}^T_2(x,t)) \cdot \frac{\partial \mathbf{h}(x,t)}{\partial t}] dx\, dt -$$

$$- \int_0^T \int_\Omega [(\mathbf{J}^T_1(x,t) - \mathbf{J}^T_2(x,t)) \cdot \mathbf{e}(x,t)] dx\, dt - \tag{4.6}$$

$$- \int_\Omega [(\mathbf{D}_1(x,0) - \mathbf{D}_2(x,0)) \cdot \mathbf{e}(x,T) - (\mathbf{B}_1(x,0) - \mathbf{B}_2(x,0)) \cdot \mathbf{h}(x,T)] dx +$$

$$+ \frac{1}{\lambda} \int_0^T \int_{\partial\Omega} (\mathbf{g}^T_1(x,t) - \mathbf{g}^T_2(x,t)) \cdot \mathbf{h}^t(x,t)\, d\sigma\, dt = 0$$

Furthermore, use of (2.7), (2.8) and equation (4.6) yields

$$\int_0^T \int_\Omega (u_1^T(x,t) - u_2^T(x,t), \Lambda^i \frac{\partial v(x,t)}{\partial x^i})\, dx\, dt -$$

$$-\int_0^T \int_\Omega (\hat{G}(u_1^T(x,t)) - \hat{G}(u_2^T(x,t), \frac{\partial v(x,t)}{\partial t})\, dx\, dt -$$

$$-\int_0^T \int_\Omega [(F_1^T(x,t) - F_2^T(x,t), v(x,t))]\, dx\, dt + \tag{4.7}$$

$$+\int_\Omega (\hat{G}(u_1(x,0)) - \hat{G}(u_2(x,0)), v(x,T))\, dx +$$

$$+\frac{1}{\lambda} \int_0^T \int_\Omega (\mathbf{g}_1^T(x,t) - \mathbf{g}_2^T(x,t)) \cdot \mathbf{h}^t(x,t) d\sigma\, dt = 0$$

Also we note that:

$$\hat{G}(u_1^T) - \hat{G}(u_2^T) = \int_0^1 \frac{d}{d\tau} \hat{G}(\tau u_1^T + (1-\tau) u_2^T)\, d\tau =$$

$$= \int_0^1 \partial_u \hat{G}(\tau \upsilon_1^T + (1-\tau) u_2^T)\, d\tau \; (u_1^T - u_2^T) \tag{4.8}$$

from which by means of definition:

$$A(x,t) = \int_0^1 \partial_u \hat{G}(\tau \upsilon_1^T + (1-\tau) u_2^T)\, d\tau \tag{4.9}$$

we have from (4.8)

$$\hat{G}(u_1^T) - \hat{G}(u_2^T) = A(x,t)\, (u_1^T - u_2^T) \tag{4.10}$$

where A is a positive and symmetric tensor.
Furthermore, the equation (4.7) can be written as

$$\int_0^T \int_\Omega [(u_1^T - u_2^T, \Lambda^i \frac{\partial v}{\partial x^i} - A\frac{\partial v}{\partial t}) - (F_1^T - F_2^T, v)]\, dx\, dt + \tag{4.11}$$

$$+ \int_\Omega \hat{G}(u_1^0(x)) - \hat{G}(u_2^0(x)), v(x,T))dx + \frac{1}{\lambda} \int_0^T \int_\Omega (\mathbf{g}_1^T - \mathbf{g}_2^T) \cdot \mathbf{h}^t\, d\sigma\, dt = 0$$

where $u_1^0(x) = (E_1^0(x), H_1^0(x))$, $u_2^0(x) = (E_2^0(x), H_2^0(x))$.

We now consider the system:

$$A(x,t)\frac{\partial v(x,t)}{\partial t} = \Lambda^i \frac{\partial v(x,t)}{\partial x^i} + f(x,t) \tag{4.12}$$

together with the initial and boundary conditions

$$v(x,0) = 0 \tag{4.13}$$

$$\mathbf{e}^t(x,t) = \lambda\, \mathbf{h}^t(x,t) \times \mathbf{n}(x) \tag{4.14}$$

The problem (4.12), (4.13), (4.14) is the same system given by the equations (3.1), (3.2), (3.3), (3.7). Therefore from Theorem 3 there exists a solution $v \in H^1(0,\infty; L^2(\Omega)) \cap L^2(0,\infty; H^1(\Omega))$ such that $v(x,0) = 0$, for any $f \in L^2(0,\infty; L^2(\Omega))$.

Hence, for every $f \in H^1(0,\infty; L^2(\Omega))$ and by (4.9) we conclude

$$\int_0^T \int_\Omega (u_1^T - u_2^T, f)\, dx\, dt =$$

$$= \int_0^T \int_\Omega (F_1^T - F_2^T, v)\, dx\, dt - \int_\Omega (\hat{G}(u_1^0(x)) - \hat{G}(u_2^0(x)), v(x,T))\, dx - \tag{4.15}$$

$$- \frac{1}{\lambda} \int_0^T \int_{\partial\Omega} (\mathbf{g}_1^T - \mathbf{g}_2^T) \cdot \mathbf{h}^t\, d\sigma\, dt$$

It follows from a - b - c hypotheses that we must have:

$F_1 - F_2 \in L^2(0,\infty; L^2(\Omega))$

$v \in H^1(0,\infty; L^2(\Omega)) \cap L^2(0,\infty; H^1(\Omega))$

$\hat{G}(u_1^0) - \hat{G}(u_2^0) \in L^2(\Omega)$

$\mathbf{g}_1^T - \mathbf{g}_2^T \in H^1(0,\infty; L^2(\partial\Omega))$

Therefore the terms on the right hand side of (4.15) are functions of T belonging to $H^1(0,\infty)$. By virtue of well known existence and uniqueness theorem on the solutions of Volterra equations (see [11]), we can prove that $u_1 - u_2 \in L^2(0,\infty; L^2(\Omega))$.

References

[1] S.I. SCHELKUNOFF, Electromagnetic waves, D. Van Nostrand Company Inc., New York (1943).

[2] G. GRAFFI, Sulle condizioni al contorno approssimate nell'elettromagnetismo, *Acc. Scienze Ist. Bologna, Rendiconti*, Serie XI Tomo 5 (1958), 581-591.

[3] D.S. JONES, The Theory of electromagnetism, Pergamon Press (1964).

[4] L. CESARI, A boundary value problem for quasilinear hyperbolic systems in Schauder's canonic form, *Ann. Scuola Morm. Pisa,* Serie IV, vol.**1** (1974), 311-355

[5] M.FABRIZIO, Théorèmes d'unicité sur un nouveau probleme aux limites relatif aux équations non lineares de Maxwell. *J. Mecanique*, **15,** (1976), 681-696.

[6] B.D. COLEMAN and E.H. DILL, Thermodynamic restrictions on the constitutive equations of electromagnetic theory, *ZAMP*, **22,** (1971), 691-702.

[7] R.L. HIGDAN, Initial-bondary value problems for linear hyperbolic systems, *SIAM Rev.*, **28** (1986), 177-217.

[8] B. LAZZARI, Decay of solution of the Maxwell's equations in bounded domains with dissipative boundary conditions, in press.

[9] A. AROSIO, Asymptotic behaviour as $t \to \infty$ of the solutions of linear hyperbolic equations with coefficients discontinuous in time (on a bounded domain), *J.Diff.Equat.,* **39** (1981), 291-309.

[10] A. AROSIO, Linear second order differential equations in Hilbert spaces. The Cauchy problem and asymptotic behaviour for large time, *Arch.Rational Mech.Anal.,* **86,** (1984), 147-180.

[11] F. RIESZ et B. SZ-NAGY, Leçons d'Analyse Functionelle, Académie des Sciences de Hangrie, Budapest, 1955.

SOME REMARKS ON A SPATIAL LATTICE MODEL

Ester Gabetta

University of Ferrara, Department of Mathematics
Via Machiavelli, 35 - 44100 FERRARA (Italy)

Abstract. A spatial lattice model connected with the Broadwell model of the discrete kinetic theory is introduced.
After assigned random initial conditions, the evolution in terms of the transition probabilities is described.

1. Introduction

The discrete velocity models (henceforth abbreviated as DVM) of the Boltzmann equation provide a substitute of the original non-linear integrodifferential equation by means of a system of semilinear coupled partial differential equations. The models are formally obtained simply by assigning to the molecules of the gas a finite number of selected velocities, chosen in such a way that momentum and energy are conserved during the binary collision. Because of this fact, only a small number of simple DVM can be created.
A complete treatment of these models is contained in the books by GATIGNOL [1] and CABANNES [2].

At this stage it is not clear whether the DVM are a kinetic limit of some dynamical systems of particles, or they are only a model.

For the Boltzmann equation, after the pioneristic paper by O. LANFORD III [3] and the subsequently written papers by ILLNER and PULVIRENTI [4], [5], the situation is clearer, in the sense that the equation is obtained in the so-called Boltzmann-Grad limit procedure, that is starting by a system of N spheres in a fixed volume, each of radius r, when $Nr^2 \rightarrow$ const. the BBGKY hyerarchy converges to the Boltzmann equation.

This is not true for the DVM , as was recently discovered by UCHYAMA [6].

In effect the derivation, in the specific case of the Broadwell model, is only formal, since, roughly speaking, the chaos does not propagate.

A different attempt to establish analogies and eventually a derivation (formal as a first step) between the Broadwell plane model and the lattice gas introduced by HARDY, DE PAZZIS and POMEAU (the HPP model) [7] is contained in [8]. In this paper the main idea relies in assigning a sequence of matrices A_n of infinite rows and columns, such that the elements of the initial matrix A_0 are given randomly, and the values of the elements of A_n depend only on that of A_{n-1} (Markov structure) in such a way to obey the collision laws of the plane Broadwell model.

In the present talk we extend the previous methodology to a spatially regular DVM, namely the six-velocities Broadwell model.

2. Space lattice model

In this paragraph we will describe briefly how the spatial lattice model connected with the 6-velocities Broadwell model is derived.

Taking into account that the spatial Broadwell model describes the evolution of a ficticious gas whose molecules possess only six admissible velocities, that is $\vec{v}_1 = v\,\vec{i}$, $\vec{v}_2 = v\,\vec{j}$, $\vec{v}_3 = v\,\vec{k}$, $\vec{v}_{i+3} = -\vec{v}_i$, $i = 1, 2, 3$, the only possible binary collisions are the head to head $\left(\vec{v}_1, \vec{v}_4\right) \leftrightarrow \left(\vec{v}_2, \vec{v}_5\right) \leftrightarrow \left(\vec{v}_3, \vec{v}_6\right)$. As a consequence, the evolution equations are

$$\frac{\partial N_i}{\partial t} + \vec{v}_i \cdot \nabla N_i = N_{i+1}\,N_{i+4} + N_{i+2}\,N_{i+5} - 2\,N_i\,N_{i+3}$$
$$\text{(mod. 6)}$$

These equations are obtained considering that the collision (v_i, v_{i+3}) gives with equal probability (v_{i+1}, v_{i+4}) and (v_{i+2}, v_{i+5}) (mod. 6) .

Consider now a space lattice with mesh $v\,\Delta t$ whose knots are described by the point $(i,\,j,\,k)$.

Initially, a three-dimensional matrix A_0 with infinite elements in any direction is given; each elements a_{ijk} is in one-to-one correspondence with the given knots of the lattice and assumes the values $0, 1, \ldots, 6$ with prescribed probability.

The elements of the matrix A_1, and therefore generally the elements of the matrix A_n, are given according to the transition probabilities of a stationary Markov process which takes into account the ***structure*** of the Broadwell model.

To be more precise, let us consider a standard probability space $(\Omega,\,\mathcal{F},\,\mathcal{P})$, in which the initial probabilities on the a^0_{ijk} are given.

Then, the event:

$$\{a^1_{i,j,k} = 1\} = \{\text{in the knot } (i,\,j,\,k) \text{ after } \Delta t \text{ instance} \\ \text{there is a molecule with velocity } \vec{v}_1\}$$

happens iff:

$$\{[(a^0_{i-1,j,k} = 1) \cap (a^0_{i+1,j,k} \neq 4)] \cup [(a^0_{i,j-1,k} = 2) \cap (a^0_{i,j+1,k} = 5)] \\ \cup [(a^0_{i,j,k-1} = 3) \cap (a^0_{i,j,k+1} = 6)]\};$$

analogously when another velocity $\vec{v}_r$ is considered.

By a repeated use of classical tools of probability theory, it follows that:

$$\begin{aligned}
&P\,\{a^n_{i,j,k} = 1\} = \\
&= \alpha_1\,P\left\{(a^{n-1}_{i-1,j,k} = 1) \cap (a^{n-1}_{i+1,j,k} \neq 4)\right\} + \\
&+ \alpha_2\,P\left\{(a^{n-1}_{i,j-1,k} = 2) \cap (a^{n-1}_{i,j,+1,k} = 5)\right\} + \\
&+ \alpha_3\,P\left\{(a^{n-1}_{i,j,k-1} = 3) \cap (a^{n-1}_{i,j,k+1} = 6)\right\} + \\
&- \beta_1\,P\left\{(a^{n-1}_{i-1,j,k} = 1) \cap (a^{n-1}_{i+1,j,k} \neq 4) \cap (a^{n-1}_{i,j-1,k} = 2) \cap (a^{n-1}_{i,j+1,k} = 5)\right\} + \\
&- \beta_2\,P\left\{(a^{n-1}_{i-1,j,k} = 1) \cap (a^{n-1}_{i+1,j,k} \neq 4) \cap (a^{n-1}_{i,j,k-1} = 3) \cap (a^{n-1}_{i,j,k+1} = 6)\right\} + \\
&- \beta_3\,P\left\{(a^{n-1}_{i,j-1,k} = 2) \cap (a^{n-1}_{i,j+1,k} = 5) \cap (a^{n-1}_{i,j,k-1} = 3) \cap (a^{n-1}_{i,j,k+1} = 6)\right\} + \\
&+ \beta_4\,P\left\{(a^{n-1}_{i-1,j,k} = 1) \cap (a^{n-1}_{i+1,j,k} \neq 4) \cap (a^{n-1}_{i,j-1,k} = 2) \cap (a^{n-1}_{i,j+1,k} = 5)\right. \\
&\qquad \left.\cap\,(a^{n-1}_{i,j,k-1} = 3)\, \cap (a^{n-1}_{i,j,k+1} = 6)\right\}
\end{aligned}$$

where

$$\alpha_1 = P\left\{\left(a^n_{i,j,k} = 1\right) / \left[\left(a^{n-1}_{i-1,j,k} = 1\right) \cap \left(a^{n-1}_{i+1,j,k} \neq 4\right)\right]\right\}$$

$$\alpha_2 = P\left\{\left(a^n_{i,j,k} = 1\right) / \left[\left(a^{n-1}_{i,j-1,k} = 2\right) \cap \left(a^{n-1}_{i,j+1,k} = 5\right)\right]\right\}$$

$$\alpha_3 = P\left\{\left(a^n_{i,j,k} = 1\right) / \left[\left(a^{n-1}_{i,j,k-1} = 3\right) \cap \left(a^{n-1}_{i,j,k+1} = 6\right)\right]\right\}$$

and where analogously the β_i represent the conditioned probabilities with respect to the intersections of more than two events.

A first step through a closer analogy to the Broadwell model is achieved by putting:

$$\alpha_1 = 1; \qquad \alpha_2 = \alpha_3 = \frac{1}{2}; \qquad \beta_i = 0, \qquad i = 1, 2, 3, 4.$$

In this case the following evolution for the probabilities is found:

$$\begin{aligned} P\left\{a^n_{i,j,k} = 1\right\} = P\left\{a^{n-1}_{i-1,j,k} = 1\right\} - P\left\{\left(a^{n-1}_{i-1,j,k,} = 1\right) \cap \left(a^{n-1}_{i+1,j,k} = 4\right)\right\} + \\ + \frac{1}{2} P\left\{\left(a^{n-1}_{i,j-1,k} = 2\right) \cap \left(a^{n-1}_{i,j+1,k} = 5\right)\right\} + \\ + \frac{1}{2} P\left\{\left(a^{n-1}_{i,j,k-1} = 3\right) \cap \left(a^{n-1}_{i,j,k+1} = 6\right)\right\} \end{aligned}$$

that seems perfectly in accord with the discrete model.

Looking deeply inside, unfortunately one notes that there is no propagation of chaos.

In fact if we consider for example the event:

$$\left\{\left(a^n_{i-1,j,k} = 1\right) \cap \left(a^n_{i+1,j,k,} = 4\right)\right\} \tag{2.1}$$

denoting by $\mathcal{A} = \left(a^n_{i-1,j,k,} = 1\right)$, $\mathcal{B} = \left(a^n_{i+1,j,k,} = 4\right)$ and

$$A_1 = \left\{\left(a^{n-1}_{i-2,j,k} = 1\right) \cap \left(a^{n-1}_{i,j,k} \neq 4\right)\right\}$$

$$B_1 = \left\{\left(a^{n-1}_{i,j,k} \neq 1\right) \cap \left(a^{n-1}_{i+2,j,k} = 4\right)\right\}$$

$$A_2 = \left\{\left(a^{n-1}_{i-1,j-1,k} = 2\right) \cap \left(a^{n-1}_{i-1,j+1,k} = 5\right)\right\}$$

$$B_2 = \left\{\left(a^{n-1}_{i+1,j-1,k} = 2\right) \cap \left(a^{n-1}_{i+1,j+1,k} = 5\right)\right\}$$

$$A_3 = \left\{\left(a^{n-1}_{i-1,j,k-1} = 3\right) \cap \left(a^{n-1}_{i-1,j,k+1} = 6\right)\right\}$$

$$B_3 = \left\{\left(a^{n-1}_{i+1,j,k-1} = 3\right) \cap \left(a^{n-1}_{i+1,j,k+1} = 6\right)\right\}$$

the formula (2.1) became:

$$\begin{aligned} P(\mathcal{A}\cap\mathcal{B}) &= P[(\mathcal{A}\cap\mathcal{B})\cap(A_1\cap B_1)\cup(\mathcal{A}\cap\mathcal{B})\cap(A_1\cap B_2)\cup\ldots \\ &\quad \cup(\mathcal{A}\cap\mathcal{B})\cap(A_3\cap B_3)] = \\ &= P[(\mathcal{A}\cap\mathcal{B})\cap(A_1\cap B_1)] + P[(\mathcal{A}\cap\mathcal{B})\cap(A_1\cap B_2)] + \ldots \\ &\quad + P[(\mathcal{A}\cap\mathcal{B})\cap(A_3\cap B_3)] \end{aligned}$$

It is evident that A_1 and B_1 are not independent events, and so:

$$P(\mathcal{A}\cap\mathcal{B}) \neq P(\mathcal{A})\,P(\mathcal{B})$$

and this essentially means that chaos (independence of events) does not propagate.

Some numerical applications suggest that, when the rarefaction is very high and the initial distribution is spatially homogeneous, the equilibrium state (global maxwellian) of the space homogeneous Broadwell model, and the steady state of the above described Markov process, are close to each other.

To conclude, the DVM seem to be related to lattice gas models, instead of to the dynamics of N bodies.

REFERENCES

1) Gatignol, R., "Théorie Cinétique des Gas à Répartition Discrète de Vitesse", *Lect. Notes in Phys.*, 36 (1975), Springer Verlag, Berlin.

2) Cabanne, H., "The Discrete Boltzmann Equation: Theory and Application", *Lect. Notes at the Univ. of California*, Berkeley, (1980).

3) Lanford III, O., "On a Derivation of the Boltzmann equation", *Astérisque*, 40, 117-137 (1976).

4) Illner, R. and Pulvirenti, M., "Global validity of the Boltzmann equation for two-dimensional rare gas in vacuum", *Comm. Math. Phys.*, 105, 189-203 (1986).

5) Pulvirenti, M., "Global validity of the Boltzmann equation for a three-dimensional rare gas in vacuum", *Comm. Math. Phys.*, 113, 79-85 (1987).

6) Uchiyama, K., "On the Boltzmann-Grad limit for the Broadwell model of the Boltzmann equation", *J. of Stat. Phys.* 52, 331-355 (1988).

7) Hardy, J., de Pazzis, O. and Pomeau, Y., "Molecular dynamics of a classical lattice gas: Transport properties and time correlation functions", *Phys. Review A*, 13, 5 (1976).

8) Gabetta, E., "From stochastic Mechanics to the Discrete Boltzmann Equation: The Broadwell model", to appear on *Mathematical and Computer Modelling.*

A CONTINUOUS STATE FEEDBACK CONTROL FOR UNCERTAIN DYNAMICAL SYSTEMS

Lucia Gambardella

Dipartimento di Matematica "G. Castelnuovo"
Università degli Studi di Roma "La Sapienza", Piazzale A. Moro 2,
00185 ROMA.

ABSTRACT. A dynamic system containing uncertain elements is considered and only the set of possible values of these uncertainties is known. Based on this information a state feedback control is proposed in order to guarantee the uniform eventual stability and the uniform actractivity in the large of the origin. This feedback control is a continuos function of the state.

INTRODUCTION.

The problem of designing a state feedback control for uncertain dynamical systems has been discussed in many papers. For instance, in |1|-|4| it is designed a control (in general discontinuous function of the state) that guarantees the uniform asymptotic stability of an equilibrium state, in |5|-|10| it is proposed a control which depends continuously on the state and guarantees the uniform ultimate boundedness of the solutions of the uncertain system. In these papers no statistical information is ever required for uncertainty but only its bound is utilized, and the system performance is described in a deterministic manner. Generally the way to generate the feedback control is based on the Lyapunov min-max design and, therefore, once a controller can be generated it guarantees the stability of the closed-loop system for all uncertainty since it is first designed based on the assumption that the "worst case" might happen.

In the present paper, by using this approach, we propose a continuous state feedback control for the uncertain system, that guarantees the uniform eventual stability and the uniform attractivity in the large of the origin, which is for the uncontrolled nominal system (i.e. the system in the absence of control and uncertainty) an uniformly asymptotically stable in the large equilibrium point. By assuming that only the sets in which the values of the uncertain quantities can range are known, we design a control which guarantees not only that every solution enters a neighborhood of the origin in finite time (uniform ultimate boundedness) but also that it happens for every neighborhood.

To illustrate the utilization of the obtained results we consider a simple pendulum subjected to a control moment and a disturbance. Only the "possible size" of the disturbance is known.

1. MAIN RESULTS

Let us consider the nonlinear uncertain dynamical system described by

$$\dot{x}(t) = f(t,x(t)) + B(t,x(t))u(t) + B(t,x(t))e(t,x(t)) \tag{1.1}$$

where $t \in I$, $I=[\tau,+\infty[$, $\tau \in R$, $x(t) \in R^n$ is the state, $u(t) \in R^m$ is the control and $e(t,x(t))$ is the "lumped" uncertain element. The "nominal" system, that is, the system without uncertainty, is described by

$$\dot{x}(t) = f(t,x(t)) + B(t,x(t))u(t) \ ,$$

where $f(.): I \times R^n \to R^n$ and $B(.) : I \times R^n \to R^{n \times m}$ are known. Concerning the system described by (1.1) we introduce the following assumptions:

i) the norm of the uncertainty is bounded by a known function $\rho(.) : I \times R^n \to R^+$, that is

$$\| e(t,x) \| \leq \rho(t,x) \qquad \forall (t,x) \in I \times R^n \tag{1.2}$$

ii) the known functions f, B, ρ as well as the unknown function e are continuous in x for all $t \in I$ and for all $x \in R^n$ are Lebesgue measurable in t;

iii) given a compact set $E \subset R^n$ and a compact interval $[a,b] \subset I$, there exist Lebesgue integrable functions $m_i(.): [a,b] \to R^+$, $i=1,2$, such that for all $(t,x) \in [a,b] \times E$

$$\|f(t,x)\| \leq m_1(t) \ , \quad \|B(t,x)\|\rho(t,x) \leq m_2(t)$$

iv) the origin $x=0$ is uniformly asymptotically stable for the uncontrolled nominal system

$$\dot{x}(t) = f(t,x(t)) \tag{1.3}$$

and, in particular, there are a function $V \in C^1 [I \times R^n, R^+]$ and continuous, strictly increasing functions $\gamma_i(.): R^+ \to R^+$, $i=1,2,3$, such that

$$\gamma_i(0) = 0 \qquad i=1,2,3 \tag{1.4}$$

$$\lim_{r \to +\infty} \gamma_i(r) = +\infty \qquad i=1,2 \tag{1.5}$$

$$\gamma_1 (\|x\|) \leq V(t,x) \leq \gamma_2(\|x\|) \qquad \forall (t,x) \in I \times R^n \tag{1.6}$$

$$\dot{V}(t,x)\big|_{(1.3)} \leq -\gamma_3(\|x\|) \qquad \forall (t,x) \in I \times R^n \tag{1.7}$$

where $\dot{V}(t,x)\big|_{(.)}$ denotes the derivative of the function V along the solutions of system (.).

In the following, corresponding to $(t_0,x_0)\in I\times R^n$, we shall denote by $x(t,t_0,x_0)$ the solution of system (1.1) corresponding to initial data (t_0,x_0), that is the solution such that $x(t_0,t_0,x_0)=x_0$.

Definition 1.1. The origin $x=0$ is eventually uniformly stable for system (1.1) if

$$\forall\, \varepsilon>0\ \exists\ \delta(\varepsilon)>0,\ T_1(\varepsilon)>0 : \|x_0\|<\delta,\ t_0>T_1 \Rightarrow \|x(t,t_0,x_0)\|<\varepsilon\ \forall t\geq t_0 .$$

Definition 1.2. The origin $x=0$ is uniformly attractive in the large for system (1.1) if $\forall\ \nu>0,\ \alpha>0\ \exists\ T_2(\nu,\alpha)>0 : \|x_0\|<\alpha,\ t_0\in I \Rightarrow \|x(t,t_0,x_0)\|<\nu\ \forall\ t\geq t_0+T_2$.

Theorem 1.3. If we suppose that system (1.1) satisfies assumptions i) - iv), than it is possible to determine a state feedback control $u(t)$ such that the following results hold

1) $\forall\ (t_0,x_0)\in I\times R^n$ there exists one and only one solution $x(t,t_0,x_0)$ defined for all $t\in[t_0,+\infty[$

2) the origin $x=0$ is eventually uniformly stable and uniformly attractive in the large.

Proof. By setting $\forall(t,x)\in I\times R^n$ $\mu(t,x) \doteq B^T(t,x)\nabla_x V(t,x)\rho(t,x)$, we consider the control function

$$u(t,x) = \frac{-\mu(t,x)\,\rho(t,x)}{\|\mu(t,x)\| + p\sigma(t)} \tag{1.8}$$

where p is a positive constant and $\sigma: I\to R^+$ is such that $0<\sigma(t)<k\ \ \forall t\in I,\ k\in]0,+\infty[$, and

$$\int_\tau^{+\infty} \sigma(t)\,dt < +\infty$$

Then, by assuming function u given by (1.8) as control in system (1.1), we have by (1.7) and (1.2) $\forall\ (t,x)\in I\times R^n$

$$\begin{aligned}\dot V(t,x)\big|_{(1.1)} &\leq -\gamma_3(\|x\|)-\nabla_x{}^T V(t,x)B(t,x)\mu(t,x)\rho(t,x)/(\|\mu(t,x)\|+\\ &\quad p\sigma(t))+\nabla_x{}^T V(t,x)B(t,x)e(t,x) \leq -\gamma_3(\|x\|) -\\ &\quad \|\mu(t,x)\|^2/(\|\mu(t,x)\|+p\sigma(t))+\|\mu(t,x)\| \leq -\gamma_3(\|x\|)+p\sigma(t)\end{aligned} \tag{1.9}$$

Then, by reasoning as in |7|, it is possible to show the uniform boundedness of the solutions of system (1.1), that is: if $x(t,t_0,x_0)$ is a solution of (1.1) defined in the interval $[t_0,t_1]\subset I$, then $\forall\ r\in]0,+\infty[$

$$\|x_0\|\leq r \Rightarrow \|x(t,t_0,x_0)\|\leq d(r) \qquad \forall t\in[t_0,t_1]$$

$$d(r)=\begin{cases}(\gamma_1{}^{-1}\circ\gamma_2)(Q) & \text{if } r\leq Q\\ (\gamma_1{}^{-1}\circ\gamma_2)(r) & \text{if } r>Q\end{cases} \qquad Q=\gamma_3{}^{-1}(pk). \tag{1.10}$$

In fact, consider a solution $x(t,t_0,x_0) : [t_0,t_1]\to R^n$ with $\|x_0\|<r$ and let $r^*\doteq\max(r,Q)$. By (1.10) $d(r)=(\gamma_1{}^{-1}\circ\gamma_2)(r^*)$ and by (1.6) $\gamma_1(r^*)\leq\gamma_2(r^*)$ so that $r^*\leq(\gamma_1{}^{-1}\circ\gamma_2)(r^*)$. Therefore $\|x(t_0,t_0,x_0)\|=\|x_0\|\leq r^*\leq d(r)$.

Now we suppose that

$$\exists\ t'\in[t_0,t_1] : \|x(t',t_0,x_0)\| > d(r). \tag{1.11}$$

Then there exists $t''\in[t_0,t_1]$ such that $\|x(t'',t_0,x_0)\|=r^*$ and $r^*\le\|x(t,t_0,x_0)\|\le d(r)$ $\forall t\in[t'',t']$ and by (1.9), (1.6) and (1.10) we have

$$\gamma_1(\|x(t',t_0,x_0)\|\le V(t',x(t',t_0,x_0))\le\gamma_2(\|x(t'',t_0,x_0)\| + \int_{t''}^{t'}(-\gamma_3(\|x(t,t_0,x_0)\|+p\sigma(t))dt\le$$
$$\le\gamma_2(r^*) + \int_{t''}^{t'}(-\gamma_3(Q) + pk)dt = \gamma_2(r^*).$$

Hence $\|x(t',t_0,x_0)\|\le(\gamma_1{}^{-1}o\gamma_2)(r^*) = d(r)$ that contradicts (1.11). Then the solutions of (1.1) are uniformly bounded and, therefore, have a continuation over $[t_0,+\infty[$. Furthermore, by Th.17.4 of |11| the origin x=0 is eventually uniformly stable.

In order to prove the uniform attractivity in the large of the origin we consider $\nu>0$, $q>0$ and, correspondently to x_0, $\|x_0\|\le q$, $t_0\in I$ the maximal solution $x(t,t_0,x_0)$.

By the uniform eventual stability of the origin there exist $\delta(\nu)>0$, $T_1(\nu)>0$: $\|x^*_0\|<\delta(\nu)$, $t^*_0 >T_1(\nu) \Rightarrow \|x(t,t^*_0,x^*_0)\|<\nu\ \forall\ t\ge t^*_0$ and by the uniform boundedness of the solutions there exists $d(q)>0$ such that $\|x(t,t_0,x_0)\|\le d(q)$ $\forall t\ge t_0$. Furthermore, by (1.6) and (1.9) there exists $t^*>T_1(\nu)$ such that $\|x(t^*,t_0,x_0)\|<\delta(\nu)$; in fact, if $\delta(\nu)\le\|x(t,t_0,x_0)\|\le d(q)$ $\forall t>T_1(\nu)$, by (1.9) we have $\lim_{t\to+\infty} V(t,x(t,t_0,x_0)) = -\infty$, an absurd by (1.6).

Then $\|x(t,t_0,x_0)\|=\|x(t,t^*,x(t^*,t_0,x_0))\|<\nu$ $\forall t>t^*$, that is the origin is uniformly attractive in the large.

2. EXAMPLE

To illustrate the utilization of Th.1.3, consider the second order nonlinear uncertain controlled system

$$\begin{aligned}\dot{x}_1 &= x_2\\ \dot{x}_2 &= -a\, f_1(x_1) + u(t) - \gamma(t)f_2(x_1)\end{aligned} \tag{2.1}$$

where $(x_1,x_2)\in R^2$, $t\in[0,+\infty[$, a is a positive constant, $f_1(.), f_2(.) : R\to R$ are such that

$$\left.\begin{aligned}&|f_1(x_1)| \le w\ \ \forall x_1\in R,\ w\in]0,+\infty[, \quad f_1(0)=0\\ &\exists\ m\in[w,+\infty[: x_1 f(x_1)>0\ \ \forall x_1\in]-m,m[\setminus\{0\}\end{aligned}\right| \tag{2.2}$$

$$\left.\begin{aligned}&f_2\in C^1(R),\ \ |f_2(x_1)|\le 1\ \ \forall x_1\in R,\ f_2(0)=1\\ &df_2(x_1)/dx_1 = -f_1(x_1)\ \ \forall x_1\in R,\\ &\exists\ \alpha\in]0,1[: |1-f_2(x_1)|\le |x_1|\ \ \forall x_1\in]-\alpha,\alpha[\end{aligned}\right| \tag{2.3}$$

and the uncertainty $\gamma :[0,+\infty[\to R$ is such that $|\gamma(t)|<\rho(t)$ $\forall t\in[0,+\infty[$, being ρ a known function.

The uncontrolled nominal system corresponding to (2.1) $(u(t)=\gamma(t)=0)$ may possesse more than one equilibrium solutions, none of which is asymptotically stable.

However, by using a suitable linear feedback u one can obtain a new uncontrolled nominal system for which $x_1 \equiv x_2 \equiv 0$ is the only equilibrium state and is uniformly asymptotically stable.

In fact if we set in (2.1)

$$u(t) = -b\, x_1 - c\, x_2 + u^*(t) \tag{2.4}$$

where b, c are positive constants, $b > a$, and u^* is a control of type (1.8), we have the system

$$\begin{aligned} \dot{x}_1 &= x_2 \\ \dot{x}_2 &= -a\, f_1(x_1) - b\, x_1 - c\, x_2 + u^*(t) - \gamma(t) f_2(x_1) \end{aligned} \tag{2.5}$$

which admits the uncontrolled nominal system

$$\begin{aligned} \dot{x}_1 &= x_2 \\ \dot{x}_2 &= -a\, f_1(x_1) - b\, x_1 - c\, x_2 \end{aligned} \tag{2.6}$$

System (2.6) has $x_1 \equiv x_2 \equiv 0$ as its only equilibrium state. Furthermore it is easy to see that the null solution is uniformly asymptotically stable by assuming as Lyapunov function

$$V(x_1,x_2) = (b + c^{2}/2)x_1^2 + c\, x_1 x_2 + x_2^2 + 2a\,(1 - f_2(x_1)) \tag{2.7}$$

In fact $\forall\, (x_1,x_2) \in R^2$

$$\gamma_1(\|(x_1,x_2)\|) \le V(x_1,x_2) \le \gamma_2(\|(x_1,x_2)\|, \qquad \dot{V}(x_1,x_2)\big|_{(2.6)} \le -\gamma_3(\|(x_1,x_2)\|)$$

where

$$\gamma_1(\|(x_1,x_2)\|) = \lambda_1 \|(x_1,x_2)\|^2,$$

$$\gamma_2(\|(x_1,x_2)\|) = \begin{cases} \lambda_2\|(x_1,x_2)\|^2 + 2a\,\|(x_1,x_2)\| & \text{if } \|(x_1,x_2)\| \le \alpha \\ \lambda_2\|(x_1,x_2)\|^2 + 2a[\alpha_1\alpha + (2-\alpha)(\|(x_1,x_2)\| - \alpha)]/\alpha_1 & \\ & \text{if } \alpha < \|(x_1,x_2)\| \le \alpha + \alpha_1, \quad \alpha_1 \in]0,+\infty[\\ \lambda_2\|(x_1,x_2)\|^2 + 4a & \text{if } \|(x_1,x_2)\| > \alpha + \alpha_1 \end{cases}$$

$$\gamma_3(\|(x_1,x_2)\|) = \lambda_3 \|(x_1,x_2)\|^2$$

$$\lambda_1 = \lambda_{max}[P], \quad \lambda_2 = \lambda_{min}[P], \quad \lambda_3 = \min\{c, cb\},$$

$$[P] = \begin{bmatrix} b + c^2/2 & c/2 \\ c/2 & 1 \end{bmatrix}$$

Then, by assuming

$$u^*(t,x_1,x_2) = -\frac{c x_1 + 2x_2}{|c\, x_1 + 2x_2| + p\sigma(t)/(\rho(t)+1)}\,\rho(t) \tag{2.8}$$

(p and σ as in (1.8)), all the hypotheses of Th.1.3 are satisfied for system (2.1) and, therefore, the origin is eventually uniformly stable and uniformly attractive in the large.

REMARK.

We observe that if in (2.1) $f_1(x_1) = \sin x_1$, $f_2(x_1) = \cos x_1$, $\gamma(t) = \gamma^*(t)/l$ $l \in [0,+\infty[$, we obtain

$$\begin{aligned} \dot{x}_1 &= x_2 \\ \dot{x}_2 &= -a \sin x_1 + u(t) + \cos x_1\ \gamma^*(t)/l \end{aligned} \tag{2.9}$$

which are the equations of a simple pendulum of lenght l subjected to a control moment u (per unit mass) and a disturbance $\gamma^*(t)$. If there exists a known function $\rho(t)$ such that $|\gamma^*(t)| \leq \rho(t)\, l$, all the results obtained for system (2.1) are available for system (2.9).

REFERENCES

|1| S.Gutman and G.Leitmann, "Stabilizing control for linear systems with bounded parameters and input uncertainty", 7th IFIP Conf. Optimization Techniques. Berlin, Germany: Springer Verlag, 1976.

|2| S.Gutman and G.Leitmann, "Stabilizing feedback control for dynamical systems with bounded uncertainty", in Proc. Conf. on Decision Contr., 1976.

|3| S.Gutman, "Uncertain dynamical systems - A Lyapunov min-max approach", IEEE Trans. Automat. Contr., vol. AC-24, 1979.

|4| G.Leitmann, "Guaranteed asymptotic stability for some linear systems with bounded uncertainties", J. Dynamic Syst., vol. 101, no. 3, 1979.

|5| G. Leitmann, " On the efficacy of nonlinear control in uncertain linear systems ", J. Dynamic Syst., vol. 103, no. 2, 1981.

|6| G.Leitmann, "Guaranteed ultimate boundedness for a class of uncertain linear dynamical systems"' in Proc. 3rd Kingston Conf. on Differential Games and Control Throry. New York: Dekker, 1979.

|7| M.J.Corless and G.Leitmann, "Continuous state feedback guaranteeing uniform ultimate boundedness for uncertain dynamic system', IEEE Trans. Autom. Contr., vol. AC-26, no. 5, 1981.

|8| B.R.Barmish, M.Corless and G.Leitmann, "A new class of stabilizing controllers for uncertain dynamical systems", SIAM J. Contr. Optimiz., vol. 21, 1986.

|9| Y.H.Chen, "On the deterministic performance of uncertain dynamical systems", Int. J. Contr., vol. 43, 1986.

|10| Y.H.Chen, "Design of robust controllers for uncertain dynamical systems", IEEE Trans. Autom. Contr., vol. 33, no. 5, 1988.

|11| T. Yoshizawa, "Stability theory by Liapunov's second method", The Mathematical Society of Japan, 1966.

STABILITY OF THE ELECTRIC FIELD IN A PARTIALLY IONIZED PLASMA UNDER SMALL PLANE WAVE PERTURBATIONS

B.D. Ganapol[+], S. Oggioni[*], and G. Spiga[o]

+ Department of Nuclear and Energy Engineering, University of Arizona, Tucson, AZ 85721, U.S.A.
* Istituto per Ricerche di Matematica Applicata del C.N.R., Campus Universitario, I-70125 Bari, Italy.
o Dipartimento di Matematica, Università di Bari, Via G. Fortunato, I-70125 Bari, Italy.

ABSTRACT

The selfconsistent electric field in a Vlasov plasma is computed, taking the effects of short range binary collisions into account. Previous analytical predictions about stability are verified and stabilizing effects, with respect to the classical description with collective electrical interactions only, are shown.

1. INTRODUCTION

The analysis of plasma oscillations, and of their stability, has been of considerable interest in plasma physics since the pioneering work of Vlasov[1)]. It is known that, in many physical situations, quite good predictions on the onset of instability are already given by a very elementary approach like the linearized Vlasov-Poisson equations[2)], in which the effects of collisions on the electron distribution function are described in terms of long-range collective interactions, accounted for by the selfconsistent electric field E. For longitudinal plane wave initial perturbations with fixed wave number k, it has been shown that the plasma is actually linearly unstable for some meaningful equilibrium distributions, like two streams and bump on tail[3]. More recently, a generalized version of the Vlasov-Poisson equations has been proposed[4], with the addition of an explicit short-range collision term, in order to take binary encounters (not negligible, especially in the presence of neutral particles) into account.

The present paper deals with the above basic stability problem. The small perturbation f with respect to the equilibrium distribution NF(v) is taken to depend only on time $t\epsilon(0,+\infty)$, longitudinal velocity $v\epsilon(-\infty,+\infty)$, and direction of propagation $z\epsilon(-\infty,+\infty)$. N denotes the constant electron density (equal to the density of ions, constituting, together with neutral particles, an immobile uniform background), and F is a differentiable function, such that $F,F'\epsilon L_1$, normalized to unity, allowing no electric current in the plasma. The governing Vlasov-Poisson equations read as

$$\begin{cases} \dfrac{\partial f}{\partial t} + v\dfrac{\partial f}{\partial x} - \dfrac{N\ e}{m} F'(v)\ E(z,t) = \left(\dfrac{\partial f}{\partial t}\right)_{coll} \\ \dfrac{\partial E}{\partial z} = -\ 4\pi e \displaystyle\int_{-\infty}^{\infty} f(z,v',t)dv' \end{cases} \tag{1}$$

where for the additional short range collision term a simple synthetic BGK model[5]

$$\left(\frac{\partial f}{\partial t}\right)_{coll} = -\ \nu f(z,v,t) + \nu F(v)\int_{-\infty}^{\infty} f(z,v',t)dv' \tag{2}$$

will be adopted, with collision frequency $\nu\geq0$ as a measure of the weight of individual collisions on the physical process. For a plane wave initial disturbance, setting $f(z,v,t)=f_k(v,t)\exp(ikz)$, the unknown $f_k(v,t)$ can be expressed analytically in terms of suitable integrals[6,7]. for any initial datum $f_{ko}(v)$ and equilibrium distribution F(v). The type of the semigroup solving the problem is equal to the spectral bound of its generator, whose spectrum is constituted by an essential part

$$\mathrm{Re}\,\lambda = -\nu \tag{3}$$

plus possibly isolated eigenvalues of finite multiplicity (due to a compact one-dimensional perturbation) which are the zeroes of the dispersion function

$$\Lambda(\lambda) = 1 - i\,\frac{\omega^2}{k}\int_{-\infty}^{\infty}\frac{F'(v)}{\lambda+\nu+ikv}\,dv - \nu\int_{-\infty}^{\infty}\frac{F(v)}{\lambda+\nu+ikv}\,dv, \tag{4}$$

where $\omega=(4\pi Ne^2/m)^{\frac{1}{2}}$ is the plasma frequency. Λ is a sectionally analytic function in the complex λ plane cut along the straight line (3). The asymptotic trend is thus determined by the rightmost eigenvalue, if any, in the half plane $\mathrm{Re}\,\lambda > -\nu$, and stability is simply related to the location of such an eigenvalue in the half plane $\mathrm{Re}\,\lambda > 0$ or in the strip $-\nu < \mathrm{Re}\,\lambda < 0$. The argument principle can be conveniently used in the investigation, and in this respect the Plemelj formulae have to be employed in order to evaluate the limiting values of Λ on the right edge of the cut. Thus, for any given F, determination of stability and number of eigenvalues is only matter of computation of single integrals of continuous functions, that can be performed numerically by standard methods for varying ν,ω,k. Actual stabilization of configurations that would be unstable for purely electrical interactions ($\nu=0$) are observed for increasing ν, and the stability threshold ν_c can be easily calculated[7].

In the following sections, attention will be focused on explicit evaluation of the most outstanding macroscopic physical quantity, namely the electric field $E(z,t)=E_k(t)\exp(ikz)$, whose behavior is crucial for all practical applications.

2. EVALUATION OF THE ELECTRIC FIELD

The analysis sketched in the Introduction gives for the Laplace transform of the selfconsistent electric field

$$\tilde{E}_k(\lambda) = -\frac{4\pi e}{ik}\frac{1}{\Lambda(\lambda)}\int_{-\infty}^{\infty}\frac{f_{ko}(v)}{\lambda+\nu+ikv}\,dv \tag{5}$$

with

$$E_{ko} = -\frac{4\pi e}{ik}\int_{-\infty}^{\infty} f_{ko}(v)\,dv \tag{6}$$

for the initial value of $E_k(t)$. The inverse Laplace transform can be performed by resorting to a recently developed numerical algorithm[8], that has proved very effective in solving neutron transport problems. It will be applied here to the simple case of a delta initial condition, $f_{ko}(v)=\delta(v)$, and of an even equilibrium distribution $F(v)$, by which the zero current condition is satisfied a priori. In addition, eigenvalues are either real or complex conjugate, and the ratio $\varphi_k(t)=E_k(t)/E_{ko}$ is always real. The method consists in recasting the electric field as

$$\varphi_k(t) = \frac{2\exp(\gamma t)}{\pi t}\left\{\int_0^{\pi/2}\mathrm{Re}[\tilde{\varphi}_k(\gamma+iu/t)]\cos u\,du + \sum_{j=1}^{\infty}(-1)^j\int_{-\pi/2}^{\pi/2}\mathrm{Re}[\tilde{\varphi}_k(\gamma+i\frac{u+j\pi}{t})]\cos u\,du\right\}, \tag{7}$$

where γ is any real number greater than the rightmost singularity of Λ. This is made possible by elementary properties of Laplace transforms, and amounts to partitioning the inversion integral in half periods of the cosine function. The finite regular integrals in (7) may be evaluated by using an efficient Romberg integration rule, and, for the final series to be summed, a very satisfactory convergence rate can be achieved by an Euler-Knopp acceleration technique.

By this procedure, not only the fundamental plasma parameter is obtained versus input data, but also analytical predictions on stability and asymptotic behavior can be verified. Moreover, initial transients (where usually the strongest and fastest fluctuations take place), as well as possible occurrence and macroscopic importance of oscillations (contributed by the cut or by imaginary parts of eigenvalues) can be seen directly. The algorithm hes been demonstrated already in Ref.9 in the limiting case of the two stream equilibrium distribution

$$F(v) = \tfrac{1}{2}\,\delta(v-v_o) + \tfrac{1}{2}\,\delta(v+v_o)\ , \tag{8}$$

in which the discontinuity across the cut disappears and the Laplace inversion can be performed also by a contour integral technique, collecting residues at isolated singularities, that provides a useful comparison test for the numerical procedure. In this paper, the electric field will be computed also for more physical equilibria, like the Maxwellian

$$F(v) = \left(\frac{m}{2\pi KT}\right)^{\frac{1}{2}} \exp\left(-\frac{mv^2}{2KT}\right) \tag{9}$$

and the symmetric double Maxwellian distribution

$$F(v) = \frac{1}{2}\left(\frac{m}{2\pi KT}\right)^{\frac{1}{2}} \left\{\exp\left[-\frac{m(v-v_o)^2}{2KT}\right] + \exp\left[-\frac{m(v+v_o)^2}{2KT}\right]\right\} . \tag{10}$$

It is known that (9) is neutrally stable already for the classical Vlasov plasma $\nu=0$, and becomes asymptotically stable for positive ν, with the fastest approach to equilibrium taking place for intermediate values of ν[6]. The distribution (10), whose limit for $T\to 0$ reproduces case (8), turns out to be electrically unstable in a certain region of parameter space, and is stabilized when ν is increased past a suitable threshold ν_c[7]. A sample of selected cases is presented in the next section.

3. RESULTS AND COMMENTS

The first example refers to the equilibrium distribution (8). In terms of dimensionless parameters

$$\tilde{\omega} = \omega/kv_o \qquad \tilde{\nu} = \nu/kv_o \qquad \tilde{t} = kv_o t \tag{11}$$

the plasma is always stable for $\tilde{\omega}<1$, but exhibits the well known two stream instability for $\tilde{\omega}\geq 1$ and $\tilde{\nu}=0$. In the latter case, it stabilizes with the addition of binary collisions, and the stability threshold is

$$\tilde{\nu}_c = [(\tilde{\omega}^2-1)/(\tilde{\omega}^2+1)]^{\frac{1}{2}}. \tag{12}$$

In the neutrally stable case $\tilde{\nu}=\tilde{\nu}_c>0$ (simple eigenvalue at $\lambda=0$), the electric field seeks a nonzero equilibrium state depending upon $\tilde{\omega}$, that can be determined on the basis of the Tauberian theorem as

$$\lim_{t\to\infty} \frac{E_k(t)}{E_{ko}} = \frac{2\tilde{\omega}^4}{(\tilde{\omega}^2-1)(\tilde{\omega}^4+3\tilde{\omega}^2+1)} . \tag{13}$$

When $\tilde{\omega}=1$ and $\tilde{\nu}$ is increased from near zero, three eigenvalues are present. At first, the rightmost ones, moving into the left half plane, are complex conjugate, then the real one takes over; it keeps decreasing until it reaches a minimum $\lambda=-kv_o$ at $\tilde{\nu}=2$, and then reverses direction and approaches again the origin with further increase in $\tilde{\nu}$. All features are confirmed by the computed electric field. In Fig.1, the ratio $E_k(\tilde{t})/E_{ko}$ is plotted versus $\tilde{t}$ when $\tilde{\omega}=2$ for $\tilde{\nu}$ varying in a neighborhood of $\tilde{\nu}_c=0.7746$; the limiting value 0.3678 for $\tilde{\nu}=\tilde{\nu}_c$ is also in agreement with Eq.(13). Analogously, in Fig.2, where $\tilde{\omega}=1$, oscillations in $E_k(\tilde{t})/E_{ko}$ actually occur for small

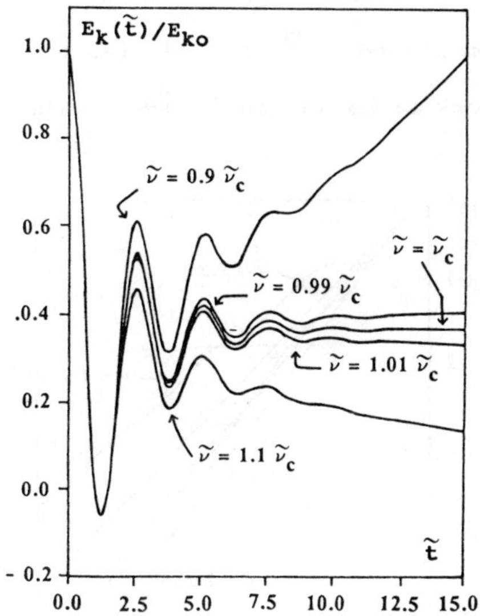

Fig.1. Two stream equilibrium: field ratio versus time when $\tilde{\omega}$=2 for different values of $\tilde{\nu}$>0.

values of $\tilde{\nu}$>0, whereas for larger values of $\tilde{\nu}$ the field exhibits a a monotonic decay, with the fastest decay rate for $\tilde{\nu}$=2.

The second example is relevant to the distribution (9), in which case it is convenient introducing dimensionless parameters

$$\hat{\omega} = \left[\frac{m}{2KT}\right]^{\frac{1}{2}} \frac{\omega}{k}$$

$$\hat{\nu} = \left[\frac{m}{2KT}\right]^{\frac{1}{2}} \frac{\nu}{k} \qquad (14)$$

$$\hat{t} = \left[\frac{2KT}{m}\right]^{\frac{1}{2}} kt.$$

The plasma is asymptotically stable for any $\hat{\nu}$>0, whereas for $\hat{\nu}$=0 the perturbation does not vanish asymptotically, but approaches the final state

$$\lim_{t\to\infty} \frac{E_k(t)}{E_{ko}} = \frac{1}{1+2\hat{\omega}^2} \; . \qquad (15)$$

The analytical prediction is that, for $\hat{\omega}>(\pi-2)^{-\frac{1}{2}}$, there are no eigenvalues when $\hat{\nu}$ is small enough, then a complex conjugate pair appears from the essential spectrum. If $\hat{\nu}$ increases further, they coalesce and split into a pair of real eigenvalues, one of which will disappear into the essential spectrum at $\hat{\nu}=(1+2\hat{\omega}^2)/\pi^{1/2}$, whereas the other survives, but is bound to tend back to the origin for $\hat{\nu}\to\infty$. When

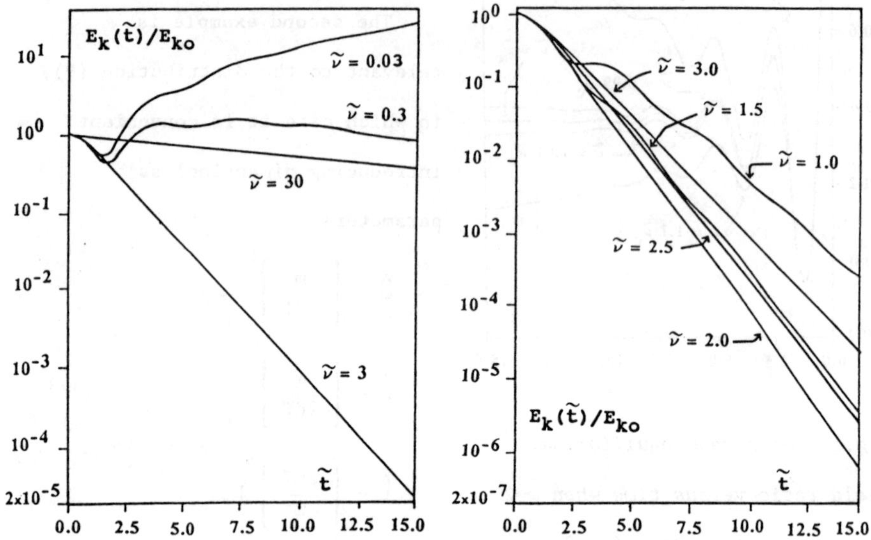

Fig.2. Two stream equilibrium: field ratio versus time when $\tilde{\omega}$=1 for different values of $\tilde{\nu}$>0.

$\hat{\omega}\leq(\pi-2)^{-1/2}$, the situation changes only in that the two initially appeared eigenvalues are absorbed by the essential spectrum at $\hat{\nu}$ = $\pi^{1/2}\hat{\omega}^2$, and only later, at $\hat{\nu}=(1+2\hat{\omega}^2)/\pi^{1/2}$, a single real eigenvalue originates again. The asymptotic trend is always a pure exponential when $\hat{\nu}$ is large enough. Once more, the computed electric field reflects all previous features. For instance, in Fig.3, where $\hat{\omega}$=1, one can observe oscillations for small $\hat{\nu}$, and realize that increasing the collision frequency beyond a certain limit slows down the approach to equilibrium. The dispersion function appearing in the inverse Laplace transform has been rewritten in terms of error functions, which in turn have been accurately evaluated by resorting to routines of the IMSL library, and/or by appropriate series expansions.

Finally the situation becomes quite hard to deal with analytically for the distribution (10). Setting

$$\hat{v}_o = \left(\frac{m}{2KT} \right)^{\frac{1}{2}} v_o \tag{16}$$

the condition for stability when $\hat{\nu}=0$ reads as

$$1 + 2\hat{\omega}^2[1 - 2\hat{v}_o D(\hat{v}_o)] > 0 \ , \tag{17}$$

where D stands for the Dawson integral. It is again a neutral stability, and the final steady state field can be shown to be

$$\lim_{t\to\infty} \frac{E_k(t)}{E_{ko}} = \frac{1}{1+2\hat{\omega}^2[1-2\hat{v}_o D(\hat{v}_o)]} \ , \tag{18}$$

which is thus always positive, like in the previous cases. All such stable configurations remain stable when binary collisions are accounted for, whereas, if Eq.(17) is not fulfilled, the electrically unstable plasma is stabilized if $\hat{\nu}$ overcomes a stability stability threshold $\hat{\nu}_c$ that can be easily determined numerically, as shown in Ref.7. At least one eigenvalue appears sooner or later for increasing $\hat{\nu}$ in all cases, and, if $\hat{\nu}$ is large enough, there exists a real dominant eigenvalue, which produces monotonic exponential asymptotics, and tends to the origin for $\hat{\nu}\to\infty$. When the parameters $\hat{v}_o, \hat{\omega}$ and $\hat{\nu}$ are varied, the number of eigenvalues can change from a minimum of zero to a maximum of four, and complex conjugates are allowed.

For illustrative purposes, the rather complicated behavior of the electric field is reported in Fig.4 for different values of $\hat{\nu}>\hat{\nu}_c=0$ when $\hat{\omega}^2=5$ and $\hat{v}_o=5$. The number of expected eigenvalues for each selected $\hat{\nu}$ is indicated in parenthesis. Effects of branch cut or

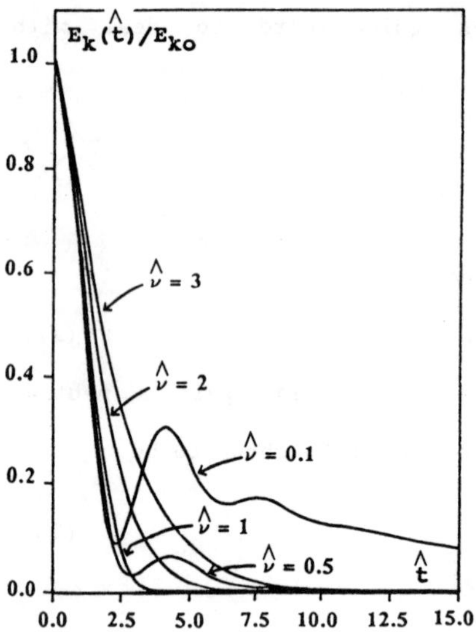

Fig.3. Maxwellian equilibrium: field ratio versus time when $\hat{\omega}=1$ for different values of $\hat{\nu}>0$.

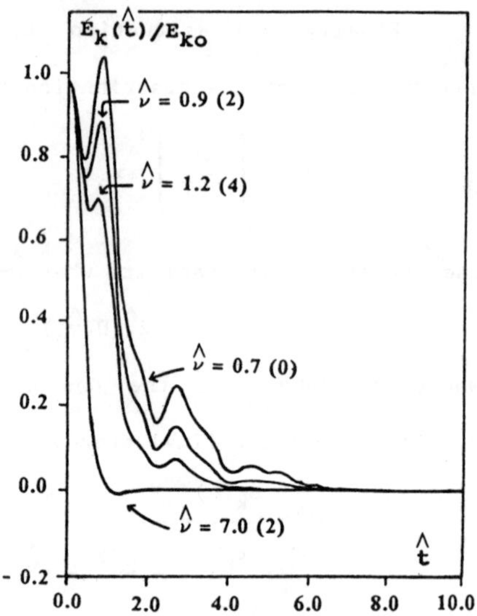

Fig.4. Double Maxwellian equilibrium: field ratio versus time when $\hat{\omega}^2=5$ and $\hat{v}_o=$ 5, for various values of $\hat{\nu}>0$.

leading complex eigenvalues are apparent for small values of $\hat{\nu}$, whereas for large $\hat{\nu}$ the damping effects of the real dominant eigenvalue are clearly shown. Numerical calculations have required here evaluation of the error functions with complex argument appearing in the dispersion function.

Acknowledgements

Work performed under the auspices of the National Group for Mathematical Physics of the Italian National Research Council (C.N.R.) in the frame of the activities PS-MMAI. In particular, one of the authors (B.D.G.) would like to thank C.N.R. for making available a visiting position at the Department of Mathematics of the University of Bari for the completion of this work. Financial support from the Italian Ministry of Public Education is also greatfully acknowledged.

REFERENCES

1. Vlasov,A.A., *J.Phys.(U.S.S.R.)* 9, 25 (1945).
2. Montgomery,D.C., Theory of the Unmagnetized Plasma, Benjamin, New York, 1973.
3. Shure,F.C., Boundary Value Problems in Plasma Oscillations, Ph.D. dissertation, Univ. Michigan, Ann Arbor, 1963.
4. Zweifel,P.F. and Ohlman,J., *Progr.Nucl.En.* 8, 145 (1981).
5. Corngold,N., Michael,P. and Wollman,W., *Nucl.Sci.Eng.* 15, 13 (1963).
6. Spiga,C. and Canapol,B.D., *Trans.Th.Stat,Phys.* 14, 595 (1985).
7. Oggioni,S. and Spiga,G., *ZAMP(J.Appl.Math.Phys.)* 39, 313 (1988).
8. Ganapol,B.D., "Application of the numerical Laplace transform inversion to neutron transport theory", G.M.Wing Conference, 1988, in press.
9. Ganapol,B.D., Spiga,G. and Oggioni,S., "The evaluation of the electric field generated by longitudinal plasma waves", *Phys.Fluids*, in press.

UNIQUENESS AND STABILITY IN LINEAR VISCOELASTICITY: SOME COUNTEREXAMPLES

CLAUDIO GIORGI (Dipartimento di Matematica Università Cattolica del Sacro Cuore
Via Trieste 17- 25121 Brescia, Italy)
BARBARA LAZZARI (Dipartimento di Automazione Industriale Università Statale
Via Valotti 9 - 25060 Brescia, Italy)

ABSTRACT: In this paper we deal with uniqueness and asymptotic stability in linear viscoelasticity. In spite of the title our goal is a little more ambitious: in essence, we do not merely exhibit some counterexamples, that is particular families of relaxation functions leading to solutions which do not have asymptotic property, but we expect to prevent them.

1. INTRODUCTION

For the sake of simplicity we restrict our attention to homogeneous one dimensional evolution problems in linear viscoelasticity by assuming a constitutive "stress - strain" equation of the following type:

$$T(x,t) = G_0\, u_x(x,t) + \int_0^\infty G'(s)\, u_x(x,t-s)\, ds \tag{1.1}$$

The *istantaneous elastic modulus* G_0 and the *Boltzmann function* G' which characterize the viscoelastic behavior are required to satisfy the following conditions:

a) Principle of Fading Memory [1]:

i) $G' \in L^1(0,\infty)$;

ii) there exists a monotonous decreasing function $h : R^+ \to R^+$ such that :

$$h(s) \le \frac{k}{(1+s)^{(2+\varepsilon)}}, \quad k,\varepsilon > 0 \quad \text{and} \quad \int_0^\infty |G'(s)|^2 h^{-1}(s)\, ds = M < +\infty . \tag{1.2}$$

b) Equilibrium Elastic Modulus Assumption

$$G_\infty = G_0 + \int_0^\infty G'(s)\, ds > 0 \tag{1.3}$$

This property characterizes viscoelastic solids.

c) Thermodynamic requirement

$$\omega \hat{G}'_s(\omega) \le 0 \qquad \forall \omega \in R \tag{1.4}$$

where $\hat{G}'_s$ is the half-range sine Fourier Transform of G'.

As shown by Fabrizio & Morro [2] the last assumption is quite equivalent to the Second Law of Thermodynamics in the form of the Clausius property for isothermal cyclic processes.

For later convenience, we shall refer to a pair (G_∞, $G'(\cdot)$) satisfying a), b), c) as the *constitutive pair* of a linear viscoelastic (one-dimensional) solid.

The usefulness of counterexamples in this field was pointed out by Fichera when attempting to investigate the "well-posedness" of the quasi-static problem of linear visco-elasticity with Dirichlet boundary conditions [3], that is:

$$\begin{cases} G_0 u_{xx}(x,t) - \int_0^\infty G'(s) u_{xx}(x,t-s)ds = f(x,t) & (x,t) \in (0,\ell)\times R \\ u(0,t) = u(\ell,t) = 0 & t \in R \end{cases} \tag{1.5}$$

The main objective is to detect what mathematical conditions guarantee the well-posedness of (1.5), but the underlying challenge is to find what physical principles (as fading memory, thermodynamical restrictions and so on) they represent, if possible.

Unfortunately, as pointed out in [4] by a family of counterexamples, conditions a), b), c), while complying with physical requirements, do not allow the quasi-static problem (1.5) to have a unique solution in the space of sinusoidal-in-time strain histories. Fabrizio & Morro suggested that nonuniqueness could ultimately be induced by the expression of the Second Law and proposed to replace c) with

c') Strong thermodynamic requirement

$$\omega \hat{G}'_s(\omega) < 0 \qquad \forall \omega \ne 0 \tag{1.6}$$

that follows from c) by assuming the equality sign refers to trivial cyclic processes only.

In [4,5] this assumption is justified by observing that the statement of the Second Law in standard textbooks requires the distinction between "reversible" and "irreversible" process and specifies that the equality sign refers to the formers and the inequality sign to the latters. Roughly speaking, in order to obtain (1.6) Fabrizio & Morro state that the process associated with a not trivial sinuosidal strain history is "irreversible".

This suggestion was very useful since assumptions a), b), c') allowed Fabrizio to prove the existence of a unique (periodic or L^1 in time) solution to (1.5) (see [5]).

As an aside we notice that the assumption c'), if joined with a.i) and b), implies very strong restrictions on the Boltzmann function G'. For istance, they turn to be stronger than the " strong dissipation principle" stated by Gurtin & Herrera [6] and lead to the invertibility of the constitutive functional from L^1 (or L^∞) into itself (see [7]). Moreover they were sucessfully applied to prove the existence and uniqueness of (periodic) solutions on the whole time axis to the dynamical boundary value problem [8] and the asymptotic stability of solutions to the initial-boundary value problem [9] of linear viscoelasticity.

However, from a mathematical point of view, we must remark that the additional condition c') is not a necessary one in order to ensure well-posedness of the quasi-static problem. For instance, any constitutive pair $(G_\infty, G'(\cdot))$ such that $G_\infty > 0$ and

$$G'(s) = -\left[s^2 - \frac{\alpha - 1}{\beta} s + \frac{\alpha^2}{8\beta^2} \right] \exp(-\beta s) \tag{1.7}$$

with $0<\alpha\leq 6$, $\beta>0$, agrees with a), b), c) but allows existence and uniqueness of the quasi-static (periodic) solution although it does not agree with c'). To see this we observe that

$$\omega \hat{G}'_s(\omega) = -\frac{\omega^2\alpha^2}{8\beta^2(\beta^2+\omega^2)^3}\left(\omega^2 - \frac{8-\alpha}{\alpha}\beta^2\right)^2 < 0 \qquad \text{for } \omega \neq 0, \pm\omega^*$$

$$\hat{G}'_s(\omega^*) = 0, \qquad \hat{G}'_c(\omega^*) = \frac{\alpha(2-\alpha)}{16\beta^3}$$

where $\omega^* = \beta\sqrt{\dfrac{8-\alpha}{\alpha}}$ and $\hat{G}'_c$ is the half-range cosine Fourier Transform of G'.

Hence it follows that:

$$G_0 + \hat{G}'_c(\omega^*) = G_\infty + \frac{(\alpha-6)(\alpha-8)}{16\beta^3} \geq G_\infty > 0$$

and summarizing

$$G_0 + \hat{G}'(\omega) = G_0 + \hat{G}'_c(\omega) - i\,\hat{G}'_s(\omega) \neq 0, \qquad \forall\omega \in R\cup\{-\infty\}\cup\{+\infty\}$$

which leads to the desidered conclusion via standard methods (see [5]).

As a consequence, at least in the quasi-static problem, the existence of a circular frequency $\omega^* > 0$ such that $\hat{G}'_s(\omega^*) = 0$ is not the crucial point in proving the non-uniqueness property of the solution.

The question that arises is the following one: in order to rule out unphysical behaviors is it really necessary to modify the statement of the Second Law of Thermodynamics, as claimed in [4,5]?

In the present paper we try to answer this question and we prove that if <u>any</u> Boltzmann function G' satisfying a), c) but not c') is given, then it is <u>always</u> possible to construct a family of constitutive pairs $(G_\infty, G'(\cdot))$, where G_∞ agrees with b), such that some "unphysical" situation does occur, namely the lack of asymptotic stability with non-zero initial data. In § 2 we start dealing with an auxiliary problem, the main theorem is proved in § 3.

2. ON COUNTEREXAMPLES TO THE UNIQUENESS THEOREM IN DYNAMICAL BOUNDARY VALUE PROBLEMS ON THE WHOLE TIME AXIS

Let u be a function defined in $(0,\ell) \times R$ which represents the displacement and satisfies the following equation and boundary conditions (*)

$$\begin{cases} u_{tt}(x,t) - G_0\, u_{xx}(x,t) - \int_0^\infty G'(s) u_{xx}(x,t-s)ds = f(x,t) & (x,t) \in (0,\ell)\times R \\ u(0,t) = u(\ell,t) = 0 & t \in R \end{cases} \tag{2.1}$$

For later convenience, we look for harmonic solutions to (2.1) only: hence, for any $\omega \neq 0$, we put

$$u(x,t) = \text{Re}\,\{\, u(x)\, e^{i\omega t} \,\}\ , \quad f(x,t) = \text{Re}\,\{\, f(x)\, e^{i\omega t} \,\}\ , \qquad u, f \in C\ ,$$

and the problem (2.1) can be rewritten in the following form:

$$\begin{cases} \omega^2\, u(x) + [G_0 + \hat{G}'(\omega)]\, u_{xx}(x) + f(x) = 0 & x \in (0,\ell) \\ u(0) = u(\ell) = 0 \end{cases} \tag{2.2}$$

where $\hat{G}'$ represents the half-range Fourier Transform of G'.

This problem was investigated in [8]) and the following result was proved there.

Theorem : Given a linear viscoelastic solid material, with constitutive pair $(G_\infty, G'(\cdot))$ satisfying the hypoteses a.i), b), c'), for any harmonic source $f(x,t) = \text{Re}\,\{f(x)\, e^{i\omega t}\}$, $f \in L^2(0,\ell)$, $\omega \in R$, there exists one and only one harmonic solution to the problem (2.1), $u(x,t) = \text{Re}\,\{u(x)\, e^{i\omega t}\}$, such that $u \in H^1_0(0,\ell) \cap H^2(0,\ell)$.

The proof of this theorem rests upon the strong ellipticity of the differential operator associated to (2.2), which easily follows from the assumptions.

It is our goal to show that if c') is replaced by the weaker assumption c) then the above result falls.

Theorem 1: *Let* G' *be any Boltzmann function which agrees with* a.i), c) *but not* c'). *This is to say,*

$$\hat{G}'_s\,(\omega^*) = 0 \tag{2.3}$$

at least for one circular frequency $\omega^* > 0$. *Then there exist two positive constants* G^*_∞ *and* ℓ^* *such that the homogeneous problem*

(*) The initial data are not given, since we study the solution u for $t \in R$.

$$\begin{cases} \omega^{*2} u(x) + [G_0^* + \hat{G}'(\omega^*)]\, u_{xx}(x) = 0 & x \in (0, \ell^*) \\ u(0) = u(\ell^*) = 0 \end{cases} \tag{2.4}$$

where $G_0^* = G_\infty^* + \int_0^\infty G'(s)\, ds$, *has infinitely many solutions.*

Proof. We put (2.3) into (2.2), so that both real and immaginary part of the solution u to (2.2) have to satisfy the problem

$$\begin{cases} \omega^{*2} w(x) + [G_0 + \hat{G}'_c(\omega^*)]\, w_{xx}(x) = 0 & x \in (0,\ell) \\ w(0) = w(\ell) = 0 \end{cases} \tag{2.5}$$

Since $G_0 + \hat{G}'_c(\omega^*) = G_\infty + \int_0^\infty G'(s)\,(\cos \omega^* s - 1) ds$ and $\int_0^\infty G'(s)\,(\cos \omega^* s - 1)\, ds$ is bounded, by virtue of a.i), we can choose a positive (in order to comply with b)) constant G_∞^* such that

$$G_0^* + \hat{G}'_c(\omega^*) = G_\infty^* + \int_0^\infty G'(s)\,(\cos \omega^* s - 1)\, ds > 0$$

Finally, by choosing

$$\ell^* = \frac{\sqrt{G_0^* + \hat{G}'_c(\omega^*)}}{\omega^*}\, \pi \tag{2.6}$$

it follows that (2.5) has more than the trivial solution, because $\dfrac{\omega^{*2}}{G_0^* + \hat{G}'_c(\omega^*)}$ turns to be an eingevalue of $-\Delta$. So we have the following family of solutions to (2.4):

$$u^*(x) = c_1 w^*(x) + i\, c_2 w^*(x) \qquad \text{where } w^*(x) = \sin\left(\frac{\pi x}{\ell^*}\right),\ c_1, c_2 \in \mathbb{R}. \tag{2.7}$$

The above feature is typical in linear elasticity: indeed, to within a frequency-dependent factor, equation (2.5) is just the same we find if we look for harmonic solutions to the wave equation. From a physical point of wiev, we may state that abandoning the restriction c') we allow the body to have an "elastic" behaviour at some prescribed frequencies. This is quite strange for a viscoelastic material. However, this nonuniqueness result is ruled out by taking under consideration initial-boundary value problems.

3. ON COUNTEREXAMPLES TO THE ASYMPTOTIC STABILITY IN BOUNDARY-INITIAL HISTORY VALUE PROBLEMS

Now we consider the boundary-initial history value problem

$$\begin{cases} u_{tt}(x,t)-G_0\, u_{xx}(x,t)-\int_0^\infty G'(s)u_{xx}(x,t-s)ds=f(x,t) & (x,t)\in(0,\ell)\times R^{++} \\ u(x,t)=u^0(x,t) & (x,t)\in(0,\ell)\times R^{-} \\ u(0,t)=u(\ell,t)=0 & t\in R^{++} \end{cases} \tag{3.1}$$

and recall the following "asymptotic stability property" which was proved in [9]:

Theorem. Given a linear viscoelastic solid material, with constitutive pair (G_∞, $G'(\cdot)$) which meets the conditions a.i), b), c'), for any source $f\in L^2(R^+,L^2(0,\ell))$ and any initial history u^0 satisfying

h_1) $u^0(\cdot,s)\in H^1_0(0,\ell)$ $\quad \dot u^0(\cdot,s)\in L^2(0,\ell)$ $\quad$ a.e. on R^-

h_2) $U^0(x,t)=\int_t^\infty G'(s)u^0_{xx}(x,t-s)\,ds\in L^2(R^+,L^2(0,\ell))$

there exists one and only one solution $u\in L^2(R^+,H^1_0(0,\ell))\cap H^1(R^+,L^2(0,\ell))$ to (3.1).

Remark: *If* a.ii) *holds, then any history* $u^0\in L^\infty(R^-, H^1_0(0,\ell)\cap H^2(0,\ell))$ *satisfies* h_2).

Proof. Let $u^0\in L^\infty(R^-,H^1_0(0,\ell)\cap H^2(0,\ell))$, putting $|||u^0|||^2_{2;\infty}=\operatorname{ess\,sup}_{t\in R^-}\|u^0(t)\|^2_2$ we have:

$$\int_0^\infty \|U^0(t)\|_2^2dt=\int_0^\ell\Big|\int_t^\infty\int_0^\infty G'(s)u^0_{xx}(x,t-s)\,dx\,ds\Big|^2dt\le$$

$$\le\int_0^\infty\Big[\int_0^\infty\frac{|G'(t+\tau)|}{\sqrt{h(t+\tau)}}\int_0^\ell\sqrt{h(t+\tau)}\,|u^0_{xx}(x,-\tau)|\,dx\,d\tau\Big]^2dt\le$$

$$\le\int_0^\infty\Big[\int_0^\infty\frac{|G'(t+\tau)|^2}{h(t+\tau)}d\tau\int_0^\infty h(t+\tau)\,\|u^0_{xx}(-\tau)\|_2^2d\tau\Big]dt$$

Hence from (1.2) it follows

$$\int_0^\infty\|U^0(t)\|_2^2dt\le M\,|||u^0_{xx}|||^2_{2;\infty}\int_0^\infty\int_0^\infty\frac{k}{(1+t+\tau)^{2+\varepsilon}}d\tau\,dt\le$$

$$\le M\,|||u^0_{xx}|||^2_{2;\infty}\int_0^\infty\int_0^\infty\frac{k}{[4(\frac12+t)(\frac12+\tau)]^{\frac{2+\varepsilon}{2}}}d\tau\,dt=\frac{k\,M\,|||u^0_{xx}|||^2_{2;\infty}}{\varepsilon^2}<+\infty \tag{3.2}$$

As in the previous section, we prove that the above theorem falls when the assumption c') is weakened into c). That is to say, the null solution of the associated homogeneous problem turns to be not asymptotically stable.

Theorem 2: *Let* G' *be any Boltzmann function which agrees with* a), c) *but not* c'), *this is to say,* $\hat{G}'_s(\omega^*) = 0$ *at least for one circular frequency* $\omega^* > 0$. *Let* G^*_∞, ℓ^* *be as in* Theorem 1 *and* $u^0(x,t) = c \sin(\frac{\pi x}{\ell^*})(\cos \omega^* t + \sin \omega^* t)$. *Then for any* $c \neq 0$:

i) u^0 *meets conditions* h_1) *and* h_2);

ii) *the homogeneous problem*

$$\begin{cases} u_{tt}(x,t) - G_0\, u_{xx}(x,t) - \int_0^\infty G'(s) u_{xx}(x,t-s)\,ds = 0 & (x,t) \in (0,\ell^*) \times R^{++} \\ u(x,t) = c \sin(\frac{\pi x}{\ell^*})(\cos \omega^* t + \sin \omega^* t) & (x,t) \in (0,\ell^*) \times R^- \\ u(0,t) = u(\ell^*,t) = 0 & t \in R^{++} \end{cases} \tag{3.3}$$

has a unique periodic (hence not belonging to $L^2(R^+, H^1_0(0,\ell^*))$) *solution.*

Proof. i) Firstly, h_1) is trivially satisfied by u^0. Secondly, the above Remark yelds h_2) because $u^0 \in L^\infty(R^-, H^1_0(0,\ell^*) \cap H^2(0,\ell^*))$.

ii) By virtue of Theorem 1 the only solution to (3.3) which agrees with the prescribed initial-history value is

$$u(x,t) = c \sin(\frac{\pi x}{\ell^*})(\cos \omega^* t + \sin \omega^* t), \quad (x,t) \in (0,\ell^*) \times R^{++}$$

and does not exhibit any decay in time.

Finally we stress that the arbitrariness of c ensures the lack of asymptotic stability of the null solution for any initial-history perturbation u^0 however small with respect to standard weighted history norms.

4. AN INTERESTING FAMILY OF RELAXATION FUNCTIONS

We consider the following three-parameter family

$$G(s) = G_\infty + \int_s^\infty \left[r^2 - \frac{\alpha - 1}{\beta} r + \frac{\alpha^2}{8\beta^2} \right] \exp(-\beta r)\, dr$$

with $\alpha \in R$, $\beta > 0$ and $G_\infty > 0$ (in order to comply with b)). It was exhibited first by Fabrizio and Morro in [4)] although with a different choice of the parameters. The associated Boltzmann functions G', given by (1.8), meet a) and c) but, if $0 < \alpha < 8$, we have

$\hat{G}'_s(\omega^*) = 0$ for $\omega^* = \beta \sqrt{\dfrac{8-\alpha}{\alpha}} \neq 0$, so that c') does not hold. Moreover it is rather surprising that relaxation functions G corresponding to $2-\sqrt{2} < \alpha < 2+\sqrt{2}$ are positive and decreasing. The following picture summarizes the previous results.

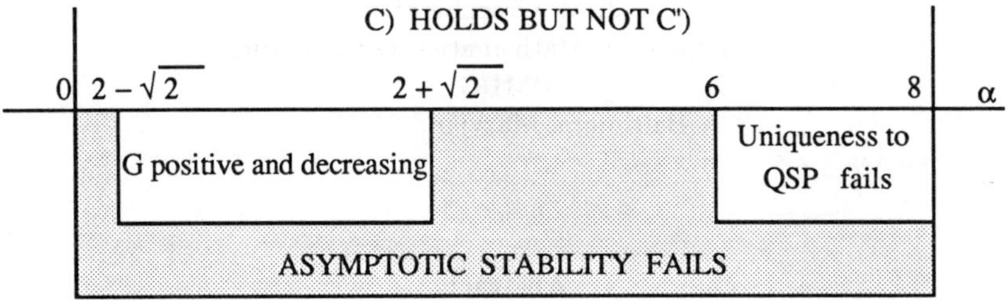

REFERENCES

[1] Coleman, B.D. and Noll,W., *Foundations of linear viscoelasticity*, Rev.Mod.Phys., **33** (1964)

[2] Fabrizio, M. and Morro, A., *Viscoelastic relaxation functions compatible with thermodynamics*, J.Elasticity, **19** (1988).

[3] Fichera,G., *Avere una memoria tenace crea gravi problemi*, Arch. Rat. Mech. Anal., **70** (1979)

[4] Fabrizio, M. and Morro, A., *On uniqueness in linear viscoelasticity: a family of counterexamples*, Quart. Appl. Math., **45** (1987)

[5] Fabrizio, M., *An existence and uniqueness theorem in quasi-static viscoelasticity*, Quart.Appl.Math., **47** (1989).

[6] Gurtin,M.E. and Herrera, I., *On dissipation inequalities and linear viscoelasticity*, Quart. Appl. Math., **23** (1965)

[7] Giorgi, C., *Alcune conseguenze delle restrizioni termodinamiche per mezzi viscoelastici lineari*, Quaderni del Dip. Matematico Univ. Cattolica S. Cuore - Brescia, **6** (1989)

[8] Lazzari, B. and Vuk, E., *Un teorema di esistenza e unicità per un problema dinamico in viscoelasticità lineare*, Atti Sem.Mat.Fis.Univ.Modena, **35** (1987).

[9] Fabrizio, M. and Lazzari, B., *On the stability of linear viscoelastic solid systems* (to appear).

EXPONENTIAL DECAY THEOREMS FOR THE FOUR VELOCITY BROADWELL EQUATIONS

J. M. Greenberg[1]
Department of Mathematics and Statistics
UMBC
Baltimore, Maryland 21228

Abstract

In this note I obtain estimates which guarantee that solutions to the Four Velocity Broadwell equations with specularly reflecting boundary conditions decay to equilibrium at an exponential rate.

1. Introduction

Given the interest at this symposia in the Broadwell equations I have decided to present a new result on the exponential stability of equilibrium solutions to the Four Velocity Broadwell equations. This result should be compared with that obtained by Slemrod [1]. For essentially arbitrary nonnegative data with finite mass Slemrod obtains the weak convergence of solutions to a purely advecting state (with the same mass as the initial data). He fails to show that the advecting state is an equilibria and fails to obtain a rate of decay estimate of the solution to the advecting state. On the other hand I am able to establish that the asymptotic state of the system is an equilibrium solution (with the same mass as the initial data) and obtain an estimate which guarantees that the solution converges (in suitably defined strong sense) to the equilibrium at an exponential rate. To establish this result I require more regularity on the initial data than required by Slemrod and further require that the initial data be close to the equilibrium (in a suitably defined strong sense); alas there is no free lunch.

The basic quantities of interest are particle densities r, l, u and d. Specifically $r(x,t)$ represents the number of particles per unit area at (x,y) at time t travelling with velocity $\mathbf{e}_1$. The densities l, u, and d have a similar interpretation except that the particles travel with velocities $-\mathbf{e}_1$, $\mathbf{e}_2$ and $-\mathbf{e}_2$ respectively[2]. The evolution

[1]This research was partially supported by the U.S. National Science Foundation and the U.S. Department of Energy.

[2]Here $\mathbf{e}_1 = \begin{pmatrix} 1 \\ 0 \end{pmatrix}$ and $\mathbf{e}_2 = \begin{pmatrix} 0 \\ 1 \end{pmatrix}$ are the standard basis vectors in R^2.

equations for the densities are

$$\left.\begin{array}{lclcl} r_t & + & r_x & = & C \\ l_t & + & l_x & = & C \\ u_t & + & u_y & = & C \\ d_t & + & d_x & = & C \end{array}\right\} \tag{1.1}$$

where the collision term C is given by

$$C = K(ud - rl). \tag{1.2}$$

For definiteness we shall consider the case where

$$K = 1. \tag{1.3}$$

My interest is in solutions which are independent of y and thus satisfy

$$\left.\begin{array}{rcl} r_t + r_x & = & (ud - rl) \\ l_t - l_x & = & (ud - rl) \\ u_t & = & (rl - ud) \\ d_t & = & (rl - ud) \end{array}\right\}. \tag{1.4}$$

I further assume that the motion takes place in the strip $0 < x < 1$ and that specular reflection takes place at the walls $x = 0$ and $x = 1$; that is

$$r(0,t) = l(0,t) \quad \text{and} \quad r(1,t) = l(1,t). \tag{1.5}$$

It is easy checked that any solution of (1.4) and (1.5) satisfies the conservation of mass identity

$$\int_0^1 (r + l + u + d)(x,t)\, dx = \int_0^1 (r_0 + l_0 + u_0 + d_0)(x)\, dx \tag{1.6}$$

where (r_0, l_0, u_0, d_0) is the nonnegative initial data for (r, l, u, d). Such solutions also satisfy the pointwise identity

$$(u - d)(x,t) = (u_0 - d_0)(x) \stackrel{\text{def}}{=} 2q_\infty(x), \quad 0 \le x \le 1. \tag{1.7}$$

It is easily checked that the only steady or equilibrium solution to (1.4) and (1.5) consistent with (1.6) and (1.7) is

$$\left.\begin{array}{rcl} r(x,t) & = & c \\ l(x,t) & = & c \\ u(x,t) & = & \sqrt{q_\infty^2(x) + c^2} + q_\infty(x) \\ d(x,t) & = & \sqrt{q_\infty^2(x) + c^2} - q_\infty(x) \end{array}\right\}. \tag{1.8}$$

Here, c is the unique positive solution of

$$2\int_0^1 (c+\sqrt{q_\infty^2(x)+c^2})\,dx = \int_0^1 (r_0+l_0+u_0+d_0)(x)\,dx. \tag{1.9}$$

Most authors, including Slemrod, deal with the case where $q_\infty \equiv 0$. Due to space limitation I shall confine myself to this case in the remainder of this paper.

I conclude this section by writing down an equivalent system for the deviations from the constant equilibrium solution. Setting

$$\left.\begin{aligned} R &= r-c \\ L &= l-c \\ 2M &= u+d-2c \end{aligned}\right\}, \tag{1.10}$$

it is easily checked that if (r,l,u,d) satisfies (1.4) and (1.5) with $q_\infty \equiv 0$, then (R,L,M) satisfies

$$\left.\begin{aligned} R_t+R_x &= c(2M-R-L)+M^2-RL \\ L_t+L_x &= c(2M-R-L)+M^2-RL \\ M_t &= c(R+L-2M)+RL-M^2 \end{aligned}\right\} \tag{1.11}$$

and

$$R(0,t)=L(0,t) \quad \text{and} \quad R(1,t)=L(1,t). \tag{1.12}$$

For solutions of (1.11) and (1.12) to be consistent with (1.9) requires that the initial data

$$(R,L,M)(x,0) \stackrel{\text{def}}{=} (R_0,L_0,M_0)(x) \tag{1.13}$$

satisfies

$$\int_0^1 (R_0+L_0+2M_0)(x)\,dx = 0. \tag{1.14}$$

This of course implies that for any $t \geq 0$

$$\int_0^1 (R+L+2M)(x,t)\,dx = 0. \tag{1.15}$$

2. A-Priori Estimates for the Forced Linear Broadwell Equations

My goal in this section is a set of a-priori estimates for the forced linear Broadwell equations:

$$\left.\begin{aligned} R_t+R_x &= c(2M-R-L)+F_0 \\ L_t+L_x &= c(2M-R-L)+F_0 \\ M_t &= c(R+L-2M)-F_0 \end{aligned}\right\} \tag{2.1}$$

which satisfy

$$R(0,t)=L(0,t) \quad \text{and} \quad R(1,t)=L(1,t). \tag{2.2}$$

I note that solving the above system is equivalent to solving

$$\left.\begin{aligned} U_t - W_t &= 0 \\ W_t - \tfrac{1}{2}(U+J)_x &= 0 \\ J_t + 4cJ - W_x &= 2F_0 \end{aligned}\right\} \tag{2.3}$$

and

$$W(0,t) = W(1,t) = 0. \tag{2.4}$$

The new unknowns (U, W, J) are given in terms of (R, L, M) via the transformation

$$\left.\begin{array}{l} 2U = R + L + 2M, \quad 2J = R + L - 2M, \quad \text{and} \quad 2W = L - R \\ \text{or equivalently} \\ R + L = U + J, \qquad R - L = -2W, \qquad \text{and} \quad M = (U-J)/2 \end{array}\right\}. \tag{2.5}$$

In what follows I shall assume the data (U_0, W_0, J_0) satisfies

$$\int_0^1 U_0(x)\,dx = 0. \tag{2.6}$$

Associated with solutions of (2.3) and (2.4) are three identities which will be of use in the sequel:

$$\frac{\partial}{\partial t}\begin{pmatrix} U^2 + 2W^2 + J^2 \\ U_t^2 + 2W_t^2 + J_t^2 \\ U_x^2 + 2W_x^2 + J_x^2 \end{pmatrix} + 8c\begin{pmatrix} J^2 \\ J_t^2 \\ J_x^2 \end{pmatrix} = 2\frac{\partial}{\partial x}\begin{pmatrix} (U+J)W \\ (U_t+J_t)W_t \\ (U_x+J_x)W_x \end{pmatrix} + 4\begin{pmatrix} F_0 J \\ F_{0t} J_t \\ F_{0x} J_x \end{pmatrix}. \tag{2.7}$$

If we now let

$$\begin{pmatrix} \Gamma_0^2 \\ \Gamma_1^2 \\ \Gamma_2^2 \end{pmatrix}(t) = \int_0^1 \begin{pmatrix} U^2 + 2W^2 + J^2 \\ U_t^2 + 2W_t^2 + J_t^2 \\ U_x^2 + 2W_x^2 + J_x^2 \end{pmatrix}(x,t)\,dx \tag{2.8}$$

and

$$\begin{pmatrix} J_0^2 \\ J_1^2 \\ J_2^2 \end{pmatrix}(t) = \int_0^1 \begin{pmatrix} J^2 \\ J_t^2 \\ J_x^2 \end{pmatrix}(x,t)\,dx, \tag{2.9}$$

then (2.7), together with the fact that W and $W_t = \frac{1}{2}(U+J)_x$ vanish at $x = 0$ and 1, implies that for any $\delta > 0$

$$e^{2\delta t}\begin{pmatrix} \Gamma_0^2 \\ \Gamma_1^2 \\ \Gamma_2^2 \end{pmatrix}(t) + 8c\int_0^1 e^{2\delta\tau}\begin{pmatrix} J_0^2 \\ J_1^2 \\ J_2^2 \end{pmatrix}(\tau)\,d\tau =$$

$$\begin{pmatrix} \Gamma_0^2 \\ \Gamma_1^2 \\ \Gamma_2^2 \end{pmatrix}(0) + 2\delta\int_0^1 e^{2\delta\tau}\begin{pmatrix} \Gamma_0^2 \\ \Gamma_1^2 \\ \Gamma_2^2 \end{pmatrix}(\tau)\,d\tau + 4\int_0^t e^{2\delta\tau}\int_0^1 \begin{pmatrix} F_0 J \\ F_{0\tau} J_\tau \\ F_{0x} J_x \end{pmatrix}(x,\tau)\,dx\,d\tau. \tag{2.10}$$

The crucial estimate of this section is the content of

Theorem 1. *For any $0 \le \delta \le \frac{1}{2}$, there are constants $C_1 - C_4$ (which may be chosen independently of δ) such that*

$$\int_0^t e^{2\delta\tau}(\Gamma_0^2+\Gamma_1^2+\Gamma_2^2)(\tau)\,d\tau \le$$
$$C_1(\Gamma_0^2+\Gamma_1^2+\Gamma_2^2)(0)+C_2\int_0^t e^{2\delta\tau}\int_0^1(F_0^2+F_{0\tau}^2+F_{0x}^2)(x,\tau)\,dx d\tau$$
$$+C_3\int_0^t e^{2\delta\tau}(J_0^2+J_1^2+J_2^2)(\tau)\,d\tau+C_4e^{2\delta t}(\Gamma_0^2+\Gamma_1^2+\Gamma_2^2)(t). \tag{2.11}$$

Proof: The boundary conditions $W(0,t)=W(1,t)=0$ guarantee that

$$\int_0^t e^{2\delta\tau}\int_0^1 W^2(x,\tau)\,dxd\tau \le \int_0^t e^{2\delta\tau}\int_0^1 W_x^2(x,\tau)\,dxd\tau = \int_0^t e^{2\delta\tau}\int_0^1 U_\tau^2(x,\tau)\,dxd\tau \tag{2.12}$$

and this, when combined with $(2.3)_3$, implies that the above three terms are bounded from above by

$$3\int_0^t e^{2\delta\tau}(16c^2J_0^2(\tau)+J_1^2(\tau)+4\int_0^1 F_0^2(x,\tau)dx)\,d\tau. \tag{2.13}$$

Noting that (2.6) implies that $\int_0^1 U(x,t)\,dx \equiv 0$, we see we are guaranteed that any continuous solution to (2.3), (2.4) and (2.6) satisfies $U(x_\#(t),t) =$ for some $x_\#(t)\in[0,1]$. This observation guarantees that

$$\int_0^1 U^2(x,t)\,dx \le \int_0^1 U_x^2(x,t)\,dx \tag{2.14}$$

and thus that it suffices to find an estimate for the latter term.

Our crucial estimate is now a consequence of the conservation law $W_t - \frac{U_x}{2} = \frac{J_x}{2}$. From this identity we obtain

$$W_t^2+\frac{U_x^2}{4}=(WU_x)_t+\frac{J_x^2}{4}-(WW_x)_x+W_x^2, \tag{2.15}$$

$$\int_0^t e^{2\delta\tau}\int_0^1(W_\tau^2+\frac{U_x^2}{4}(x,\tau))\,dxd\tau =$$
$$e^{2\delta t}\int_0^1(WU_x)(x,t)\,dx-\int_0^1(WU_x)(x,0)\,dx$$
$$-2\delta\int_0^t e^{2\delta\tau}\int_0^1(WU_x)(x,\tau)\,dxd\tau+\int_0^t e^{2\delta\tau}\int_0^1\frac{J_x^2}{4}(x,\tau)\,dxd\tau$$
$$+\int_0^t e^{2\delta\tau}\int_0^1 W_x^2(x,\tau)\,dxd\tau \tag{2.16}$$

and finally the inequality

$$\int_0^t e^{2\delta\tau}\int_0^1 (W_\tau^2+\frac{U_x^2}{4}(x,\tau)\,dxd\tau \le$$
$$e^{2\delta t}\int_0^1 (2W^2+\frac{U_x^2}{4})(x,t)\,dx+\frac{1}{2}\int_0^1 (W^2+U_x^2)(x,0)\,dx$$
$$+\delta\int_0^t e^{2\delta\tau}\int_0^1 (4W^2+\frac{U_x^2}{4})(x,\tau)\,dxd\tau$$
$$+\int_0^t e^{2\delta\tau}\int_0^1 \frac{J_x^2}{4}(x,\tau)\,dxd\tau$$
$$+\int_0^t e^{2\delta\tau}\int_0^1 W_x^2(x,\tau)\,dxd\tau. \qquad (2.17)$$

The Theorem now follows from (2.12)-(2.17) and the definitions (2.8) and (2.9).

□

If we now exploit the fact that for any $0<\alpha$

$$4\int_0^t e^{2\delta\tau}\int_0^1 (F_0J+F_{0\tau}J_\tau+F_{0x}J_x)(x,\tau)dx \le$$
$$2\alpha\int_0^t e^{2\delta\tau}(J_0^2+J_1^2+J_2^2)(\tau)d\tau+\frac{2}{\alpha}\int_0^t e^{2\delta\tau}\int_0^1 (F_0^2+F_{0\tau}^2+F_{0x}^2)(x,\tau)dxd\tau, \quad (2.18)$$

we find that (2.10), (2.11) and (2.18) can be combined to yield

Theorem 2. *There exists a $0<\delta_0=\mathcal{O}(c)$ and constants D_1-D_6 such that*

$$2e^{2\delta t}(\Gamma_0^2+\Gamma_1^2+\Gamma_2^2)(t) \le$$
$$D_1(\Gamma_0^2+\Gamma_1^2+\Gamma_2^2)(0)+D_2\int_0^t e^{2\delta\tau}\int_0^1 (F_0^2+F_{0\tau}^2+F_{0x}^2)(x,\tau)\,dxd\tau, \quad (2.19)$$

$$\int_0^t e^{2\delta\tau}(\Gamma_0^2+\Gamma_1^2+\Gamma_2^2)(\tau)\,d\tau \le$$
$$D_3(\Gamma_0^2+\Gamma_1^2+\Gamma_2^2)(0)+D_4\int_0^t e^{2\delta\tau}\int_0^1 (F_0^2+F_{0\tau}^2+F_{0x}^2)(x,\tau)\,dxd\tau \quad (2.20)$$

and

$$\int_0^t e^{2\delta\tau}(J_0^2+J_1^2+J_2^2)(\tau)\,d\tau \le$$
$$D_5(\Gamma_0^2+\Gamma_1^2+\Gamma_2^2)(0)+D_6\int_0^t e^{2\delta\tau}\int_0^1 (F_0^2+F_{0\tau}^2+F_{0x}^2)(x,\tau)\,dxd\tau. \quad (2.21)$$

Similar arguments may be used to obtain L_∞ estimates for U, W and J. The estimates for U and W follow from the fact that solutions of (2.3), (2.4) and (2.6) of necessity satisfy

$$U^2(x,t)\le\int_0^1 U_x^2(x,t)\,dx \quad\text{and}\quad W^2(x,t)\le\int_0^1 W_x^2(x,t)\,dx. \qquad (2.22)$$

The estimate for J is a simple consequence of the identity

$$\begin{aligned} J(x,t) &= U(x,t) + (J-U)(x,0)e^{-4ct} - 4c\int_0^t e^{-4c(t-s)}U(x,s)\,ds \\ &\quad + 2\int_0^t e^{-4c(t-s)}F_0(x,s)\,ds \end{aligned} \tag{2.23}$$

and the result is

$$\begin{aligned} &e^{\delta t} \mid J(\cdot,t) \mid_\infty \leq \\ &\quad e^{\delta t} \mid U(\cdot,t) \mid_\infty + e^{-(4c-\delta)t}(\mid U(\cdot,0) \mid_\infty + \mid J(\cdot,0) \mid_\infty) \\ &\quad + \frac{1}{4c-\delta}(4c \sup_{0\leq s\leq t} e^{\delta s} \mid U(\cdot,s) \mid_\infty + 2 \sup_{0\leq s\leq t} e^{\delta s} \mid F_0(\cdot,s) \mid_\infty). \end{aligned} \tag{2.24}$$

where $\mid f(\cdot) \mid_\infty = \sup_{0\leq x\leq 1} \mid f(x) \mid$.

Equipped with the estimates (2.19)-(2.24) it is now a relatively simple matter to show that the following iteration scheme

$$\left.\begin{aligned} &\frac{\partial U^n}{\partial t} - \frac{\partial W^n}{\partial x} = 0 \\ &\frac{\partial W^n}{\partial t} - \frac{1}{2}\frac{\partial}{\partial x}(U^n + J^n) = 0 \\ &\frac{\partial J^n}{\partial t} + 4cJ^n - \frac{\partial W^n}{\partial x} = 2((M^{n-1})^2 - R^{n-1}L^{n-1}) = 2((W^{n-1})^2 - U^{n-1}J^{n-1}) \end{aligned}\right\} \tag{2.25}$$

$$W^n(0,t) = W^n(1,t) = 0, \tag{2.26}$$

and

$$(U^n, W^n, J^n)(x,0) = (U_0, W_0, J_0)(x), \quad 0 \leq x \leq 1 \tag{2.27}$$

is convergent provided the initial data (U_0, W_0, J_0) is small enough; that is that the L_∞ norms of these quantities are small as well as the quantities $\Gamma_i^2(0)$, $i = 0,1,2$. One of course obtains the solution in a Banach space whose norm is modelled on the norms appearing in (2.19)-(2.24). The exponential decay result quoted in the introduction follows from the above local existence Theorem and the exponential weights appearing in the Banach space norm[3].

3. Concluding Remarks

It should be noted that many of the arguments used here are similar to ones used by the author in [2] where he sought a-priori estimates for the equations motion of a nonlinear viscoelastic solid. In hindsight this result is not surprising since the Broadwell equations can be cast in a form where they strongly resemble the equations of viscoelasticity.

[3]The solution so obtained is locally unique, that is it is the only small solution. This fact follows from the a-priori estimates and the contraction mapping principle.

If one takes as starting point the original system (1.1)-(1.3) or the equivalent system presented below for the moments (U, W, J) (see (2.5) for the connection between (R, L, M) and (U, W, J))

$$\left.\begin{aligned} U_t \quad - \quad & W_x = 0 \\ W_t \quad - \quad & \tfrac{1}{2}(U+J)_x = 0 \\ J_t + 4cJ - W_x \quad = \quad & 2(W^2 - UJ) \end{aligned}\right\} \tag{3.1}$$

and if one now lets $\gamma = -U$ be the longitudinal strains, $V = -W$ be the particle velocity, and $\Sigma = -\frac{1}{2}(U+J)$ be the longitudinal stress, then a simple calculation shows that if (U, W, J) satisfies (3.1), then (γ, V, Σ) satisfies

$$\left.\begin{aligned} \gamma_t - V_x \quad & = 0 \\ V_t - \Sigma_x \quad & = 0 \\ (\Sigma - \gamma)_t + 4c(\Sigma - \tfrac{\gamma}{2}) \quad & - 2\Sigma\gamma + \gamma^2 + V^2 = 0 \end{aligned}\right\}. \tag{3.2}$$

The linearized version of (3.2) are equations of motion of a linear-Maxwell material and the nonlinear system corresponds to a viscoelastic material whose instantaneous elastic response is given by $\Sigma = \gamma$, whose equilibrium elastic response is given by $\Sigma = \frac{\gamma}{2} + \frac{V^2}{(4c-2\gamma)}$, and whose relaxation time is $4c - 2\gamma$.

Finally, it should be noted that the results of the preceding section are conservative. The author has done a great deal of numerical experimentation with the full system (1.1)-(1.3) and at no time have the solutions failed to converge to the unique equilibrium solution consistent with the conservation of mass identity (1.9). The author has tried to expand Slemrod's estimates by producing global functionals other than the entropy, $H = \int_0^1 (r \ln r + l \ln l + u \ln u + d \ln d)(x,t)dx$, which decay on trajectories. At present he has not succeeded.

References

1. Slemrod, M., "Large Time Behavior of the Broadwell Model of Discrete Velocity Gas with Specularly Reflective Boundary Conditions", this volume.

2. Greenberg, J. M., "A-Priori Estimates for Flows in Dissipative Materials", Jour. Math. Anal. and Appl. 60, 617-630(1977).

PERTURBATIVE METHOD FOR A NONLINEAR INTEGRAL EQUATION OF THE PARTICLE TRANSPORT THEORY

Gabriele Guerriero
Dipartimento di Matematica e Applicazioni
R. Caccioppoli dell'Universita' di Napoli
via Mezzocannone, 8 80134 Napoli (Italy)

ABSTRACT. The existence and asymptotic stability of the steady solution to a nonlinear integral equation of the particle transport theory in the stochastic models is proved.

1. INTRODUCTION.

In the frame of the scattering kernel formulation the space independent nonlinear integro-differential Boltzmann problem, governing the one particle distribution function $f(\mathbf{v},t)$ in a gas of particles diffusing in an infinite homogeneous medium through binary collisions and in absence of removal and external forces, reads as[1]:

$$\begin{cases} f_t+kQf=k\int_{\mathbb{R}_3}\int_{\mathbb{R}_3}\pi(\mathbf{v}',\mathbf{v}'',\mathbf{v})f(\mathbf{v}',t)f(\mathbf{v}'',t)d\mathbf{v}'d\mathbf{v}'' \\ f(\mathbf{v},0)=QS(\mathbf{v})\ ; \quad k>0 \quad \mathbf{v},\mathbf{v}',\mathbf{v}''\in\mathbb{R}_3\ ,\quad t\in[0,\infty)\ ,\ \int_{\mathbb{R}_3}S(\mathbf{v})d\mathbf{v}=1 \end{cases} \tag{1}$$

In (1)
$$Q=\int_{\mathbb{R}_3} f(\mathbf{v},t)d\mathbf{v}$$
represent the number of particles injected in the medium while $\mathbf{v}',\mathbf{v}'',\mathbf{v}$ are the velocities before and after the collisions respectively. The function $\pi(\mathbf{v}',\mathbf{v}'',\mathbf{v})$ is the scattering probability distribution which is a non negative function on $R_3\times R_3\times R_3$, is summable respect to $\mathbf{v}$ and obeys respectively the normalization and symmetry conditions:

$$\int_{\mathbb{R}_3}\pi(\mathbf{v}',\mathbf{v}'',\mathbf{v})d\mathbf{v}=1 \quad , \quad \pi(\mathbf{v}',\mathbf{v}'',\mathbf{v})=\pi(\mathbf{v}'',\mathbf{v}',\mathbf{v}) \tag{2}$$

The problem considered, has been, in the last years object of several investigations aimed at focusing the mathematical problems connected with the existence, uniqueness and stability of the solutions. We recall that in [2-3-5] the global existence and uniqueness of solutions to equation (1) has been given in the natural space $L^1(R_3)\times[0,\infty)$. In [6-7] has been studied essentially the stability of the solutions to equation (1) (and some implications) respectively in the space $L^2\times[0,\infty)$ and in the natural space $L^1\times[0,\infty)$.

In the present paper - in the framework of stochastic models for which π is unspecified - we assume that the scattering probability distribution π is expanding as power series of a parameter ε and we define an iterative constructive scheme leading to explicit solution of the form of power series of ε. We then prove the existence of a steady solution to equation (1) and its L^1-asymptotic stability.

2. SOLUTIONS.

It is easily seen that when the scattering probability distribution is independent of the velocities before the collision i. e.

$$\pi(\mathbf{v}',\mathbf{v}'',\mathbf{v})=\pi_o(\mathbf{v}) \quad , \qquad \int_{\mathbb{R}_3}\pi_o(\mathbf{v})d\mathbf{v}=1 \tag{3}$$

The solution to equation (1) is [2]:

$$f_o(\mathbf{v},t)=Q[\pi_o(\mathbf{v})+(\ S(\mathbf{v})-\pi_o(\mathbf{v})\ )e^{-kQt}] \tag{4}$$

In the sequel we shall assume:

Hypothesis I. The scattering probability distribution is expanding about $\pi_o(\mathbf{v})$ as power series of a parameter ε, not necessarily small, according to

$$\pi(\mathbf{v}',\mathbf{v}'',\mathbf{v})=\pi_o(\mathbf{v})+\sum_{i=1}^{\infty}\varepsilon^i\pi_i(\mathbf{v}',\mathbf{v}'',\mathbf{v}) \tag{5}$$

In order to satisfy equations (2) we require

$$\int_{\mathbb{R}_3}\pi_i(\mathbf{v},\mathbf{v}'',\mathbf{v})d\mathbf{v}=0 \quad , \quad \pi_i(\mathbf{v}',\mathbf{v}'',\mathbf{v})=\pi_i(\mathbf{v}'',\mathbf{v}',\mathbf{v}) \tag{6}$$

It is then quite natural to seek a solution of equation (1) in the form

$$f(\mathbf{v},t)=f_o(\mathbf{v},t)+\sum_{i=1}^{\infty}\varepsilon^i f_i(\mathbf{v},t) \tag{7}$$

with:

$$\int_{\mathbb{R}_3} f_i(\mathbf{v},t)=0 \qquad , \qquad f_i(\mathbf{v},0)=0 \tag{8}$$

It is easy to verify that consequently the evolution equations for f_i $(i=1,2,..)$ are

$$\frac{\partial f_1}{\partial t} + kQf_1 = k\int_{\mathbb{R}_3}\int_{\mathbb{R}_3} \pi_1(\mathbf{v}',\mathbf{v}'',\mathbf{v})f_0(\mathbf{v}',t)f_0(\mathbf{v}'',t)d\mathbf{v}'d\mathbf{v}'' \tag{9}$$

$$\frac{\partial f_2}{\partial t} + kQf_2 = k\int_{\mathbb{R}_3}\int_{\mathbb{R}_3} [\pi_2(\mathbf{v}',\mathbf{v}'',\mathbf{v})f_0(\mathbf{v}',t)f_0(\mathbf{v}'',t) + \tag{10}$$

$$+ 2\pi_1(\mathbf{v}',\mathbf{v}'',\mathbf{v})f_0(\mathbf{v}',t)f_1(\mathbf{v}'',t)]d\mathbf{v}'d\mathbf{v}''$$

$$\frac{\partial f_3}{\partial t} + kQf_3 = k\int_{\mathbb{R}_3}\int_{\mathbb{R}_3} [\ \pi_3(\mathbf{v}',\mathbf{v}'',\mathbf{v})f_0(\mathbf{v}',t)f_0(\mathbf{v}'',t)+ \tag{11}$$

$$+2\pi_1 f_0(\mathbf{v}',t)f_2(\mathbf{v}'',t)+2\pi_2(\mathbf{v}',\mathbf{v}'',\mathbf{v})f_0(\mathbf{v}',t)f_1(\mathbf{v}'',t)+$$

$$+\pi_1(\mathbf{v}',\mathbf{v}'',\mathbf{v})f_1(\mathbf{v}',t)f_1(\mathbf{v}'',t)]d\mathbf{v}'d\mathbf{v}''$$

- -

i.e., accounting for the $(6)_2$

$$\frac{\partial f_i}{\partial t} + kQf_i = k\sum_{s+k+l=i}^{s,k,l\,\in\,N} \int_{\mathbb{R}_3}\int_{\mathbb{R}_3} \pi_s(\mathbf{v}',\mathbf{v}'',\mathbf{v})f_k(\mathbf{v}',t)\times \tag{12}$$

$$f_l(\mathbf{v}'',t)d\mathbf{v}'d\mathbf{v}''. \qquad (i=1,2,3,...)$$

Recalling that π_i $(i=0,1,2..)$ are known, from equation (9) one obtains $f_1(\mathbf{v},t)$. Inserting $f_1(\mathbf{v},t)$ in (10) one obtain $f_2(\mathbf{v},t)$ and so on.

We are now in position to prove the following theorem:

Theorem 1. If $\exists$ M>0 such that:

$$\left|\int_{\mathbb{R}_3} \pi_n(\mathbf{v}',\mathbf{v}'',\mathbf{v})\varphi_1(\mathbf{v}')\varphi_2(\mathbf{v}'')d\mathbf{v}'d\mathbf{v}''\right| \le M^n \tag{13}$$

$$\forall\ \varphi_i(\mathbf{v})\ :\ \int_{\mathbb{R}_3}\varphi_i(\mathbf{v})d\mathbf{v} \le Q \qquad i=1,2$$

and moreover holds (5) with $\varepsilon<1/M$, then the series

$$f(\mathbf{v},t)=f_o(\mathbf{v},t)+\sum_{i=1}^{\infty}\varepsilon^i f_i(\mathbf{v},t) \quad , \ \varepsilon<1/M \tag{14}$$

(where $f_i(\mathbf{v},t)$ are L^1-solutions to the initial value problem (12)-$(8)_2$), is a.e. uniformly convergent in $\mathbb{R}_3\times[0,\infty]$. Moreover (14) can be differentiated (respect to t) term by term, is termwise L^1-summable (respect to **v**) and is solution to equations (1).

Proof. By equations (12) and the hypothesis of the theorem we obtain:

$$|f_n(\mathbf{v},t)| \le n^2M^n/Q \tag{15}$$

Then for $\varepsilon<1/M$ series (14), accordingly to the Weierstrass' test, is a.e. uniformly convergent in $\mathbb{R}_3\times[0,\infty]$. Moreover by equation (12) we obtain

$$\left|\frac{\partial f_n(\mathbf{v},t)}{\partial t}\right| \le 2kn^2M^n \tag{16}$$

hence derivative series is uniformly convergent and (14) can be differentiated, respect to t, term by term. Finally one verify easily that this series is termwise L^1-summable (respect to **v**) and is solution to equation (1).

3. EXISTENCE OF THE STEADY SOLUTION.

Let us consider now the steady version of equation (1):

$$f(\mathbf{v})=(1/Q)\int_{\mathbb{R}_3}\int_{\mathbb{R}_3}\pi(\mathbf{v}',\mathbf{v}'',\mathbf{v})f(\mathbf{v}')f(\mathbf{v}'')d\mathbf{v}'d\mathbf{v}'' \tag{17}$$

The following theorem holds:

Theorem 2. Let assumptions of theorem 1 hold. Then exists a solution to equation (17).

Proof. Let us consider the function

$$M(\mathbf{v})=M_o(\mathbf{v})+\sum_{i=1}^{\infty}\varepsilon^i M_i(\mathbf{v}) \tag{18}$$

where

$$M_o(\mathbf{v})=Q\pi_o(\mathbf{v}) \tag{19}$$

$$M_1(\mathbf{v})=(1/Q)\int_{\mathbb{R}_3}\int_{\mathbb{R}_3}\pi_1(\mathbf{v}',\mathbf{v}'',\mathbf{v})M_o(\mathbf{v}')M_o(\mathbf{v}'')d\mathbf{v}'d\mathbf{v}'' \tag{20}$$

$$M_2(\mathbf{v})=(1/Q)\int_{\mathbb{R}_3}\int_{\mathbb{R}_3}[\pi_2(\mathbf{v'},\mathbf{v''},\mathbf{v})M_o(\mathbf{v'},t)M_o(\mathbf{v''},t)\ + \tag{21}$$

$$+\ 2\pi_1(\mathbf{v'},\mathbf{v''},\mathbf{v})M_o(\mathbf{v'},t)M_1(\mathbf{v''},t)]d\mathbf{v'}d\mathbf{v''}$$

Substituting (18) in (17) and taking into account theorem 1 follows that series (18) is a.e. uniformly convergent in $\mathbb{R}_3\times[0,\infty]$, termwise L^1-summable (respect to $\mathbf{v}$) and is solution to equation (17).

4. L^1-ASYMPTOTIC STABILITY OF THE STEADY SOLUTION.

Let $f(\mathbf{v},t)$ and $g(\mathbf{v},t)=f+u$ be two solutions to equation (1) corresponding respectively to the initial data $(1)_2$ and

$$g(\mathbf{v},0)=QS(\mathbf{v})+u_o(\mathbf{v}). \tag{22}$$

Indicated by

$$\|u\|=\int_{\mathbb{R}_3}|u(\mathbf{v},t)|d\mathbf{v} \tag{23}$$

the L^1-norm of the perturbation $u(\mathbf{v},t)$ to the basic solution, it is possible to prove the following theorems:[4)]

Theorem 3. Let us assume the validity of theorem 1. If $\exists$ $\Phi_i(\mathbf{v'},\mathbf{v''},\mathbf{v})\in L^1(R_6)$ $(\forall\mathbf{v})$, such that:

$$|\pi_i(\mathbf{v'},\mathbf{v''},\mathbf{v})\Psi_1(\mathbf{v'})\Psi_2(\mathbf{v''})|\le\Phi_i(\mathbf{v},\mathbf{v''},\mathbf{v}) \tag{24}$$

$$\forall\ \Psi_i(\mathbf{v})\ :\ \int_{\mathbb{R}_3}\Psi_i(\mathbf{v})d\mathbf{v}\le Q$$

then indicated by $f(\mathbf{v},t)$ a solution to equation (1) exists the limit:

$$\lim_{t\to\infty} f(\mathbf{v},t)=M(\mathbf{v})\qquad \text{a.e.} \tag{25}$$

and does not depend on the initial data.

Moreover if exists a function $\phi(\mathbf{v})=L^1(R_3)$ such that:

$$f(\mathbf{v},t)\le\phi(\mathbf{v},t)\qquad \text{a.e.} \tag{26}$$

then the steady solution to equation (1) is unconditionally attractive in the L^1-norm.

Theorem 4. Suppose that assumptions of theorem 3 are verified. Then the solution $M(\mathbf{v})$ to equation (1) is uniformly asymptotically stable in

L^1-norm.

Proof of Theorem 4. Since the steady solution is unconditionally L^1-attractive we have:

(i) $\forall\ \varepsilon>0\ \exists\ T>0\ :\ t>T\ \Rightarrow\ \|u(\mathbf{v},t)\|<\varepsilon$.

Moreover the solutions to equation (1) depend continuously on the initial data in the L^1-norm [6] i.e.:

(ii) $\forall\ \varepsilon>0, T>0,\ \exists\ \delta>0\ :\ \|u_0\|<\delta\ \Rightarrow\ \|u(\mathbf{v},t)\|<\varepsilon\quad t\in[0,\infty)$

by (i) and (ii) and accounting for theorem 3, theorem 4 follows.

Acknowledgments.

This research has been performed under the auspices of the G.N.F.M.of C.N.R. and has been partially supported by the Italian Ministry for Education (M.P.I.) under 40% and 60% contracts.

REFERENCES

[1] Boffi V.C.,Molinari V.G.,Nonlinear Transport Problems by Factorization of the Scattering Probability, Il Nuovo Cimento Vol. 65 B, N.1, p. 29-44 (1981).

[2] Boffi V.C.,Spiga G.,Rigorous iterated solutions to a nonlinear integral evolution problem in particle transport theory, J. Math. Phys., **23**, 1859,1982, (1982).

[3] Boffi V.C.,Spiga G.,Global solution to a nonlinear integral evolution problem in particle transport theory, J. Math. Phys., **23**, 2299,2303, (1982).

[4] Guerriero G.,On The existence and stability of a steady solution to a nonlinear integral equation of the particle transport theory,Rendiconti dell'Accademia di Scienze Fisiche e Matematiche di Napoli (to appear).

[5] Rionero S.,Global solutions to a nonlinear integral evolution problem in particle transport theory, Atti Accademia Scienze Torino, Suppl. Vol. 120 p.199-209,(1986).

[6] Rionero S.,Guerriero G.,On the stability of the solutions to a nonlinear integral evolution problem arising in particle transport theory, Meccanica **22**,p.59-64, (1987).

[7] Rionero S.,Guerriero G.,L^1-stability of the solutions to an integral evolution equation of the nonlinear particle transport theory, Meccanica 24,p.191-199, (1989).

INVESTIGATION OF THE STABILITY OF TRANSVERSAL VIBRATIONS OF A NON-LINEAR STRING BY THE METHOD OF DIFFERENTIAL INEQUALITIES

GUTOWSKI ROMAN
WARSAW UNIVERSITY OF TECHNOLOGY

ABSTRACT

Stability in the Liapunov sense of the zero-solution of the vibrations of a non-linear string is investigated. The stability is tested directly on the basis of the definition with respect to an appropriate selected distance between the perturbed solution of the problem belonging to the Sobolev space $W^{2,2}(\Omega)$ and the zero solution. The method consist in construction of some differential inequality satisfied by the mentioned distance. The investigation of this inequality allows to obtain the conditions of stability and moreover the estimation of the perturbed solutions. As an example the stability and an effective estimation of the solution of the considered problem with the nonlinearity of the power function type is constructed.

1. METHOD OF DIFFERENTIAL INEQUALITIES OF INVESTIGATION THE STABILITY OF A NON-LINEAR STRING.

Let us consider the differential equation in the form

$$u_{tt} + 2bu_t + cu + f(x,t,u,u_x,u_t) = au_{xx} \tag{1}$$

$$x \in (0,1) \quad , \quad t \in (0,\infty)$$

where a,b,c are positive constants.

This equation may be considered as a mathematical model describing the transversal vibrations of a string supported on the non-linear elastic base with linear exterior damping.

Let the initial and boundary conditions have the form

$$u(x,0) = u_0(x) \quad , \quad u_t(x,0) = u_1(x) \tag{2}$$

$$u(0,t) = 0 \quad , \quad u(1,t) = 0 \; . \tag{3}$$

Let us assume, that Eq.(1) has a zero-solution $u = 0$. We shall investigate the stability in Liapunov sense of the unperturbed solution $u = 0$.

Suppose the regularity of function f assuring the existence and uniqueness of the solution to the investigated initial-boundary problem in the Sobolev spaces $u \in W^{2,2}(\Omega)$,

$u_t \in W^{1,2}(\Omega)$, $\Omega = (0,1)$ restricted to a set of functions satisfying the boundary conditions.

Let us introduce the distances:

$\rho(u)$ – between the perturbed solutions $u(x,t)$ and the zero solution $u = 0$.

$\rho_i(0)$ – between the initial values (2) of the perturbed solutions and the zero solution $u = 0$, $u_t = 0$,

in the form

$$
\left.\begin{aligned}
\rho(u) &= \int_0^1 \left(u_t^2 + a u_x^2 + 2 b u u_t + c u^2 \right) dx = \\
&= \int_0^1 \left[(c-b^2) u^2 + a u_x^2 + (bu+u_t)^2 \right] dx \\
\rho_i &= \rho \Big|_{t=0}
\end{aligned}\right\} \tag{4}
$$

where is assumed that

$$
c - b^2 > 0 \,. \tag{5}
$$

In this paper we shall use the following definition of stability.

The unperturbed solution $u = 0$ is called stable in the Liapunov sense with respect to the two distances ρ and ρ_i in the interval $t \in [0,\infty)$ if:

a. all perturbed solutions $u(x,t)$ are definite in the interval $t \in [0,\infty)$

b. for every $\varepsilon > 0$ there exist such a $\eta(\varepsilon) > 0$ that for arbitrary solution $u(x,t)$ the inequality $\rho_i \le \eta(\varepsilon)$ implies the inequality $\rho \le \varepsilon$ for all $t \ge 0$.

The time derivative of the distance ρ given by formula (4), using Eq.(1) and taking into account the boundary conditions (3) may be expressed in the form:

$$
\dot{\rho} = -2b\rho - 2 \int_0^1 (bu + u_t) \, f \, dx. \tag{6}
$$

The method of investigation used in the sequel is based on obtaining an estimation of the second term in the right-hand side of the formula (6) by a function depending on ρ (and eventually explicit of time) in the form:

$$-2\int_0^1 (bu + u_t) f\, dx \leq \beta(t)\rho + \omega(t,\rho) \tag{7}$$

where $\beta(t)$ (in particular we may have $\beta =$ const) is a known continuous function of time and $\omega(t,\rho)$ is a known continuous function of variables t and ρ.

In view of (7) the relation (6) takes the form of a differential inequality

$$\dot{\rho} \leq w(t)\rho + \omega(t,\rho) \tag{8}$$

where

$$w(t) = \beta(t) - 2b.$$

In the sequel we shall use the following method to obtain the inequality (8). Let us apply to the second term in the right-hand side of the formula (6) the inequality of Cauchy

$$-2\int_0^1 \tau^{1/2}(bu + u_t)\frac{1}{\tau^{1/2}} f\, dx \leq$$

$$\leq \int_0^1 \tau(bu + u_t)^2 dx + \int_0^1 \frac{1}{\tau} f^2 dx \leq$$

$$\leq \tau \int_0^1 \left[(c-b^2)u^2 + au_x^2 + (bu+u_t)^2\right] dx + \int_0^1 \frac{1}{\tau} f^2 dx =$$

$$= \tau\rho + \frac{1}{\tau}\int_0^1 f^2 dx$$

where τ = const > 0. If we may determine such a continuous non-negative function $\omega(t,\rho)$ that the inequality

$$\frac{1}{\tau}\int_0^1 f^2\,dx \le \omega(t,\rho) \tag{9}$$

holds, then the relation (6) takes the form of the differential inequality (8)

$$\dot{\rho} \le (\tau-2b)\rho + \frac{1}{\tau}\int_0^1 f^2\,dx \le (\tau-2b)\rho + \omega(t,\rho) \tag{10}$$

where in view of (7) and(8)

$$\beta(t) = \tau = \text{const} > 0 \quad , \quad w(t) = \tau - 2b = \mu = \text{const}. \tag{11}$$

Therefore the problem is reduced to the investigation of differential inequality (10). This inequality while solved allows to obtain the following estimation (see (4))

$$\rho = \int_0^1 \left[(c-b^2)u^2 + au_x^2 + (bu+u_t)^2\right] dx \le \Phi(t) \tag{12}$$

where $\Phi(t)$ is a known function. On the basis of this inequality we may investigate the stability in Liapunov sense and moreover, obtain the estimation

$$u^2(x,t) \le \frac{1}{a}\Phi(t). \tag{13}$$

The main result on basis of which one can obtain the estimation (12) is the following theorem [1,2].

THEOREM. Let us assume that there exist such two functions:

$v(t)$ - continuous and non-negative for all $t \ge 0$,

$g(z)$ - continuous, non-negative and non-decreasing with respect to $z \ge 0$

that for $t \geq 0$, $z \geq 0$ the following inequality

$$\omega\left[t, z \exp\left(\int_0^t w(s)ds\right)\right] \exp\left(-\int_0^t w(s)ds\right) \leq v(t)g(z) \qquad (14)$$

is satisfied. If the inequality (8) holds then the function $\rho(t)$ satisfies in the interval $[0,T)$ the inequality

$$\rho \leq G^{-1}\left[G(c) + \int_0^t w(s)ds\right] \exp\left(\int_0^t w(s)ds\right) = \Phi(t) \qquad (15)$$

where G^{-1} is the function inverse to

$$G(u) = \int_{u_o}^{u} \frac{ds}{g(s)} \quad , \quad 0 < u_0 \leq c \leq u \qquad (16)$$

and

$$c = \rho_i = \rho\Big|_{t=0} \;. \qquad (17)$$

The value of T is determined by the relation

$$\int_{u_o}^{\infty} \frac{ds}{g(s)} = G(c) + \int_0^T v(s)\, ds \;. \qquad (18)$$

Furthermore, if there exist such a constant $\tilde{c} > 0$, that the following inequality

$$\int_0^{\infty} v(t)\, dt \leq \int_{\tilde{c}}^{\infty} \frac{ds}{g(s)} \qquad (19)$$

is satisfied, then the inequality (15) holds for $t \in [0,\infty)$ that means for $T = \infty$ and $c \in (0,\tilde{c})$.

2. EXAMPLE.

The case when the non-linear function f appearing in Eq.(1) has the form

$$f = \gamma u^{\frac{1}{2}R} = \gamma u^{r}, \quad \gamma = \gamma(x) > 0, \quad r = \frac{1}{2} R = \text{const} > 0 \qquad (20)$$

where $\gamma(x)$ is a known 2k-integrable ($k = \text{const} > 0$) function shall be investigated as an example.

Let us apply in the considered case to the integral in the formula (9) the Holder's inequality

$$\int_0^1 f^2 dx = \int_0^1 \gamma^2(x) u^R dx \le \left(\int_0^1 \left(\gamma^2(x)\right)^k dx \right)^{1/k} \left(\int_0^1 |u^R|^l \, dx \right)^{1/l} . \qquad (21)$$

Let us denote

$$M_1 = \left(\int_0^1 \gamma^{2k}(x) \, dx \right)^{1/k} . \qquad (22)$$

Hence we have

$$\int_0^1 f^2 \, dx = \int_0^1 \gamma^2(x) \, u^R \, dx \le M_1 \left(\int_0^1 |u|^{Rl} \, dx \right)^{1/l} . \qquad (23)$$

If we are able to estimate the integral in the right-hand side of the inequality (23) by a function depending on ρ, then we obtain the function $\omega(t,\rho)$ and therefore the inequality (10) with the known function $\omega(t,\rho)$.

Now we make use of the assumption that the solution belongs to Sobolev space. We are interested in the space

$$W^{m,p}(\Omega) = W^{1,2}(\Omega)$$

that means $m = 1$, $p = 2$.

In the theory of Sobolev spaces the following imbedding holds [3)]

$$W^{m,p}(\Omega) \hookrightarrow L^s(\Omega), \quad 1 < p \le s < \infty, \quad mp \ge n$$

where n denotes the dimension of the space, which means

$$W^{m,p}(\Omega) \subset L^s(\Omega), \quad \|u\|_s \le M_2 \|u\|_{m,p}$$

where $M_2 = \text{const} > 0$ denotes the so called imbedding constant and

$$\|u\|_s = \left(\int_0^1 |u|^s \, dx \right)^{1/s}, \quad \|u\|_{m,p} = \left(\sum_{0 \le \alpha \le m} \|D^\alpha u\|_p^p \right)^{1/p}.$$

In the case under consideration we have $m = 1$, $p = 2$, $n = 1$ therefore $mp > n$ and

$$\left. \begin{aligned} & W^{1,2}(\Omega) \hookrightarrow L^s(\Omega), \quad 1 < 2 \le s < \infty, \\ & \|u\|_s \le M_2 \|u\|_{1,2} \end{aligned} \right\} \tag{24}$$

In view of (24) we have

$$\left(\int_0^1 |u|^s \, dx \right)^{1/s} \le M_2 \left(\int_0^1 u^2 \, dx + \int_0^1 u_x^2 \, dx \right)^{1/2}. \tag{25}$$

Let us assume that

$$s = R\iota \ge 2 . \tag{26}$$

Hence we have

$$\left(\int_0^1 |u|^{R\iota} \, dx \right)^{1/R\iota} \le M_2 \left(\int_0^1 u^2 \, dx + \int_0^1 u_x^2 \, dx \right)^{1/2}$$

and taking the power $R = 2r$ (see(20)) of both sides we get

$$\left(\int_0^1 |u|^{R\iota}dx\right)^{1/\iota} \le M_2^R\left(\int_0^1 u^2\,dx + \int_0^1 u_x^2\,dx\right)^{R/2} \le$$

$$\le M_2^{2r}\left[\frac{1}{c-b^2}\int_0^1\Big((c-b^2)u^2 + au_x^2 + (bu+u_t)^2\Big)\,dx + \right.$$

$$\left. + \frac{1}{a}\int_0^1\Big((c-b^2)u^2 + au_x^2 + (bu+u_t)^2\Big)\,dx\right]^r .$$

Therefore in view of (4) we obtain

$$\left(\int_0^1 |u|^{R\iota}\,dx\right)^{1/\iota} \le M_2^{2r}\left(\frac{1}{c-b^2} + \frac{1}{a}\right)^r \rho^r. \tag{27}$$

Let us denote by M the following constant

$$M = M_1\, M_2^{2r}\left(\frac{1}{c-b^2} + \frac{1}{a}\right)^r . \tag{28}$$

Hence in view of (23) the inequality (9) takes the form

$$\frac{1}{\tau}\int_0^1 f^2\,dx \le \frac{M}{\tau}\rho^r = \omega(t,\rho).$$

Therefore in the case of the considered non-linearity, the differential inequality (10) may be written in the form

$$\mathring{\rho} \le \mu\rho + \frac{M}{\tau}\rho^r , \qquad \mu = \tau - 2b = \text{const.} \tag{29}$$

Since in view of (26) we have

$$s = R\iota, \quad \tfrac{1}{2}R\iota \ge 1 , \quad r\iota \ge 1, \quad r \ge \frac{1}{\iota} , \quad 1 < \iota < \infty$$

therefore

$$r \in (0,\infty) . \tag{30}$$

Determining the constant M is a separate problem. The constant M_1 is known and given by the formula (22). The determination of the imbedding constant M_2 is possible, however laborious. In this paper we omit the determination of this constant.

The differential inequality in the form (29) may be solved explicit on the basis of the theorem presented above in the point 1. We shall give the explicit estimation of the distance ρ satisfying the differential inequality (29).

Let us consider the case $r > 1$, $\mu = \tau - 2b < 0$. In this case we have the following estimation

$$\rho \leq \frac{e^{\mu t}}{\left[\frac{1}{c^{r-1}} + \frac{M}{\mu\tau}\left(1 - e^{\mu(r-1)t} \right) \right]^{\frac{1}{r-1}}} = \Phi(t) \qquad (31)$$

where $c = \rho\big|_{t=0} = \rho_i$.

From this formula follows that inequality (31) holds for $t \in [0,\infty)$ but only for these $c \in (0,\tilde{c})$ for which the expression within the square brackets in the denominator is non- -negative for arbitrarily large t.

The value of $\tilde{c}$ may be determined from the condition

$$\frac{1}{\tilde{c}^{\,r-1}} + \frac{M}{\mu\tau} \geq 0 \qquad \mu = \tau - 2b < 0$$

and hence

$$\tilde{c} \leq \left[\frac{(2b-\tau)\tau}{M} \right]^{\frac{1}{r-1}} . \qquad (32)$$

The same value may be obtained using the general formula (15) given in the theorem mentioned in point 1.

Since the greatest possible value of $\tilde{c}$ is desired, therefore we calculate for which value of τ the function $\tilde{c}(\tau)$ takes the maximal value. We obtain

$$\tilde{c}_{max} = \left[\frac{b^2}{M} \right]^{\frac{1}{r-1}}, \qquad \text{for } \tau = b. \tag{33}$$

Therefore the function $\Phi(t)$ in the formula (31) being an estimation of the distance ρ is given for $t \in [0,\infty)$ and for $c \in (0,\tilde{c}_{max})$ which means

$$\rho_i \in \left(0 \ , \left[\frac{b^2}{M} \right]^{\frac{1}{r-1}} \right). \tag{34}$$

Since the denominator in the formula (31) tends to a positive constant for $t \longrightarrow \infty$ when $c \in (0,\tilde{c}_{max})$ and $\mu = \tau - 2b = b - 2b = -b < 0$ therefore the function $\Phi(t)$ given by the formula (31) tends to zero for $t \longrightarrow \infty$. In view of (13) the perturbed solutions $u(x,t)$ have in the case $r > 1$ the following result.

For

$$\rho_i = c \le \eta(\varepsilon) = \varepsilon \qquad \rho_i \in \left(0 \ , \left[\frac{b^2}{M} \right]^{\frac{1}{r-1}} \right) \tag{35}$$

we have

$$\rho(t) \le \rho_i = \Phi(0) = c \le \varepsilon. \tag{36}$$

Therefore the unperturbed solution $u = 0$ of the considered initial-boundary problem is stable with respect to admitted distances ρ and ρ_i. Since moreover $\rho \longrightarrow 0$ for $t \longrightarrow \infty$ exponentially, this solution is asymptotically and exponentially stable.

REMARK 1. If $\mu = \tau - 2b > 0$ in the considered case then the expression within the square brackets beginning from a certain time-moment shall be negative and the inequality (31) loses sense, that means it is not true for $t \in [0,\infty)$.

REMARK 2. If $\mu = \tau - 2b = 0$ then the estimation of the distance has the form

$$\rho \le \frac{1}{\left[\frac{1}{c^{r-1}} + \frac{M}{\tau}(r-1)t\right]^{\frac{1}{r-1}}} = \Phi(t) . \qquad (37)$$

Unfortunately beginning from a certain time-moment

$$t^{*} = \frac{\tau}{M(r-1)} \frac{1}{c^{r-1}}$$

the expression within the square brackets shall be negative, therefore the inequality (37) loses sense and is not of use for investigation the stability in the interval $t \in [0,\infty)$.

REFERENCES

1) R.GUTOWSKI, B.RADZISZEWSKI : "The behaviour of the solutions of equations of motion of the mechanical system with program constraints"; Bulletin de l'Academie Polonaise des Sciences, Serie des sciences techniques, Vol.XVII, Nr.2. (1969)

2) R.GUTOWSKI : "On the stability of some material systems" I-Vol.211, II-Vol.212, III-Vol.213; Conferenze del Seminaro di Matematica dell'Università di Bari (1986).

3) R. Adams : Sobolev Spaces, 1975 Academic Press, New York, San Francisco, London (1975).

THE GROWTH OF CRYSTALS

by

Donatella Iannece

Universitá degli Studi di Napoli
Dipartimento di Matematica e Applicazioni "R. Caccioppoli"
Via Mezzocannone, 8 - 80134 Napoli (Italy)

Abstract. A nonlocal continuum theory is employed to describe the process of growth of a crystal with plane faces in its melt. In particular the crystal evolution is determined as a function of the face orientation.

1. INTRODUCTION.

There are many experimental proofs (see, for instance, [1–5]) that in the solidification process of a small polyhedric crystal the effects of surface tension are relevant. In the ordinary theory of continuum thermodynamics these effects are closely related to the curvature of the interface between the solid and its melt so that they disappear in crystals with planar faces. In [6] it has been proved that a nonlocal continuum theory provides for the surface tension effects in the equilibrium problem of a small crystal having planar faces. In [7–8] we propose a dynamic nonlocal theory of the crystal growth which leads us to the experimental result that the growing velocity of a crystal face depends on its orientation with respect to the elementary cell. However such dependence cannot be brought back to a unique growth mechanism which prescinds from the crystal size. In fact in agreement with the most suitable theory to describe the evolution of a macroscopic crystal in its melt (having linear dimensions greather than 1 mm.) the growth is produced by the birth of spirals on faces [9] and it is evident that such a mechanism is not appropriate for crystal germs (10^{-5}–10^{-2} cm.).

In our theory a suitable interpretation of localization residuals permits us to introduce two different growth factors that alternatively prevail according to the crystal size: mechanical effects are dominant for large crystals (spiral growth) whereas thermal effects are dominant for very small crystals.

2. NONLOCAL BALANCE LAWS.

Let $\mathcal{S}$ be a two-phase system consisting of a small polyhedric crystal in the volume V_S and its melt in the volume V_L and let σ be the surface which separates the bulk phases; σ will be a nonregular surface containing a line c_σ on which the exterior* unit normal $\mathbf{n}$ to σ suffers a jump.

Now, our two-phase system will be characterized by the following geometrical and physical assumptions:

a) the crystal is a convex polyhedron with n faces σ_i and m edges c_σ^k so that we have:

$$\sigma = \bigcup_{i=1}^{n} \sigma_i, \qquad c_\sigma = \bigcup_{k=1}^{m} c_\sigma^k;$$

b) the crystal V_S and its melt V_L are rigid and moreover their mass densities ρ_S, ρ_L are almost equal, so that the effects of mass transport can be neglected;
c) each crystal face σ_i has characteristic uniform surface energy E_{σ_i} and entropy S_{σ_i}, i. e. we do not take into account the effects of stretching in the faces during their growth;
d) the temperature θ is uniform inside the crystal V_S and continuous across ∂V_S owing to the small size of the crystal.

Under these assumptions the global balance laws of energy and entropy we supposed to be valid for the whole system $\mathcal{S}$ lead to the following local ones

$$\begin{aligned} &E_S = \text{constant}, \\ &\dot{S}_S \geq 0, \end{aligned} \quad \text{in } V_S \qquad\qquad \begin{aligned} &\dot{E}_L = -\nabla \cdot \mathbf{q}, \\ &\dot{S}_L \geq -\nabla \cdot \frac{\mathbf{q}}{\theta}, \end{aligned} \quad \text{in } V_L \tag{1}$$

$$\begin{aligned} &[\![E]\!] c_{n_i} + \mathbf{q}_i^L \cdot \mathbf{n}_i = e_{\sigma_i}, \\ &-\theta [\![S]\!] c_{n_i} + \mathbf{q}_i^L \cdot \mathbf{n}_i \geq \theta s_{\sigma_i}, \end{aligned} \qquad \text{on } \sigma_i - {\bigcup_k}' c_\sigma^k, \tag{2}$$

$$\begin{aligned} &E_\sigma^i \mathbf{w}_{ij} \cdot \boldsymbol{\nu}_{ij} + E_\sigma^j \mathbf{w}_{ij} \cdot \boldsymbol{\nu}_{ij} = e_{c_{ij(k)}}, \\ &S_\sigma^i \mathbf{w}_{ij} \cdot \boldsymbol{\nu}_{ij} + S_\sigma^j \mathbf{w}_{ij} \cdot \boldsymbol{\nu}_{ij} = s_{c_{ij(k)}}, \end{aligned} \qquad \text{on } c_\sigma^k, \tag{3}$$

where $\mathbf{q}$ is the heat flux vector, c_{n_i} the geometrical velocity of σ_i along the unit normal vector $\mathbf{n}_i$, $\mathbf{w}_{ij}$ the geometrical velocity of the edge between σ_i and σ_j, $\boldsymbol{\nu}_{ij}$ the exterior normal vector to this edge in the tanget plane of σ_i. Finally e_{σ_i}, c_σ^k, $e_{c_{ij(k)}}$, $s_{c_{ij(k)}}$ represent the localization residuals which have to verify (see [7]) the

* Here exterior means that $\mathbf{n}$ lies in V_L.

global conditions

$$\begin{aligned}\sum_{i=1}^{n}\int_{\sigma_i} e_{\sigma_i}\,d\sigma+\sum_{k=1}^{m}\int_{c_k} e_{c_k}\,dl=0,\\ \sum_{i=1}^{n}\int_{\sigma_i} s_{\sigma_i}\,d\sigma+\sum_{k=1}^{m}\int_{c_k} s_{c_k}\,dl=0.\end{aligned}\tag{4}$$

We stress that in (1) volume localization residuals do not appear because in our theory these residuals are related to surface and edge effects; moreover we observe that $(3)_2$ is an equality because we neglect thermal conduction and, consequently, dissipation on the edges.

Finally, if h_i denotes the distance between the face σ_i and an internal point O of the crystal, α_{ij} the angle between σ_i and σ_j, A_{ij} the quantities

$$A_{ij}=\frac{E_\sigma^j}{\sin\alpha_{ij}}-E_\sigma^i\cot\alpha_{ij},\tag{5}$$

simple geometrical considerations allow us to write (4) in form

$$\sum_{i=1}^{n}\Big([\![E]\!]-\frac{1}{\sigma_i}\sum_j{}'_{(i)}A_{ij}l_{ij}\Big)\dot h_i=\sum_{i=1}^{n}\frac{1}{\sigma_i}\int_{\sigma_i}\mathbf{q}_i^L\cdot\mathbf{n}_i\,d\sigma.\tag{6}$$

Here $\sum_j{}'_{(i)}$ is a summation on the edges forming the boundary of σ_i and l_{ij} is the length of each c_{ij}.

3. CRYSTAL GERMS.

Up to this point the developed theory is suitable for both germs and macroscopic crystals. However different constitutive assumptions will characterize from now on the different growth processes.

In this section we consider a germ and, as a consequence of its very small size, we assume

a) $\mathbf{q}_i^L\cdot\mathbf{n}_i=k^L\dfrac{d\theta}{dn_i}=\text{constant}\qquad\text{on }\sigma_i;$ (7)

b) the velocity $\dot h_i$ of a face σ_i is affected by the heat flux $\mathbf{q}_i^L\cdot\mathbf{n}_i\sigma_i$ and vanishes with it.

This means that we are neglecting the fast variations of $\mathbf{q}_L$ in the neighbourhood of the edges and moreover the face transition has thermal origin: if there is a negative temperature gradient normal to the face σ_i, this face advances in the melt with velocity $\dot h_i$ which depends on $d\theta/dn_i$ and is independent of the thermal condition of other faces. In other words the temperature effects are primary but local.

Owing to assumptions a) and b) equation (7) gives us on each σ_i

$$\left([\![E]\!] - \frac{1}{\sigma_i}\sum_j{}'_{(i)} A_{ij} l_{ij}\right)\dot{h}_i = -k^L \frac{d\theta}{dn_i} \qquad (i = 1, \ldots, n), \tag{8}$$

so that:

1) from (2) and (8) we derive the surface localization residuals

$$e_{\sigma_i} = -\frac{1}{\sigma_i}\sum_j{}'_{(i)} A_{ij} l_{ij} \dot{h}_i \qquad (i = 1, \ldots, n), \tag{9}$$

as well as their physical interpretation. In fact, owing to (5), on each σ_i, the residual represents that part of the power $E_L c_{n_i}$ which the volume loses during the surface changes accompanying the volume growth;

2) the latent heat per unit volume of the small crystal is the sum of the latent heat $\lambda_\infty = [\![E(\theta)]\!]$ of a macroscopic crystal and a quantity e_{σ_i} which depends on the whole structure of the crystal and varies with the face;
3) the higher the surface energy E^i_σ of σ_i with respect to those of contiguous faces the higher the velocity $\dot{h}_i$ of σ_i;
4) the quantities l_{ij} are linear combinations of h_i's, whereas σ_i is quadratic in h_i's (see [6]). This implies that (8) reduces to the well known Stefan condition on each face when the crystal is sufficiently large to neglect the surface effects.

A futher result, experimentally well known, can be derived from the entropy inequality: in growing the crystal is at temperature θ less than the solidification temperature of macroscopic crystals (supercooling) and the greather the supercooling the greather the growth velocities.

4. MACROSCOPIC CRYSTALS.

The constitutive equations assumed in the previous section led to correct results for small crystals, however for larger crystals they imply that the growth velocities depend on the faces if only the thermal conductivity tensor is anysotropic. This is in evident disagreement with experimental results. Therefore we substitute them with the more general ones

$$e_{\sigma_i} = -\frac{1}{\sigma_i}\sum_j{}'_{(i)} A_{ij} l_{ij} \dot{h}_i + e'_{\sigma_i} \qquad (i = 1, \ldots, n), \tag{10}$$

where

$$\sum_{i=1}^{n} e'_{\sigma_i}\sigma_i = 0$$

and moreover*

$$e'_{\sigma_i} = \sum_{j=1}^{n} a_{ij}(\theta_\sigma, p - p_e, h_1, \ldots, h_n)\dot{h}_j, \qquad \lim_{h_i \to 0} a_{ij} = 0. \tag{11}$$

Thus e'_{σ_i} describes the cooperative action of the crystal face velocities.

Under these assumptions the evolution equations (8) are substituted by the following ones

$$\left([\![E]\!] - \frac{1}{\sigma_i} {\sum_j}'_{(i)} A_{ij} l_{ij} + a_{ii} \right) \dot{h}_i = -k^L \frac{d\theta}{dn_i} - \sum_{j \neq i} a_{ij} \dot{h}_j \qquad (i = 1, \ldots, n). \tag{12}$$

Thus we obtain that

— for large crystals $(h_i \to \infty)$ $e_{\sigma_i} \to e'_{\sigma_i}$ so that the crystal growth is essentially governed by the velocities of all faces;

— for small crystals $(h_i \to 0)$ the growth of each face depend s on the form of the whole crystal and the velocity of that particular face.

Finally in agreement with "spiral growth" theory we suggest putting

$$e'_{\sigma_i} = \frac{1}{\sigma_i} \left(\int_{\sigma_i} \nabla_S \mathbf{j}_{\sigma_i} \, d\sigma - {\sum_j}'_{(i)} \int_{c_{ij}} \mathbf{j}_{\sigma_i} \cdot \boldsymbol{\nu}_{ij} \, ds \right).$$

In other words we suppose that a surface energy flux vector $\mathbf{j}_{\sigma_i}$ is associated with spiral expansion on σ_i and this vector gives rise to quantities e'_{σ_i} we axiomatically introduced.

REFERENCES

[1] Gibbs, J. W.: The scientific Papers of Willard Gibbs, Ph. D. L. L. D. vol.III, Dover Publications, New York (1961).

[2] Landau, L. D.: The equilibrium form of crystals, Collected Papers, Gordon and Breach Science Publisher, 540 (1950).

[3] Defay, R. and Prigogine, I.: Tension surficielle et Adsorption, Editions Desoer, Liege (1951).

[4] Strickland-Constable, R. F.: Kinetics and Mechanism of Cristallization, Academic Press, London and New York (1968).

[5] Vainshtein, B. K.: Modern Crystallography, Formation of Crystals, Springer-Verlag, Berlin, Heidelberg, New York (1981).

[6] Romano, A. and Suhubi, E.S.: On Wulff's law about Equilibrium Configurations of Crystals, *Int. J. Engng. Sci.*, **27**, 1135 (1989).

---

* Here θ_σ coincides with solidification temperature and the crystal volume is at uniform temperature θ_σ (large crystals) and p_e is the equilibrium pressure.

[7] Iannece, D. and Romano, A.: Solidification of small crystals and nonlocal theories, *Int. J. Engng. Sci.* (in press).
[8] Iannece, D. and Romano, A.: Growth of macroscopic crystals and nonlocal theories, *Int. J. Engng. Sci.* (in press).
[9] Verma, A. R.: Crystal growth and dislocations, Butterworths, London (1953).

LOCALIZATION OF PLANE WAVES IN RANDOMLY LAYERED INHOMOGENEOUS MEDIA

V.I.Klyatskin

Pacific Oceanological Institute,
Far East Branch, USSR Academy of Sciences,
Vladivostok, 690032, USSR

ABSTRACT

There are considered problems: the plane-wave incidence on the half-space of the randomly inhomogeneous medium, the problem of the plane-wave source and the problem of the time-pulse in the unbounded space in the limit case of the attenuation absence in the medium. The space localization is shown to exist for the mean intensity of the source wave field and the pulse.

At present, the question of the wave-field localization in the layered randomly inhomogeneous media in the absence of the absorption in the media is seriously discussed[1-3]. In spite of this, the assumption of the exponential attenuation of the wave-field characteristics deep into the medium is taken as the initial information. This does not take place for the intensity statistical characteristics of the field in the case of plane-wave incidence on the half-space[4-6]. The absolutely different situation can exist for the source of plane waves in the unbounded space and for the nonstationary problems.

1. GENERAL FORMULATION OF THE PROBLEM

Let the layer of the randomly inhomogeneous medium to occupy a part of the space $L_0 \leq x \leq L$. The source of plane waves is placed in the point x_0 inside the layer. Then, the wave field in the layer of medium is described by the equation

$$\frac{d^2}{dx^2} G(x,x_0)+k^2 [1+\varepsilon(x)] G(x,x_0)=2ik\,\delta(x-x_0) \tag{1}$$

with boundary conditions

$$\frac{d}{dx} G(L_0,x_0)=-ik\, G(L_0,x_0),\quad \frac{d}{dx} G(L,x_0)=ikG(L,x_0). \tag{2}$$

We suppose here $\varepsilon(x)=\varepsilon_1(x)+i\gamma$ ($\varepsilon_1(x)=\varepsilon_1^*(x)$, $\gamma \ll 1$ - describes the absorption of the wave in the layer of medium), outside the layer $\varepsilon=0$.

If the source of plane waves is placed on the layer boundary $x_0=L$, the boundary problem (1),(2) corresponds to the problem of the unit amplitude plane-wave incidence on the layer of medium from the region $x \geq L$ .It is described by the boundary problem for the function $u(x)=G(x;L)$

$$\frac{d^2}{dx^2} u(x)+k^2 [1+\varepsilon(x)] u(x)=0, \tag{3}$$

$$\frac{d}{dx} u(L_0)=-ik\, u(L_0),\quad \frac{d}{dx} u(L)=ik[u(L)-2].$$

In this case the field on the layer boundary $x=L$ is $u(L)=1+R_1(L)$, where $R_1(L)$ is the complex coefficient of the wave reflection from the layer of medium.

The solution of the boundary problem (1),(2) has the structure[4)]

$$G(x,x_0)=\begin{cases} G(x_0,x_0)\exp[ik\int_x^{x_0} d\xi\, \psi_1(\xi)], & x_0 \geq x \\ G(x_0,x_0)\exp[ik\int_{x_0}^{x} d\xi\, \psi_2(\xi)], & x_0 \leq x \end{cases}, \tag{4}$$

where $G(x_0,x_0)=2[\psi_1(x_0)+\psi_2(x_0)]^{-1}$ and functions $\psi_i(x)$ satisfy the Riccati equations

$$\frac{d}{dx} \psi_{1,2}(x)=\pm ik[\psi_{1,2}^2-1-\varepsilon(x)],\ \psi_1(L_0)=1,\ \psi_2(L)=1. \tag{5}$$

Let introduce the functions $R_1(x)$, $R_2(x)$ instead the functions $\psi_i(x)$, using the equality $\psi_i(x)=[1-R_i(x)][1+R_i(x)]^{-1}$.
Then, for example the function $R_1(x)$ will be described by the Riccati equation

$$\frac{d}{dx} R_1(x)=2ikR_1(x)+\frac{ik}{2}\varepsilon(x)[1+R_1(x)]^2,\ R_1(L_0)=0. \tag{6}$$

Note, that the wave equation (1), when $x<x_0$, gives the equality

$$\frac{d}{dx} S(x,x_0)=k\gamma I(x,x_0),\ I(x,x_0)=|G(x,x_0)|^2 , \tag{7}$$

where the energy flow density in the region $x<x_0$ is defined by the relation

$$S(x,x_0)=\frac{1}{2ik}\left[G(x,x_0)\frac{d}{dx}G^*(x,x_0)-G^*(x,x_0)\frac{d}{dx}G(x,x_0)\right] \quad (8)$$

and, therefore, integrating (7) over the half-space $(-\infty,x_0)$, we obtain the equality

$$k\gamma\int_{-\infty}^{x_0}dx\, I(x,x_0)=S(x_0,x_0). \quad (9)$$

From the solution (4) we have

$$I(x,x_0)=\frac{|1+R_1(x)|^2|1+R_2(x_0)|^2}{|1-R_1(x_0)R_2(x_0)|^2}\,\frac{1-|R_1(x_0)|^2}{1-|R_1(x)|^2}\,e^{-k\gamma\int_x^{x_0}d\xi\frac{|1+R_1(\xi)|^2}{1-|R_1(\xi)|^2}}$$

$$S(x_0,x_0)=\frac{[1-|R_1(x_0)|^2]\,|1+R_2(x_0)|^2}{|1-R_1(x_0)R_2(x_0)|^2}.$$

If the source is placed on the layer boundary $x_0=L$ (the wave incidence on the layer of medium) $R_2(L)=0$, and all formulae become simpler. Let pass now to the statistical description of the wave field.

2. THE WAVE INCIDENCE ON THE HALF-SPACE

For the simplicity let consider the function $\varepsilon_1(x)$ to be the Gaussian delta-correlated random function with parameters $\langle\varepsilon_1(x)\rangle=0$, $\langle\varepsilon_1(x)\varepsilon_1(x')\rangle=2\sigma^2\ell\,\delta(x-x')$. Averaging (9) we obtain $k\gamma\int_{-\infty}^{L}dx\langle I(x,L)\rangle=1-\langle|R_1(L)|^2\rangle$, where the mean value $|R_1(L)|^2$ is counted using the probability distribution for $W=|R_1(L)|^2$ (see, for example[4])

$$P_\infty(W)=\frac{2\beta}{(1-W)^2}\exp\left\{-2\beta\frac{W}{1-W}\right\},\quad \beta=k\gamma/D,$$

where the diffusion coefficient is $D=k^2\sigma^2\ell/2$.

In the limit of the small attenuation, when $\beta\to 0$, we have $D\int_{-\infty}^{L}dx\langle I(x;L)\rangle=2\ln\frac{1}{\beta}$, $\langle I(L;L)\rangle=2$, and we see that the mean part of energy is visible to be concentrated in the region of the space $\ell_\beta\sim\frac{1}{D}\ln\frac{1}{\beta}$, that is the localization, linked to the attenuation, takes place. The quantity $\ell_\beta\to\infty$, and the localization is absent as $\beta=0$.

3. THE SOURCE LOCATION IN THE UNBOUNDED SPACE

Let return now to the problem of the field of the point source, placed in the unbounded space ($L_0\to-\infty$,

$L\to\infty$). Accomplishing the averaging in formulae (9), we obtain $< I(x_0,x_0)> = 1+\frac{1}{\beta}$, $< S(x_0, x_0)> = 1$; that is the mean flow of the energy density does not depend on the fluctuations of the medium parameters. Hence, $\int_{-\infty}^{x_0} dx < I(x,x_0)> =$ $= 1/\beta D$, as in the case of the free space. The main part of energy becomes localized in the region of space $\ell \sim 1/D$.

The localization curve in the limit $\beta\to 0$ is defined by means of equation

$$\frac{\partial}{\partial \xi}\Phi_\xi(u) = -u\Phi_\xi(u) + \frac{\partial}{\partial u}u^2\Phi_\xi(u) + \frac{\partial}{\partial u}u^2\frac{\partial}{\partial u}\Phi_\xi(u), \quad \Phi_0(u) = e^{-u} \tag{10}$$

and the equality

$$\Phi_{loc}(\xi) = \lim_{\beta\to 0}\beta < I(x,x_0)> = \int_0^\infty du\, u\,\Phi_\xi(u) \quad (\xi = D|x-x_0|). \tag{11}$$

We had examined above the problem of statistical characteristics of the point source intensity in the unbounded space in the limit of the small attenuation. We have shown, that

$$\lim_{\beta\to 0}\lim_{L_0\to -\infty} < I(x,x_0)>/< I(x_0,x_0)> = \Phi_{loc}(\xi)\,(\xi = D|x-x_0|). \tag{12}$$

However, it is easy to see, that the interchange of limits in (12) leads to the result

$$\lim_{L_0\to -\infty}\lim_{\beta\to 0} < I(x,x_0)>/< I(x_0,x_0)> = \frac{2}{3} \quad (D|x-x_0| \gg 1),$$

which does not depend on $(x-x_0)$. Hence, the space localization is absent here. The similar situation takes place also in the case of the source location on the reflective boundary. From the physical point of view, the passage to the limit (12) appears to be correct.

4. THE NONSTATIONARY PROBLEM

Let examine now the nonstationary problem of the field of the time-pulse, which is generated in the randomly inhomogeneous medium. The wave field is described in this case by the wave equation

$$\left[\frac{\partial^2}{\partial x^2} - \frac{1}{c^2(x)}\frac{\partial}{\partial t}\left(\frac{\partial}{\partial t} + 2\tilde{\gamma}\right)\right]u(x,x_0;t) = \frac{2}{c_0}\delta(x-x_0)\frac{\partial}{\partial t}f(t).$$

Its solution can be written in the form of the Fourier integral (the parameter $\tilde{\gamma}$ is supposed to be small)

$$u(x,x_0;t)=\frac{1}{2\pi}\int_{-\infty}^{\infty}d\omega\, G_\omega(x,x_0)f(\omega)e^{-i\omega t},\quad G_\omega(x,x_0)=\int_{-\infty}^{\infty}dt\, G(x,x_0;t)e^{i\omega t}.$$

The function $G_\omega(x,x_0)$ is the solution of the stationary problem in the field of the plane-waves point source

$$\frac{d^2}{dx^2}G_\omega(x,x_0)+\kappa^2[1+\varepsilon(x)]G_\omega(x,x_0)=2i\kappa\,\delta(x-x_0),$$

where $c^{-2}(x)=c_0^{-2}[1+\varepsilon(x)]$, $\kappa=\omega/c_0$, $\varepsilon(x)=\varepsilon_1(x)+2i\tilde{\gamma}/\omega$.

We shall be interested in limit values of the mean magnitude of the wave field intensity $I(x,x_0,t)=u^2(x,x_0,t)$, which is determined by the quantity

$$\langle I(x,x_0;t)\rangle=\frac{1}{(2\pi)^2}\iint_{-\infty}^{\infty}d\omega\, d\psi\,\langle I_{\omega,\psi}(x,x_0)\rangle\, f(\omega+\tfrac{\psi}{2})f^*(\omega-\tfrac{\psi}{2})e^{-i\psi t}.$$

When $t\to\infty$, the value of this integral is defined by the integrand behaviour at small values of ψ, i.e. by the quantity

$$\langle I(x,x_0;t)\rangle_{t\to\infty}=\frac{1}{(2\pi)^2}\int_{-\infty}^{\infty}d\omega\,|f(\omega)|^2\int_{-\infty}^{\infty}d\psi\,\langle I_{\omega,\psi}(x,x_0)\rangle e^{-i\psi t}, \quad (13)$$

where the two-frequency analogue of the plane-waves intensity is introduced

$$I_{\omega,\psi}(x,x_0)=G_{\omega+\frac{\psi}{2}}(x,x_0)\,G^*_{\omega-\frac{\psi}{2}}(x,x_0).$$

As it was shown in the work[5)], all statistical characteristics, received in the stationary problem, being prolongated analytically with respect to the parameter $\tilde{\gamma}$, define the corresponding two-frequency statistical characteristics. So, if the damping is absent ($\tilde{\gamma}=0$), it is necessary to replace the parameter $\tilde{\gamma}$ by $0-i\psi$ in the corresponding statistical characteristics of the plane-wave stationary problem.

As a result we obtain for the mean intensity localization curve, as

$$\langle I(x,x_0;\infty)\rangle=\frac{1}{2\pi}\int_{-\infty}^{\infty}d\omega\, D(\omega)|f(\omega)|^2\Phi_{loc.}(\xi)\quad\left(\xi=D(\omega)|x-x_0|,\ D(\omega)=\frac{\sigma^2 l\omega^2}{2c_0^2}\right),$$

where $\Phi_{loc.}(\xi)$ is localization curve for the stationary problem.

In the case of the time-pulse incidence on the

half-space of the randomly inhomogeneous medium $x<L$, we have for $t\to\infty$

$$< I(L;L;t)> = \frac{c_0}{2\pi}\int_{-\infty}^{\infty} d\omega\, |f(\omega)|^2 \frac{D(\omega)}{[2+D(\omega)c_0 t/2]^2},$$

$$\mathcal{E}(t) = \int_{-\infty}^{L} dx < I(x,L;t)> = \frac{c_0}{\pi}\int_{-\infty}^{\infty} d\omega\, |f(\omega)|^2 \frac{1}{2+D(\omega)c_0 t/2}.$$

Note, that the first formulae for $<I>$ have been received before in the work[6)] and the particular case in the work[5)].

REFERENCES

1. Sheng P., White B., Zhang Z., Papanicolaou G., Phys.Rev.B34,4757(1986).
2. Sheng P., White B., Zhang Z., Papanicolaou G., In:"Scattering and localization of classical waves in Random media",World Sc.Publ.Co.Singapure (1989).
3. Gradeschool S.A., Freilicher V.D., Izv.Vuzov, Radiophysika, 31, 1210(1968) (in Russian).
4. Klyatskin V.I."Stochastic equations and waves in the randomly inhomogeneous media",Moscow, Nauka (1980) (in Russian), (1985) (in French).
5. Shevtsov B.M., Izv.Vuzov,Radiophysika,25,1032 (1982) (in Russian).
6. Burridge R., Papanicolaou G., Sheng P., White B., SIAM J.Appl.Math.,49,582(1989).

SPATIAL DECAY ESTIMATES IN THE VIBRATING ANISOTROPIC ELASTIC BEAM

by

R J Knops
Department of Mathematics
Heriot-Watt University
Riccarton
Edinburgh EH14 4AS

ABSTRACT

This paper, which is mainly expository and based on [6], shows that in an elastic beam subject to a dynamically applied end-displacement the energy is bounded by a constant and by a exponentially decaying spatial term. A similarity thus exists with Saint-Venant's Principle for the corresponding equilibrium problem.

1. INTRODUCTION

This paper seeks to examine the nature of edge-effects in a semi-infinite elastic beam subjected to an applied dynamical end-load. It might be reasonable to expect a similarity with the edge-effects encountered in the corresponding equilibrium problem where, in accordance with Saint-Venant's Principle, it is well-known that there is exponential decay with respect to distance from the loaded base of the beam. (Cp. [1,3-5,7-10,12,13].) A moment's reflection, however, reveals that the similarity cannot be exact.

For a beam initially at rest, a suddenly applied dynamic end-load creates a disturbance that propagates along the beam with a definite wave speed. Ahead of the wave-front the material is undisturbed, while behind the wave-front the disturbed motion may consist of a mixture of dilation, shear, surface and other waves, interacting with shocks and other effects due, for example, to resonance. (Cp [2,15.)

Nevertheless, provided the end-load is such as to avoid the generation of shocks and resonance effects, examination of some averaged quantity of the motion may simplify the complexity and moreover indicate the same exponential decay found in the corresponding equilibrium problem. It is the aim of this paper to confirm this conjecture and to describe in some detail the averaged structure behind the wave-front as measured by the energy.

The problem of the vibrating beam has a long history, which commenced with the studies of Saint-Venant (Cp Love [11]). The subject is still one of active interest and is of importance to acoustics, nonlinear optics, structural vibrations, wave-guides and many other fields.

Our analysis closely follows that of Flavin, Knops and Payne [6] and depends upon showing that the total work done over a cross-section of the beam satisfies a pair of differential inequalities. Integration then establishes how the total energy in the beam evolves with respect to both space and time. For simplicity, we consider only a prismatic beam with fixed lateral sides. The derivation of the basic differential inequalities is carried out for a nonlinear elastic beam subject to certain constitutive assumptions which, together with the basic equations, are set down in Section 2. It is not possible within the context of the nonlinear theory to describe any detailed behaviour and accordingly in Section 4 attention is confined to the linearised anisotropic theory of small deformations superposed upon large. The assumption of a specified sinusoidal end-displacement enables a decomposition of the displacement into two terms, the sum of whose associated energies is shown to be bounded by a constant and by an exponentially spatially decaying factor. Thus, in these circumstances at least, the conjecture is clarified and proved. For a related study, see [14].

Throughout we assume the existence of a smooth solution. We omit several of the detailed arguments since these may be found in [6] which also contains results not included here. Our discussion, however, is presented in the slightly more general context of linear anisotropy.

2. BASIC NOTATION, EQUATIONS AND ASSUMPTIONS

We consider a semi-infinite prismatic cylinder of uniform cross-section $D \subset \mathbb{R}^2$ whose boundary is sufficiently smooth to admit application of the divergence theorem. The origin of a rectangular cartesian system is located at an interior point of the base cross-section with the x_3 - axis parallel to the generators of the cylinder. The region B occupied by the cylinder is therefore given by $B = D \times [0,\infty)$ and its lateral surface by $\partial B \times [0,\infty)$. The notation $D(x_3)$ signifies that respective quantities are evaluated over the cross-section at axial distance x_3 from the origin so that, for example, the integral of a function over D may be written as follows:

$$\int_{D(x_3)} h\,dA = \int_D h(x_1,x_2,x_3)\,dx_1 dx_2, \tag{2.1}$$

where the area element of integration is denoted by dA. The summation convention is also adopted, with latin and greek suffixes having the ranges 1,2,3 and 1,2 respectively, and a comma is used to indicate partial differentiation with respect to spatial variables.

It is now supposed that a nonlinear homogeneous elastic body, which in its stress-free reference configuration occupies the region B, is set in motion by assigning throughout B the initial values of the displacement and velocity. Subsequently the lateral boundary is held fixed at zero displacement, whereas over the base the displacement is prescribed as a function of space and time. The body-force is assumed to be zero throughout the motion, and all data is supposed compatible with the existence of a smooth (classical) solution.

We let $u : B \times \mathbb{R}_+ \to \mathbb{R}^3$ be the smooth displacement field and we let $W : M^+ \to \mathbb{R}$ be the strain-energy density of the deformation, where M^+ is the set of 3×3 - matrices F that satisfy the condition

$$\det(F + I) > 0 \tag{2.2}$$

with I denoting the unit matrix. It is well-known that the strain-energy is a function of the displacement gradient and that the

Piola-Kirchhoff stress is given by

$$\sigma_{ji}(u) = \frac{\partial W}{\partial u_{i,j}}. \tag{2.3}$$

We assume that $\rho : B \to \mathbb{R}$ is the uniform mass density, taken to be strictly positive. The equations governing the motion of the elastic body then become

$$\sigma_{ji,j} = \rho \ddot{u}_i, \quad x \in B \times \mathbb{R}_+ \tag{2.4}$$

where a superposed dot indicates differentiation with respect to time. The lateral boundary conditions are specified by

$$u_i(x_j,t) = 0, \quad (x,t) \in \partial D \times \mathbb{R}_+ \times \mathbb{R}_+, \tag{2.5}$$

while the initial conditions are

$$u_i(x_j,0) = f_i(x_j), \quad \dot{u}_i(x_j,0) = g_i(x_j), \quad x \in B \tag{2.6}$$

where f_i, g_i are assigned functions having bounded initial total energy:

$$E(0) \equiv \frac{1}{2}\int_B \left[\rho g_i g_i + 2W(f_{i,j})\right] dx \le M_1^2 \tag{2.7}$$

for some positive constant M_1. Note that (2.7) specifies the spatial asymptotic behaviour of the initial data.

In the absence of any generally agreed constitutive hypotheses on the strain-energy density function W, apart from those imposed by invariance arguments, we introduce the following assumptions:-

(A.1) The strain-energy satisfies the inequality

$$\int_D \sigma_{i3}(v)\sigma_{i3}(v)dA \le c\int_D W(v_{k,l})dA, \tag{2.8}$$

uniformly in x_3 and for all sufficiently smooth vector fields v, where c is a positive constant.

(A.2) The strain-energy satisfies the lower bound

$$\int_D W(v_{k,l})dA \ge 0, \tag{2.9}$$

uniformly in x_3.

Assumption (A.1) implies unique inversion of the constitutive relations (2.3) and hence precludes dynamic buckling of the beam.

The sequel requires the linearised version of our initial-boundary value problem. It is convenient to treat the theory of small deformations upon large, when the formulation of the problem continues to be given by (2.4) - (2.7), but with the constitutive relations (2.3) replaced by

$$\sigma_{ij} = c_{ijkl}u_{k,l}. \tag{2.10}$$

The elasticities, c_{ijkl}, are functions of the "large deformation", assumed to be in equilibrium, and by virtue of the existence of the strain-energy density function satisfy the symmetry condition

$$c_{ijkl} = c_{klij}. \tag{2.11}$$

We also suppose that the elasticities are positive-definite in the sense that there holds

$$c_{ijkl}\psi_{ij}\psi_{kl} \geq c_1\psi_{ij}\psi_{ij} \tag{2.12}$$

for every second order tensor ψ_{ij} and some positive constant c_1. (In Section 4 the additional assumption of boundedness is introduced.) It then may easily be shown that assumptions (A1) and (A2) are satisfied with the strain-energy density given by

$$W(u_{k,l}) = \frac{1}{2}c_{ijkl}u_{i,j}u_{k,l}. \tag{2.13}$$

We remark that the method of treatment described in this paper does not immediately apply to the classical linear theory, for which the elasticities additionally possess the minor symmetries

$$c_{ijkl} = c_{jikl} = c_{ijlk}, \tag{2.14}$$

Such an extension requires the introduction of Korn-type inequalities and is beyond the scope of the present discussion.

3. EVOLUTION OF A CROSS-SECTIONAL WORK FUNCTION

The main purpose of this section is to establish differential inequalities that are satisfied by the total work done in a cross-section of the beam during the motion. Integration of these inequalities provides some general information on how the total energy fed by the dynamic end-loading propagates along the beam. Not surprisingly, these calculations show that the disturbance propagates into the beam with a definite speed but the analysis is too general to describe in adequate detail the structure of the solution behind the travelling front of the disturbance. To achieve the requisite refinement we consider in the next section the linearised version of the theory and further restrict the end-loading to be periodic. Then a more complete description of the behaviour of the solution becomes possible and reveals a transient exponential decay term reminiscent of the spatial decay exhibited in the standard Saint-Venant's Principle.

We commence the discussion by defining the total cross-sectional work function $\Phi(z,t)$:

$$\Phi(z,t) = -\int_0^t\int_{D(z)} \sigma_{i3}(x_j,t)u_{i,\eta}(x_j,\eta)dAd\eta. \tag{3.1}$$

In terms of this function we have the following basic result:

Theorem 3.1

The cross-sectional work function $\Phi(z,t)$, defined in (3.1), satisfies the pair of differential inequalities

$$\frac{\partial\Phi}{\partial t}(z,t) + \beta\frac{\partial\phi}{\partial z}(z,t) \le \beta H(z), \tag{3.2}$$

$$\frac{\partial\phi}{\partial t}(z,t) - \beta\frac{\partial\phi}{\partial z}(z,t) \ge -\beta H(z), \tag{3.3}$$

for $z \ge 0$, $t \ge 0$, where

$$H(z) \equiv \frac{1}{2}\int_{D(z)}\left[\rho g_i g_i + 2W(f_{i,j})\right]dA \tag{3.4}$$

and $\beta = (c\rho^{-1})^{1/2}$.

We note that the function $H(z)$ is related to the initial total energy by means of the expression

$$E(0) = \int_0^\infty H(z)dz. \tag{3.5}$$

Proof of theorem Upon introducing the function $H(z,t)$ defined by

$$H(z,t) = \frac{1}{2}\int_{D(z)} \left[\rho\dot{u}_i(x_j,t)\dot{u}_i(x_j,t) + 2W(u_{i,j}(x,t))\right]dA, \tag{3.6}$$

so that $H(z,0) = H(z)$, we may integrate by parts in (3.1) to easily obtain

$$\Phi(z,t) = \Phi(0,t) - \int_0^z H(y,t)dy + \int_0^z H(y)dy, \tag{3.7}$$

and hence

$$\frac{\partial\Phi}{\partial z}(z,t) = - H(z,t) + H(z). \tag{3.8}$$

Again, from (3.1) we have

$$\left|\frac{\partial\phi}{\partial t}\right| = \int_{D(z)} \sigma_{i3}\dot{u}_i dA, \tag{3.9}$$

which by Schwarz's inequality and assumption (A.1) leads to

$$\left|\frac{\partial\phi}{\partial t}\right| \le \beta H(z,t). \tag{3.10}$$

Elimination of $H(z,t)$ between (3.8) and (3.10) yields the desired inequalities (3.2), (3.3). ■

In order to extract from (3.2), (3.3) information on the evolutionary behaviour of the total energy in the beam, we proceed first to derive the following Corollary:

Corollary 3.1 The cross-sectional work function $\Phi(z,t)$ satisfies the asymptotic behaviour

$$\lim_{z\to\infty} \Phi(z,t) = 0 \qquad \forall t. \tag{3.11}$$

Proof: Let $z > z_0 \ge 0$, for fixed z_0. Along the line segment

$\beta t = z - z_0$ we see that (3.2) becomes $d\Phi \le H(z)dz$. However, we have $\Phi(z_0,0) = 0$ so that by integration we obtain

$$\Phi(z,\beta^{-1}(z - z_0)) \le \int_{z_0}^{z} H(y)dy$$

which by (2.7) yields for fixed t:

$$\lim_{z\to\infty} \Phi(z,t) \le 0.$$

Again, on letting $0 \le z < z_1$, for fixed z_1, and integrating along the line segment $\beta t = z_1 - z$ we similarly obtain for fixed t

$$\lim_{z\to\infty} \Phi(z,t) \ge 0. \qquad \blacksquare$$

The next corollary confirms that the disturbances due to the dynamically applied end loads do not propagate faster than some speed β. The proof follows easily from Corollary 1 and the identity (3.7) and hence is omitted.

Corollary 2. The total energy contained in that part of the beam given by $D \times [z,\infty)$, viz.,

$$E(z,t) = \int_{z}^{\infty} H(y,t)dy, \tag{3.12}$$

satisfies, for $\beta t = z - z_0$, $0 \le z_0 < z$, the bound

$$E(z,t) \le E(z_0,0), \tag{3.13}$$

and for $\beta t = z_1 - z$, $0 \le z < z$, the bound

$$E(z,t) \ge E(z_1,0). \tag{3.14}$$

The last inequality partially describes the disturbed behaviour in the region $z < \beta t$. Refinement is afforded by integrating (3.2) along the segment $z = \beta(t - t_0)$ from $(0,t_0)$ to (z,t), where $t \ge t_0 \ge 0$. This yields

$$E(z,t) \le \Phi(0,t_0) + E(0), \quad z \le \beta t. \tag{3.15}$$

Of course, neither (3.14) nor (3.15) establish spatial decay of the solution in the region $z \le \beta t$. Demonstration of this property is undertaken for the special case of a sinusoidal end-displacement in the next Section. For generalisations, see [6].

4. PERIODIC END DISPLACEMENT

We continue to study the initial boundary value problem specified by (2.4) - (2.7), (2.10), but subject to the end displacement

$$u_i(x_\alpha, 0, t) = h_i(x_\alpha)\sin \omega t, \tag{4.1}$$

where h_i is a given function and ω is a real positive constant. Resonance can therefore be expected for critical values of ω.

The displacement $u_i(x,t)$ is decomposed as follows

$$u_i(x,t) = v_i(x)\sin \omega t + w_i(x,t)$$

where in terms of the operator L, defined by

$$(L\varphi)_i = (c_{ijkl}\varphi_{k,l})_{,j},$$

the functions v_i and w_i satisfy

$$\left.\begin{aligned} (Lv)_i + \rho\omega^2 v_i &= 0, && x \in B \times \mathbb{R}_+ \\ v_i(x) &= 0, && x \in \partial D \times \mathbb{R} \times \mathbb{R}_+ \\ v_i(x_\alpha) &= h_i(x_\alpha), && x \in D(0) \end{aligned}\right\} \tag{4.2}$$

$$\left.\begin{aligned} (Lw)_i &= \rho\ddot{w}_i && x \in B \times \mathbb{R}_+ \\ w_i(x,t) &= 0, && x \in \partial D \times \mathbb{R} \times \mathbb{R}_+ \\ w_i(x_\alpha, 0, t) &= 0, && x \in D(0) \times \mathbb{R}_+ \\ w_i(\cdot,0) &= f_i, && \dot{w}_i(\cdot,0) = g_i - \omega v_i, x \in B. \end{aligned}\right\} \tag{4.3}$$

Upon introducing the definitions

$$(\varphi,\varphi) = \int_D \varphi_i\varphi_i dx_1 dx_2 \qquad (L\varphi,\varphi) = \frac{1}{2}\int_D c_{ijkl}\varphi_{i,j}\varphi_{k,l} dx_1 dx_2,$$

we easily see from the triangle inequality that

$$[E(z,t)]^{1/2} \le E_w^{1/2}(z,t) + Q^{1/2}(z,t) \tag{4.4}$$

where $E_w(z,t)$ is given by (3.12) with u replaced by w, and $Q(z,t)$ is defined to be

$$Q(z,t) = \frac{1}{2}\rho\omega^2 \cos^2\omega t\int_z^\infty (v,v)dx_3 + \sin^2\omega t\int_z^\infty (Lv,v)dx_3. \tag{4.5}$$

In order to preclude resonance, we assume that

$$\Omega^2 \equiv (1 - \rho\omega^2/c_1\lambda) > 0, \tag{4.6}$$

where λ is the constant in the Poincaré inequality for the cross-section D. Hence on recalling (2.12) we immediately deduce that

$$Q(z,t) \le \int_z^\infty (Lv,v)dx_3. \tag{4.7}$$

In order to accomplish our next task, which is to establish a decay estimate for the right-side of (4.7), we assume that the elasticities additionally satisfy the boundedness condition

$$\text{(A3)} \qquad c_{ijkl}\xi_i\xi_k\eta_j\eta_l \le c_2\xi_i\xi_i\eta_j\eta_j, \qquad \forall\xi,\ \eta \in \mathbb{R}^3, \tag{4.8}$$

where c_2 is some positive constant. Then, by means of the Schwarz and Poincaré inequalities, it may be shown that the function

$$I(x_3) = \int_D c_{i3kl}v_{k,l}v_i dA \tag{4.9}$$

satisfies the differential inequality

$$\kappa|I(x_3)| \le I'(x_3), \qquad \kappa = (c_2/c_1\lambda)^{1/2}\Omega^{-2}. \tag{4.10}$$

A complete discussion of the last inequality is presented in [6] where, in particular, is proved that the boundedness of $\int_z^\infty (Lv,v)dx_3$ leads to the inequalities

$$\int_z^{\infty}(Lv,v)dx_3 \le - I(x_3) \le \Omega^{-1}e^{-\kappa z}\int_0^{\infty}(Lv,v)dx_3, \tag{4.11}$$

which is the required estimate. We note that $\int_0^{\infty}(Lv,v)dx_3$ may be further bounded in terms of the given function h and its transverse derivations. (See [6] for the isotropic case.)

Finally, we may insert (4.11) and (4.7) into (4.4) and use (3.15) to bound $E_w(s,t)$ in terms of initial data (recall $(4.3)_3$) in order to conclude that for $z < \beta t$,

$$[E(z,t)]^{1/2} \le [E(0)]^{1/2} + \left[\omega\left(\frac{\rho}{c_1\lambda}\right)^{1/2} + \Omega^{-1}e^{-\kappa z/2}\right]\left[\int_0^{\infty}(Lv,v)dz\right]^{1/2} \tag{4.12}$$

We have thus demonstrated that within the disturbed region $0 < z < \beta t$ the energy is bounded above by the sum of a constant term and a term that decays exponentially with distance from the base of the beam. Hence, the aim of the paper is achieved.

REFERENCES

1. Biollay, Y., First boundary value problem in elasticity: Bounds for the displacements in Saint-Venant's Principle, ZAMP. 31, 556-567 (1980).
2. Boley, B.A., Application of Saint-Venant's Principle in dynamical problems, J. Appl. Mechs. 22 (2), 204-206 (1955).
3. Fichera, G., Il principio di Saint-Venant: intuizione dell'ingegnere e rigore del matematico. Rend. di Mat., 10 (Ser VI), 1-24 (1977).
4. Fichera, G., Remarks on Saint-Venant's Principle. Complex Analysis and its Applications, I.N. Vekua Anniversary Volume, 543-557 (Moscow 1978).
5. Flavin, J.N., Knops R.J and Payne L.E., Decay estimates for the constrained elastic cylinder of variable cross section. Q. Appl. Maths. 47, 325-350 (1989).

6. Flavin, J.N., Knops R.J and Payne L.E., Energy bounds in dynamical problems for a semi-infinite elastic beam. Elasticity. Mathematical Methods and Applications. The Ian N. Sneddon 70th Birthday Volume. (Ed. by G. Eason and R.W Ogden), Ellis Horwood. 101-112 (1990).

7. Galdi, G.P., Knops, R.J. and Rionero, S., Asymptotic behaviour in the nonlinear elastic beam, Arch. Rat. Mech. Anal. 87, 305-318 (1985).

8. Horgan, C.O. and Knowles, J.K., Recent developments concerning Saint-Venant's Principle. Advances in Applied Mechanics (J.W. Hutchison, ed.) 23, 179-269. Acad. Press. New York (1983).

9. Knops, R.J., Payne, L.E. and Rionero, S., Saint-Venant's Principle on unbounded domains. Proc. Roy. Soc. Edinburgh A (to appear).

10. Knowles, J.K., On Saint-Venant's Principle in the two-dimensional linear theory of elasticity. Arch. Rational Mech. Anal. 21, 1-22 (1966).

11. Love, A.E.H., Theory of Elasticity, C.U.P. (1952).

12. Oleinik, O.A. and Yosifian, G.A., On the asymptotic behaviour at infinity of solutions in linear elasticity. Arch. Rat. Mech. Anal. 78, 29-53 (1982).

13. Toupin, R.A., Saint-Venant's Principle. Arch. Rational Mech. Anal. 18, 83-96 (1965).

14. Novozhilov, V.V. and Slepian, L.I., On Saint-Venant's Principle in the dynamics of beams. J. Appl. Maths and Mechs (PMM) 29,(2), 261-281 (1965).

15. Werner, P., Resonance phenomena in cylindrical waveguides. J. Math. Anal. Appl. 121, 173-214 (1987).

A Brief History of Solitons

Alan Jeffrey

Department of Engineering Mathematics,
University of Newcastle upon Tyne,
NE1 7RU, England.

Introduction

When I was invited to speak before this learned Society it was suggested that I should choose for my theme something which illustrated the effect history has had on the development of the study of wave propagation in applied mathematics. After some reflection, and because the study of wave phenomena is so large, I decided to concentrate my discussion on the very special type of waves called **solitons** which have recently become so important.

To appreciate how unusual are the properties of solitons, it is first necessary to understand something about waves in general. In its simplest form, a progressive wave may be considered to be a disturbance which propagates at a finite speed through space as time evolves. If, for simplicity, we consider one–dimensional motion, and plot the magnitude u of the disturbance in space at any given fixed time we obtain a curve which we will call a **wave profile**. A typical wave profile is shown in Fig. 1(a) which could represent, for example, the amplitude of a water wave or perhaps the transverse displacement of a long stretched string.

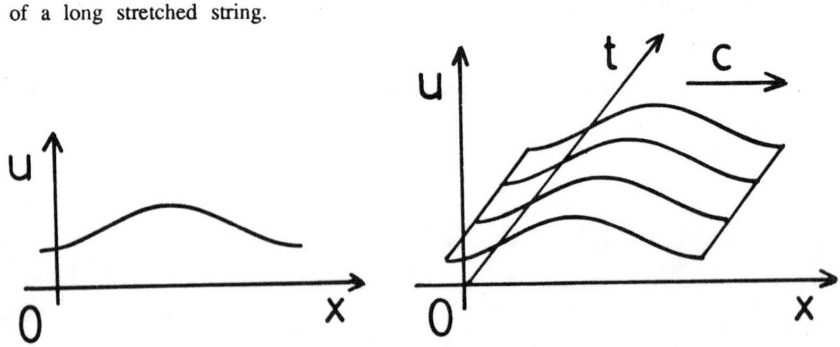

(a) (b)

Fig. 1

In the simplest linear case, each part of the initial wave profile moves in space with speed c, either to the left or right, but the **shape** of the wave remains unchanged. This process is illustrated in Fig. 1(b) in which the wave profile is shown at different times t.

A progressive wave of this type is of the form

$$u = g(x - ct) \ , \tag{1}$$

where g(x) is the shape of the wave profile when $t = 0$. We see from this that $u = \text{const.}$ along the lines $x - ct = \text{const.}$, which is in agreement with the fact that the initial wave profile moves with speed c to the right without change of shape. The straight lines

$$x - ct = \xi \qquad\qquad (\xi = \text{constant})$$

are called the **characteristic curves** of the equation satisfied by u (a **first order partial differential equation**, which has the form

$$u_t + c\ u_x = 0 \ , \tag{2}$$

where the subscripts signify partial differentiation.

A more general case arises when the solution is of the form

$$u = g(x - ct) + h(x + ct) \ . \tag{3}$$

Here one wave with initial shape h(x) moves to the left with speed c and another with initial shape g(x) moves to the right, also with speed c, though neither changes its shape. The solution u is thus the sum of these two simple solutions. This addition of solutions which occurs in linear equations is a special property only possessed by such equations and it is called the **linear superposition** of solutions. The equation satisfied by (3) is the second order linear **wave equation**

$$u_{xx} = \frac{1}{c^2} u_{tt} \ . \tag{4}$$

The situation becomes more complex when nonlinearity is involved, for then the wave can change its nature entirely at some stage of its evolution. Here, by nonlinearity, I mean that the magnitude of the disturbance u occurs in the governing equation in a form which is not linear, like u^2 or $f(u)u_x$, with f(u) some function of u which is not identically constant.

The first order prototype equation to be considered in the nonlinear case is

$$u_t + \underbrace{f(u)u_x}_{\text{Nonlinearity}} = 0 , \tag{5}$$

where the initial shape of the wave profile is

$$u(x,0) = g(x) , \tag{6}$$

with g(x) some given function.

Without going into details, let it suffice to say that the characteristic curves of (5) are obtained by integrating

$$\frac{dx}{dt} = f(u) , \tag{7}$$

and that along each characteristic curve u = const., with the constant value associated with a characteristic equal to the value of u determined by the initial wave profile g(x) at the point at which the characteristic meets the x–axis. It follows directly from this and (7) that in this case the characterstics are non–parallel straight lines. If they *diverge*, then the solution of (5) with the initial condition (6) is determined for all t. However, if they *converge* the solution becomes **non–unique** when the characteristics first intersect (they form an envelope). This is because each characteristic carries along it a *different* constant value of u, and the intersection of characteristics corresponds to there being more than one solution possible at the point of intersection.

The form taken by the solution in this nonlinear case turns out to be

$$u = g(x - t\ f(u)) , \tag{8}$$

which should be compared with the simple linear solution in (1).

The implications of this solution are not difficult to see. Result (7) tells us that the speed of propagation of any part of the wavefront depends on the value of u at that point. Now u varies over the wavefront, so this means that different parts of the wavefront move at different speeds, and so the *shape* of this nonlinear wave *changes* with

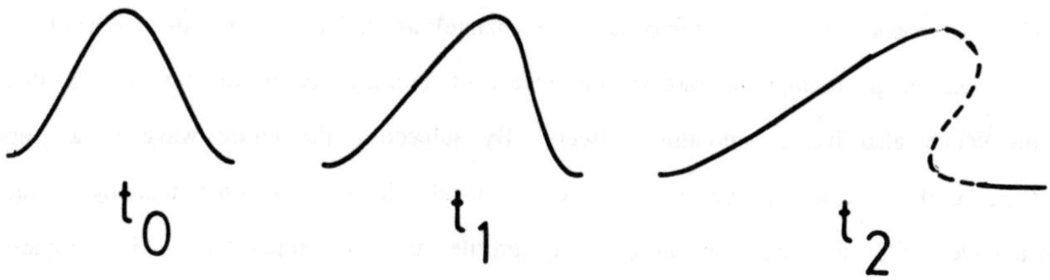

Fig. 2

time. A typical change of shape of the wave profile is shown in Fig. 2 for $t_0 < t_1 < t_2$. It is seen from this that the wave profile steepens as it advances, and at time $t = t_2$ the solution becomes non–unique in the way we have already mentioned (there is more than one value of u for a given value of x). If, for example, the profile in Fig. 2 represented a water wave, the start of non–uniqueness in the solution would correspond to the breaking of the wave, since the top of the wave (the dotted portion) would be unsupported. Thus, once non–uniqueness occurs, the nature of the solution changes completely. Behaviour such as this is typical of first order equations describing nonlinear waves, so that waves described by such equations cannot retain their initial shape.

We have seen that the effect of nonlinearity is, in general, to steepen wave profiles and eventually lead to non–uniqueness. It is thus natural to enquire whether the addition of some smoothing term to equation (5) can modify this behaviour and lead to the propagation of a constant shaped wave profile.

It can be shown that this is indeed possible if a term is added which corresponds to a dissipative effect (like viscosity in a fluid). However, when this is done no form of superposition of solutions is possible and the wave is such that a disturbance to any local part of the wave eventually influences the entire wave shape.. A dissipative effect only proves to be important when the wave profile steepens dramatically, for then it prevents non–uniqueness, but elsewhere it has little effect on the wave shape. The prototype equation involved when dissipation is involved is Burgers' equation

$$u_t + uu_x = \mu u_{xx} . \qquad (\mu > 0) \qquad (9)$$

This equation serves to model the transition of gas pressure and density across a thin layer of gas in a shock wave. Waves of this type are associated with supersonic flight and they produce a bang (a rapid pressure change) as the aircraft passes overhead.

Our next attempt to modify the effect of nonlinearity is to introduce a **dispersive** term which also has a smoothing effect. By subjecting the entire wave to a smoothing effect, unlike the dissipative effect which is localised, it is possible that by including a dispersive effect a more interesting steady profile may be propagated. The simplest such term is u_{xxx} , and when we include this term we arrive at the celebrated Korteweg–de Vries (KdV) equation

$$u_t + uu_x + \mu u_{xxx} = 0 \qquad (\mu = \text{const.}). \qquad (10)$$

This equation does, indeed have wave solutions of constant form which propagate to the right at a finite speed, but the properties of these waves turn out to be quite remarkable.

The constant form pulse–like solution of this equation, originally called a **solitary wave**, though now called a **soliton**, propagates at a speed proportional to its height h as shown in Fig. 3(a). This means that if two such waves are widely separated, with the larger one to the left, the larger one will overtake the smaller wave as indicated in Fig. 3(b). As the KdV equation is nonlinear, superposition of solutions is not possible, so it is reasonable to expect that after the overtaking has taken place both waves will have lost their separate identities and some general oscillatory solution will result. It is this issue which lies at the heart of the study of all equations possessing solitary wave solutions, and so is central to the study of the KdV equation which is the prototype equation of its kind.

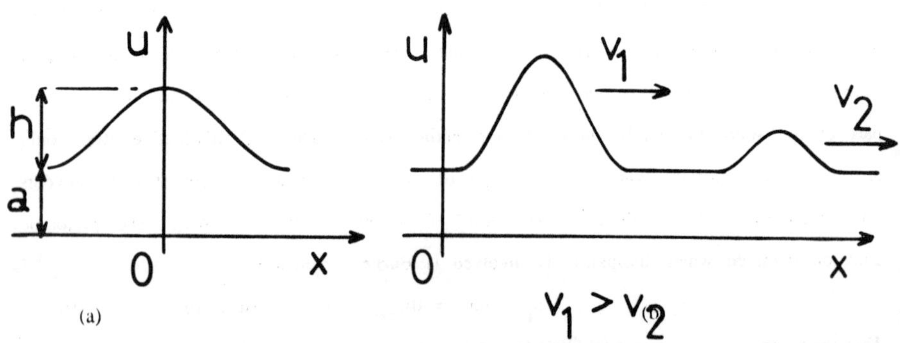

Fig. 3

We now move on from this outline of the main mathematical issues involved in the study of wave propagation and turn to the history of the soliton.

The Discovery of Solitary Waves – The Pre–History of Solitons

The start of the study of solitons must, without question, be credited to J. Scott Russell the English engineer who observed this phenomenon in the form of a solitary wave on the surface of the water in the Edinburgh to Glasgow canal in 1834. In his own words he described such a wave as "a great wave of translation" and he reported it to the British Association in 1844 in a communication entitled "Report on Waves". I can do no better than quote his own words:

> "I believe I shall best introduce the phenomenon by describing the circumstances of my own first acquaintance with it. I was observing the motion of a boat which was rapidly drawn along a narrow channel by a pair of horses, when the boat suddenly stopped – not so the mass of water in the channel which it had put in motion; it accumulated round the prow of the vessel in a state of violent agitation, then suddenly leaving it behind, rolled forward with great velocity, assuming the form of a large solitary elevation, a rounded, smooth and well–defined heap of water, which continued its course along the channel apparently without change of form or diminution of speed. I followed it on horseback, and overtook it still rolling on at a rate of some eight or nine miles an hour, preserving its original figure some thirty feet long and a foot to a foot and a half in height. Its height gradually diminished, and after a chase of one or two miles I lost it in the windings of the channel."

Let me now describe experiments which were performed by Scott Russell and summarise his general observations, the full significance of which only became apparent comparatively recently. He generated solitary waves in a long narrow shallow tank of water by dropping a large weight into one end of the tank, as illustrated in Fig. 4.

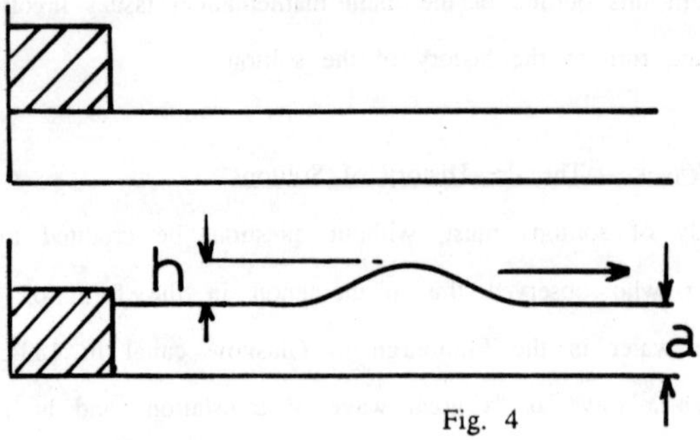

Fig. 4

His experimental findings were as follows:

1. the speed c of propagation of the solitary wave is

$$c^2 = g(a + h) \,,$$

where g is the acceleration due to gravity,

2. the water displaced by the weight is localised and confined within the volume represented by the solitary wave and not throughout the entire tank,

3. a solitary wave is a gravity wave in the sense that the acceleration due to gravity g enters into the expression determining the propagation speed c,

4. he deduced that large amplitude waves move faster than small amplitude waves, so that overtaking can occur,

5. he found that depression waves generated by withdrawing the weight from the water when it was in equilibrium did *not* produce solitary waves, but instead gave rise to a propagating oscillatory wave motion.

Less conspicuous in his report to the British Association were some observations concerning the overtaking process. These gave the first indication (overlooked until some two decades ago of the remarkable properties of such waves. He called the interaction process a **compound wave**, and indicated by his diagrams based on observations that a compound wave seeming to contain two humps revolved itself into two separate solitary waves once the interaction was finished. This is, indeed, remarkable, and certainly not to be expected in view of the properties of nonlinearity we have commented on so far. The essence of his observations are reproduced in Fig. 5, and his description of the phenomenon was as follows

> "This represents the genesis by a large low column of fluid of a compound or double wave of the first order, which immediately breaks down by spontaneous analysis into two, the greater moving faster and altogether leaving the latter".

Fig. 5

To emphasise the international nature of developments in science, and in applied mathematics in particular, I mention that J. Boussinesq (1871), and later Lord Rayleigh (1876), deduced on theoretical grounds both the speed of propagation c and the precise wave profile without first finding the differential equation governing the motion. This last and vital step was taken by D.J. Korteweg and G. de Vries whose paper entitled "On the change of form of long waves advancing in a rectangular channel, and on a new type of long stationary wave" which was published in London in 1895. We must now turn to more recent developments in the subject and establish the link with waves called solitons.

Recent History of Solitons

In the paper by Korteweg and de Vries, it was shown that the equation governing such water wave motion is

$$u_t + uu_x + \mu u_{xxx} = 0 .$$

This is now called the KdV equation in their honour. However, after the publication of this paper no further interest was taken in the equation until 1965, but to see how this came about we need to consider the 1955 publication by E. Fermi, J. Pasta and S.M. Ulam who when working at Los Alamos encountered a related problem, though failed to recognise the connection with the KdV equation.

In brief, they were trying to model the behaviour of certain crystals which show anomalous heat conduction properties. The mathematical model they used represented the crystal as a one–dimensional system of point masses (the molecules) joined together by springs which exhibited a nonlinear elastic behaviour (the inter–molecular forces). The response of the crystal to an input of energy was then interpreted in terms of the longitudinal oscillations of this system.

Their idea was to start the oscillation of the system in the simplest possible manner (in its lowest linearised mode) and then to see how long it took for there to be approximate equi–partition of the energy between the most important lower order modes, since this could provide some measure of the rate at which energy was transmitted through such a crystal. They performed their analysis numerically using an early computer and, much to their surprise, found the energy was very slow to leave the lowest mode, and that after many oscillations it all appeared to return to the original mode. This **recurrence phenomenon**, in which the solution of a nonlinear problem returned to its initial state, was totally unfamiliar at the time and their results were left unexplained.

Next on the scene with a crucially important paper were N.J. Zabusky and M.D. Kruskal, who again examined the Fermi, Pasta and Ulam problem, but this time replaced the discrete system by a continuum approximation in the form of the nonlinear wave equation

$$\frac{1}{c^2} u_{tt} = (1 + k\, u_x)u_{xx} .$$

However, after careful study, it was discovered that the solutions of this equation always became non–unique after a suitable interval of time, so this equation could *not* describe the recurrence phenomenon. Then, in 1965, Zabusky and Kruskal derived a better continuum approximation which led to the KdV equation

$$u_t + uu_x + \mu u_{xxx} = 0 .$$

They examined the Fermi, Pasta and Ulam problem using this model with the help of numerical computation and found that it did, indeed, exhibit the required recurrence phenomenon. Thus the vital connection between the KdV equation and recurrence phenomenon had been established.

On examining the problem more closely using numerical methods they found that the recurrence was explained by the initial disturbance decomposing into solitary waves which then moved back and forth, interacting with one another, but always *returning* to their *original form* after each interaction. Eventually, after many oscillations, these solitary waves re–combined into the initial wave form. This totally unexpected behaviour suggested to them that the solitary wave solutions of the KdV equation were of fundamental importance and behaved in some sense like particles. As a result they invented the name **soliton** to describe such a special wave. A typical interaction process is shown in Fig. 6, and it should be emphasised that the interaction does not involve linear superposition. Thus we have an example of a nonlinear equation whose solutions exhibit certain of the aspects associated with linear equations, yet there is no linear superposition involved.

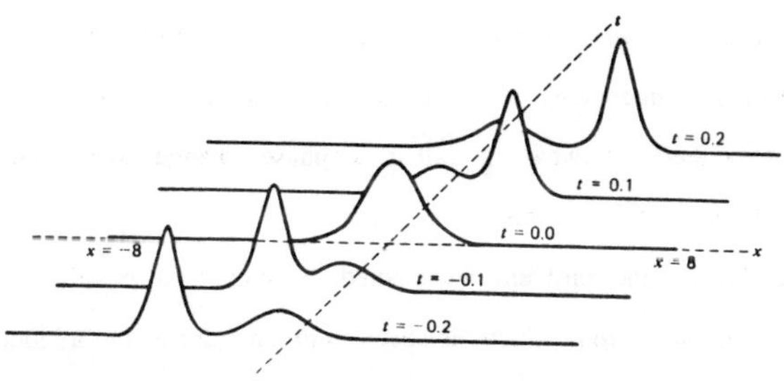

Fig. 6

Since this early fundamental paper by Zabusky and Kruskal much has been published in this area and many equations are now known to have constant profile travelling wave solutions with the soliton property. However, by no means all equations with constant profile travelling wave solutions have the soliton property though some approximate this behaviour.

The applications of soliton theory, which is now well understood in many respects, are many and varied, and numerous people throughout the world have contributed to its development. Applications range from water waves in canals and tidal waves (tsunami waves) in the oceans, through to wave propagation in certain semiconductors, atmospheric waves, surface and internal waves in the Andaman sea with their consequences for off–shore drilling rigs, and waves in nonlinear elastic solids. There has also been the suggestion that soliton behaviour may be a contributing factor to the permanence of the great red spot observed on Jupiter.

An Application to a Nonlinear Elastic Rod

I now offer a brief description of a contribution of my own, made together with a Japanese colleague K. Konno in 1984. This is of some technical interest because the method developed to solve the problem involved making an extension of an existing method which until then was limited in its application. However, rather than discussing the technical details, I will instead give the graphical results obtained from our analytical solution. These are interesting in their form, and they also provide an easily understood example of why there is what is called a **phase change** associated with soliton inter–actions. I will describe what this means later.

Let me explain the problem we studied as simply as possible. Suppose we have a very flexible long thin rod which is bent into the form of a loop and that the rod together with its loop is held in tension. Then a constant profile wave solution exists for this system in the form of the loop which moves along the rod at a constant speed without change of shape. Our question was "is this loop solution a soliton and how can we obtain an analytical solution which describes the interaction of two or more such loops".

We were able to solve the problem exactly and to show that these loops are, indeed, solitons. We called them **loop solitons** and showed that after any number of interaction the two loops always returned to their initial forms. We also found the exact solution and so were able to study the interaction process in detail. To understand what is to follow, it is necessary to know that the speed of propagation of a loop soliton is inversely proportional to the size of the loop, so a small loop travels faster than a large loop and so can overtake it. A typical loop soliton is shown in Fig. 7.

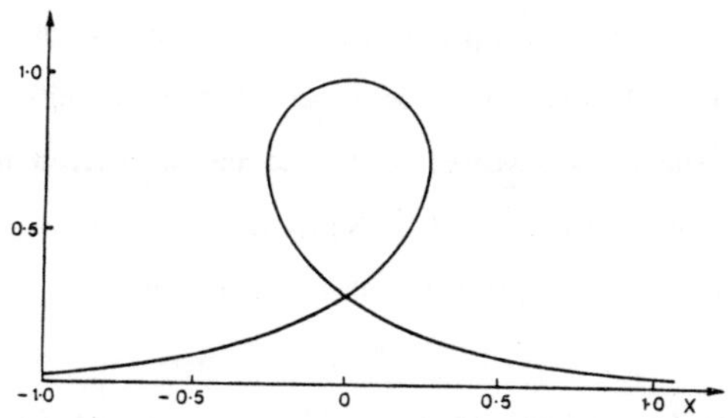

Fig. 7

The interaction process is shown in detail in Fig. 8, which was obtained from the exact analytical solution of this problem.

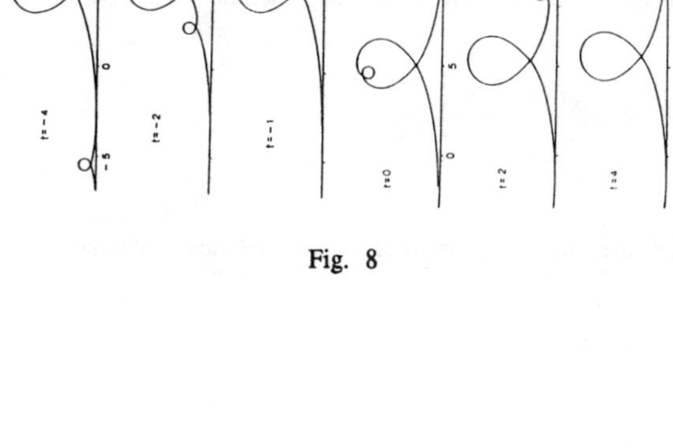

Fig. 8

I referred earlier to a **phase change** which always accompanies soliton interaction. Let me now make clear the precise meaning of this term. Consider two solitons, initially well separated one from the other, which will subsequently overtake and interact, and let us start our timing and measurement at this stage. Confine attention to either one of these solitons, and let T be some time by which the interaction process has been completed and the two solitons are again well separated. Then the distance travelled by the soliton we are considering will be d, say. Now suppose there had only been this one soliton and that it had travelled undisturbed at its own constant speed for this same time T, during which time it travelled a distance D, say. Then the **phase change** of this soliton is the distance d − D, which may be either positive or negative. If the phase change is positive the position has advanced relative to the undisturbed motion, while if it is negative the position has been retarded. Now, although this is a well established mathematical fact for all soliton interactions, there is not usually any simple physical explanation for the effect.

In the case of the loop solitons shown in Fig. 8 the reason for the phase change is easy to see. Until the small loop approaches close to the large loop they each move independently at their own characteristic speeds. However, while the interaction takes place, the small loop is delayed running around the large loop, so that after the interaction it is no longer as far advanced as it would have been had the interaction not occurred. Thus in this case the phase change is easily understood on physical grounds.

A phase change also occurs with the large loop which, it can be shown, experiences a phase advance after the interaction. For the sake of interest, I give below the partial differential equation governing loop soliton behaviour to show its extreme nonlinear nature

$$u_{xt} + \operatorname{sgn}\left(\frac{dx}{dt}\right)\left[\frac{u_{xx}}{(1 + u_x^2)^{3/2}}\right] = 0 \ .$$

I should add that we have obtained exact analytical solutions for the interaction of N loop solitons.

The Future – What Next?

Aside from the continuing flow of papers reporting new equations with soliton solutions, and their associated applications, the most important developments to be expected in the immediate future are likely to be technical in nature. Some of these are listed below;

1. So far equations with soliton solutions have been identified by first finding that they have a constant profile progressive wave solution and then showing that such solutions are indeed solitons. An important advance would be to be able to show *a priori* from the equation itself, without solving it, that it does or does not possess soliton solutions. Some progress has been made in this area but the question remains open.
2. The development of new and more general methods for finding exact solutions.
3. As soliton interaction is a form of **nonlinear superposition** of solutions, can the notion of superposition be generalised? Encouraging progress has already been made in this area by means of Bäcklund transformations. These interesting but very difficult transformations were first introduced by A.V. Bäcklund in 1876 in the context of geometry and only comparatively recently applied to soliton theory.
4. The study of soliton propagation and interaction needs to be extended systematically to two or more space–dimensions. Some progress in this direction has been made, but it has been slow.
5. All mathematical models of physical situations are idealisations of the real world and, in particular, one of the ways in which our mathematical representation of a wave is deficient is due to the fact that waves in the real world are not smooth. This suggests that the study of soliton behaviour should be extended to include wave profiles which are described in statistical terms. A connection would then need to be established with the very recently developed rigorous mathematical theory of chaotic behaviour associated with the solution of nonlinear dynamical systems.

Conclusion

I have tried to trace a path in the development of soliton theory, starting with the observation of Scott Russell while on horseback in Edinburgh in 1834, through important numerical and theoretical contributions of many people to more recent work. In any general talk of this nature it is impossible to acknowledge everyone who has made important contributions or, indeed, to make any real reference to the technical aspects of the subject. My hope is that I will be forgiven by the many people whose names I have omitted to mention, but that I will have succeeded in conveying something of the "spirit of the hunt" in mathematics and of its international nature.

WATER LUBRICATED PIPELINING—A REVIEW OF THEORY AND EXPERIMENTS

Daniel D. Joseph[1]

[1] *Department of Aerospace Engineering and Mechanics, University of Minnesota, Minneapolis, MN, U.S.A.*

ABSTRACT This is a largely nontechnical review of nearly everything known about theory and experiments on water lubricated pipelining. The object of the review is to give readers a broad view of what has been done and where the subject is going, together with a relatively complete list of references. A complimentary but more technical paper by Preziosi, et al also appears in this collection.

1. POSITION OF THE PROBLEM

There is a strong tendency for two fluids to arrange themselves so that the low-viscosity constituent is in the region of high shear (Joseph, Nguyen and Beavers, [1984]). This gives rise to a kind of gift of nature in which the lubricated flows are stable, and it opens up very interesting possibilities for technological applications in which one fluid is used to lubricate another.

We can imagine that it may be possible to introduce a beneficial effect in any flow of very viscous liquid important in applications by introducing small amounts of lubricating fluid. Nature's gift is evidently such that the lubricating fluid will migrate to the right places so as to do the desired job.

There are significant reserves of heavy viscous crude oils in the United States, Canada, Venezuela and Europe. Heavy crudes may have viscosities of 1000 P at room temperature. These viscous crudes cannot be transported by the usual pipeline methods. It is customary to reduce the viscosity of the oil either through the addition of a hydrocarbon diluent or through the installation of heating equipment at short intervals along the pipeline. The former method can only be used in the unusual case in which there is an abundant supply of light oil in the same region as the heavy oil; heating is inconvenient and costly.

Another method which has been proposed to facilitate the transport of viscous crudes is the addition of an immiscible lubricating liquid, usually water. Experiment to examine this possibility have been carried out by Russell and Charles [1959], Russell, Hodgson and Govier [1959], Charles, Govier and Hodgson [1961], Gemmell and Epstein [1962], Charles and Lilleleht [1966], and Bai, Chen and Joseph [1990]. Oliemans and Ooms [1986] have written a comprehensive review of pipe flows of oil and water, theory and experiment, prior to 1984.

Various arrangement of the oil and water occur in the aforementioned experiments. This type of non-uniqueness is typical of flowing bicomponent fluids. The arrangements that appear in horizontal pipes are: (a) stratified flow with heavy fluid below; (b) concentric oil in water (core-annular) flow; (c) water drops in oil; (d) oil drops (bubbles) in water (these include large bubbles and slugs of oil lubricated by water).

The measured pressure drops indicated that the addition of water can greatly reduce the pressure gradient. There is a powerful tendency for the water to migrate to the pipe walls where the shearing is greatest, lubricating the flow.

It was found that the core-annular flow had the greatest volume flux for a given pressure drop among all of the realized rearrangements. The pressure drop over the pipe could be even smaller than the pressure drop in water along at the same value of the volume flux. Lubrication of oil by water can reduce the drag on the oil by a factor proportional to the viscosity ratio, a thousandfold or more.

Prior to 1987 there were a few papers on this problem and a larger but still not large secondary literature. Frenkel [1988] found a condition in which the motion of the core could be coupled into the Kuramoto-Sivashinsky system, leading to an additional linear term which disperses waves. A clear exposition of this work based on a systematic expansion in powers of a small parameter, together with numerical solutions of Kuramoto-Sivashinsky-Frenkel equations have been given by Papageorgiou, Maldarelli, and Rumschitzki [1989]. The conditions under which this equation may be applied are rather severe.

The first stability study of water lubricated pipelining was by Joseph, Renardy and Renardy [1984]. They neglected density differences and surface tension and found that lubricated transport was stable if the water fraction was not too great. Earlier, Hickox [1971] had shown the core-annular flow with less viscous liquid in the core was unstable to long waves.

The secondary literature will not be reviewed here since it is collected in references given by Preziosi, Chen and Joseph [1989].

Several studies of thin film lubrication, using lubrication theory, have led to nonlinear amplitude equations of the KS type. The first such study by Frenkel, Babchin, Levich, Shlang and Sivashinsky [1987] neglected the effects of the core.

D. D. JOSEPH, M. RENARDY AND Y. RENARDY. INSTABILITY OF THE FLOW OF TWO IMMISCIBLE LIQUIDS WITH DIFFERENT VISCOSITIES IN A PIPE. *J. FLUID MECH.* (1984), VOL. **141**, PP. 309–317

We study the flow of two immiscible fluids of different viscosities and equal density through a pipe under a pressure gradient. This problem has a continuum of solutions corresponding to arbitrarily prescribed interface shapes. The question therefore arises as to which of these solutions are stable and thus observable. Experiments have shown a tendency for the thinner fluid to encapsulate the thicker one. This has been "explained" by the viscous-dissipation principle, which postulates that the amount of viscous dissipation is minimized for a given flow rate. For a circular pipe, this predicts a concentric configuration with the more viscous fluid located at the core. A linear stability analysis, which is carried out numerically, shows that while this configuration is stable when the more viscous fluid occupies most of the pipe, it is not stable when there is more of the thin fluid. Therefore the dissipation principle does not always hold, and the volume ratio is a critical factor.

L. PREZIOSI, K. CHEN AND D. D. JOSEPH. LUBRICATED PIPELINING: STABILITY OF CORE ANNULAR FLOW. *J. FLUID MECH.* (1989), VOL. **201**, PP. 323–356

The stability of core-annular flow (CAF) in pipes is analyzed using the linear theory of stability. Attention is confined to the potentially stable case of lubricated pipelining with the less viscous liquid, say water, in the annulus. The effects of surface tension and density are included, but gravity is excluded. We find upper and lower branches of the neutral curve in a Reynolds number ($\mathbb{R}$) *vs.* wavenumber (α) plane. A window of parameters is identified in which CAF is stable to small disturbances. When $\mathbb{R}$ is below the lower critical value, CAF is destabilized by surface tension and long waves break up into slugs and bubbles. The sizes of slugs and bubbles of oil in water observed by Charles, Govier and Hodgson [1961] are given by the wavelength of the fastest growing long wave. This long-wave instability is a capillary instability, modified by shear, which reduces to Rayleigh's instability in the appropriate limit. At higher $\mathbb{R}$, the capillary instability is stabilized by shear. At yet higher $\mathbb{R}$, above the upper critical value, the flow is unstable to generally shorter waves which leads to emulsification, water droplets in oil. The theory agrees with experiments. The analysis seems to be applicable to the design of lubricated pipelines; for example, there is an optimum viscosity ratio for stability, greater stability can be obtained by using heavy liquid as a lubricant when the flow is unstable to capillary modes on the lower branch and by using light liquids when the flow is unstable to emulsifying disturbances on the upper branch.

H. H. HU AND D. D. JOSEPH. STABILITY OF CORE-ANNULAR FLOW IN A ROTATING PIPE. *PHYS. FLUIDS A* (1989), VOL. **1**(10), PP. 1677–1685

The linear stability of core-annular flow in rotating pipes is analyzed. Attention is focused on the effects of rotating the pipe and the difference in density of the two fluids. Both axisymmetric and nonaxisymmetric disturbances are considered. Major effects of the viscosity ratio, interfacial tension, radius ratio, and Reynolds number are included. It is found that for two fluids of equal density the rotation of the pipe stabilizes the axisymmetric (n=0) modes of disturbances and destabilizes the nonaxisymmetric modes. Except for small $\mathbb{R}$, where the axisymmetric capillary instability is dominant, the first azimuthal mode of disturbance $|n|=1$ is the most unstable. When the heavier fluid is outside centripetal acceleration of the fluid in the rotating pipe is stabilizing; there exists a critical rotating speed above which the flow is stabilized against capillary instability for certain range of small $\mathbb{R}$. When the lighter fluid is outside the flow is always unstable.

H. H. HU AND D. D. JOSEPH. LUBRICATED PIPELINING: STABILITY OF CORE-ANNULAR FLOW, PART 2. *J. FLUID MECH.* (1989), VOL. **205**, PP. 359–396

In this paper, we study the linearized stability of three symmetric arrangements of two liquids in core-annular Poiseulle flow in round pipes. Deferring to one important application, we say oil and water when we mean more viscous and less viscous liquids. The three arrangements are (i) oil is in the core and water on the wall, (ii) water is in the core and oil is outside and (iii) three layers, oil inside and outside with water in between. The arrangement in (iii) is our model for lubricated pipelining when the pipe walls are hydrophobic and it has not been studied before. The arrangement in (ii) was studied by Hickox [1971] who treated the problem as a perturbation of long waves, effectively suppressing surface tension and other essential effects which are necessary to explain the flows observed, say, in recent experiments of W. L. Olbricht and R. W. Aul. The arrangement in (i) was studied in Part 1 of this paper (Preziosi, Chen and Joseph [1987]). We have confirmed and extended their pseudo-spectral calculation by introducing a more efficient finite-element code. We have calculated neutral curves, growth rates, maximum growth rate, wavenumbers for maximum growth and the various terms which enter into the analysis of the equation for the evolution of the energy of a small disturbance. The energy analysis allows us to identify the three competing mechanisms underway: interfacial tension, interfacial friction and Reynolds stress. Many results are presented.

K. CHEN, R. BAI AND D. D. JOSEPH. LUBRICATED PIPELINING III: STABILITY OF CORE-ANNULAR FLOW IN VERTICAL PIPES. *J. FLUID MECH.* (TO APPEAR, 1990)

The stability of core-annular flow in vertical pipes is analyzed using the linearized theory of stability. In previous studies instabilities due to interfacial fraction, interfacial tension and Reynolds stresses in the bulk fluid were identified and associated with observed instabilities. In this study we include and analyze the effects of gravity. In one case gravity opposes and in the other aids the applied pressure gradient. Some preliminary results from our experiments are also presented. The prediction of stability of perfect core-annular flow in a carefully selected window of parameters is verified for the case of free fall in which the applied pressure gradient vanishes.

The abstract for this paper is only a partial description of the results given in the paper. The effects of the density in vertical flow were shown to introduce an always destabilizing term $(\rho_2-\rho_1)g$ in the balance equation for the shear stress at the interface. To achieve stable core-annular slow flow, use a heavy lubricant for down flow and a light one for up flow. There is an optimizing density ratio which gives the largest interval of Reynolds numbers for stability.

The next paper by Chen and Joseph is very important for the computation of bifurcation parameters of Ginzburg-Landau equations which arise in the nonlinear theory of stability of core-annular flow. We believe that this singular value decomposition is the state of the art and will become standard in a few years.

K. CHEN AND D. D. JOSEPH. APPLICATION OF THE SINGULAR VALUE DECOMPOSITION TO THE NUMERICAL COMPUTATION OF THE COEFFICIENTS OF AMPLITUDE EQUATIONS AND NORMAL FORMS. *APPL. NUMERICAL MATHEMATICS* (TO APPEAR, 1990) (AVAILABLE AS A UMSI REPORT 89/94, JUNE 1989)

The Fredholm alternative is a standard procedure by which one generates the coefficients of amplitude equations and normal forms. The alternative requires that the inhomogeneous terms in the underlying system of differential equations, which contain the unknown coefficients, be orthogonal to the independent eigenvectors spanning the null space of the adjoint system of differential equations. The numerical computation of the adjoint eigenvectors and the application to solvability is frequently difficult and inefficient. Typically the underlying system of the inhomogeneous differential equation is discretized and solved as an inhomogeneous matrix-valued eigenvalue problem. We find that the solvability conditions which lead to values of the unknown coefficients are conveniently and economically computed by application of the singular value decomposition directly to the matrix formulation, avoiding completely the computation of an adjoint system of differential equations.

2. EXPERIMENTS

We have built three devices to study lubricated pipelining. The first setup is a free fall apparatus. The second apparatus is an inverse U loop whose long legs are pipes for up and down flows. A third apparatus is a horizontal pipeline designed for water lubricated pipelining of oil and coal-oil mixtures (COMs). We have collected a nearly complete set of data on up and down flows of a 600 weight Mobil cylinder oil with a nominal viscosity of 4 poise and density of 0.881 gm/cm^3. A complete set of data means that we have the values of the pressure drop as a function of three parameters: the flow rates Q_o and Q_w of oil and water and the holdup ratio

$$h = \frac{\text{Flow rate ratio}}{\text{Volume ratio}} = \left(\frac{Q_o}{Q_w}\right) \Big/ \left(\frac{V_o}{V_w}\right)$$

where V_o and V_w are the volumes of oil and water in a length ℓ of pipe which is sufficiently long so that h is independent of ℓ.

We have taken enough data to distinguish all the major trends. We have only to fill in the places where the data may be ambiguous or to emphasize data which are crucial to a major conclusion.

One major conclusion is that the holdup ratio h in *upflow* does not depend on the flow rates Q_0 and Q_w of oil and water or on the ratio of flow rates. The holdup ratio in downflow depends strongly on these parameters. When the flow rates are sufficiently large the effects of gravity disappear; the difference between downflow and upflow diappear and the holdup ratio in both flows is independent of flow rate. This result disagrees with Charles, et al [1961], who found that when the density of oil was matched to water by additives, so that gravity was not important, the holdup ratio in a horizontal pipe does depend on the input ratio and flow velocity, even at large flow rates.

3. COMPARISON OF EXPERIMENTS WITH LINEAR STABILITY THEORY

There is an important regime of flow called wavy core flow which is greatly important in the technology of lubricated transport. We took nine arbitrary cases of such flows, measured average wave speeds and wave lengths and compared the measured values with values computed using prescribed input data from the experiments and got the nearly perfect results you see exhibited in Table 1. The flow has huge waves, nowhere near to the linear case, but the linear theory works anyway. It really does work and the agreements we are reporting are not fortuitious. There are also some regimes of flow where the linear theory does not work well and we want to carefully delineate these.

	Experiments		Computations	
Exp. no.	λ (cm)	c (cm/sec)	λ (cm)	c (cm/sec)
1	1.21	57.70	1.04	50.84
2	1.31	43.28	1.16	40.64
3	1.41	35.65	1.23	32.12
4	1.22	27.81	1.17	27.51
5	1.3739	19.16	1.3692	17.40
6	1.79	22.90	1.37	21.82
7	1.34	28.22	1.32	26.18
8	1.17	31.06	1.16	30.88
9	0.90	36.25	1.05	35.40

Table 1. Comparison of wave speed, wave length between experiments and numerical computations (the computational results are for the most unstable modes)

We are preparing a paper: "Lubricated pipelining: stability of core-annular flow. Part IV: Comparison of linear stability theory and experiments" by R. Bai, K. Chen and D. D. Joseph.

4. NONLINEAR THEORY. GINZBURG-LANDAU EQUATIONS FOR LUBRICATED PIPELINING

We have a good understanding of the predictive power of linear theory for the nonlinear flows we see in experiments. We also have in place good numerical codes to calculate eigenvalues and eigenfunctions for the various problems which may arise in the future. Now we are looking at nonlinear effects and we have made progress on the regimes of flow which lend themselves to analysis in terms of Ginzburg-Landau equations.

We are able to obtain Ginzburg-Landau equations whenever the parameters are such as to allow a stable interval of Reynolds numbers with upper and lower branches of the neutral curve as in figure 1. In this situation it is possible to compute different solutions which bifurcate from the upper and lower branches, respectively.

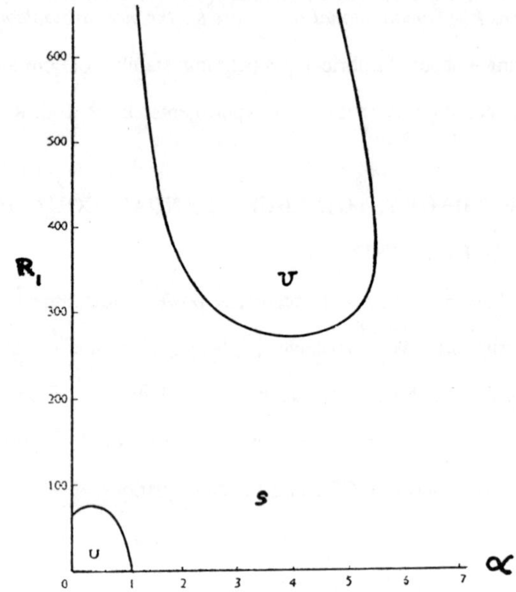

Figure 1. Neutral curve for core-annular flow when the viscosity ratio, density ratio, water fraction and interfacial tension allow stable, perfect core-annular flow.

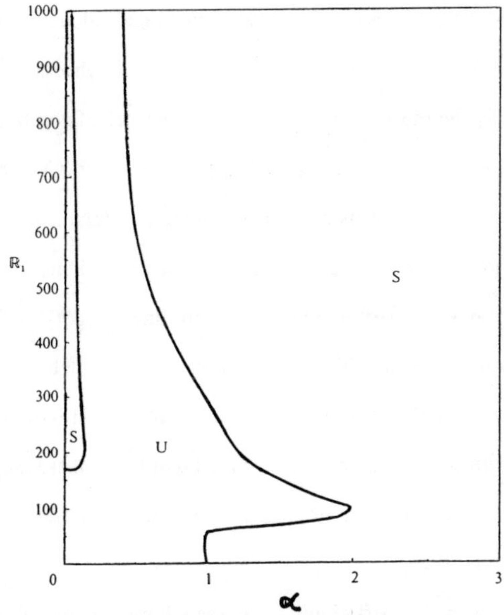

Figure 2. One of many cases in which perfect core flow is not possible. We don't have amplitude equations for these cases.

In the Ginzburg-Landau case, we are restricted to small neighborhoods near the minimum point m_u of the upper branch and the maximum M_L of the lower branch of the neutral curve.

The linear theory shows that axisymmetric disturbances are always most dangerous so that non-axisymmetric modes are not active in the smallest regions hugging criticality. In this case we get the usual one space dimension (x along the axis of the pipe) Ginzburg-Landau equation

$$\frac{\partial A}{\partial t} - \gamma \frac{\partial^2 A}{\partial x^2} = \mu A - \beta |A|^2 A$$

where $\beta=\beta_r+i\beta_i$ and $\gamma=\gamma_r+i\gamma_i$ are complex and μ the bifurcation parameter, proportional to a Reynolds number, is real. $A=0$ is stable when $\mu<0$. If $\beta_r<0$, there is a subcritical bifurcating solution with $A\neq0$ which is unstable. If $\beta_r>0$, then $A=0$ is unstable when $\mu>0$ and there is a supercritcal bifurcation solution $A\neq0$ which can be stable.

The computation of the coefficients γ, μ, β in the present problem is very complicated. The equations are very involved and the adjoint problem needed for solvability is equally complicated.* All the analytic expression for these complicated calculations have been derived by Kangping Chen, checked against known limiting cases and checked again using symbol manipulators. Fortunately the most complicated part of the work of computing the coefficients can be removed by application of the singular value decomposition (Chen and Joseph [1990]). This is an entirely numerical recipe using an off-the-shelf program of Businger and Golub [1969]. Using this method, Chen has already computed the coefficients for about 100 different cases. We have not done a detailed comparison of nonlinear theory with experiments yet. However, most of the solutions which bifurcate from the lower branch of the neutral curve are unstable. This is consistent with the observation that instability on the lower branch leads to slugs and bubbles far from core-annular flow. On the other hand, there are cases in which the solutions which bifurcate from the upper branch are stable and are observed as wavy core flow close to perfect core-annular flow.

We are computing more coefficients for cases which we control completely; that is, cases where we have done the experiments and have good data for prescribed operating conditions and have also computed stability results corresponding to these conditions. These results are being collected into a paper "Lubricated pipelining: Stability of core-annular flow. V: Ginzburg-Landau equations" by Chen and Joseph.

The astonishing agreements between linear stability theory and experiments in the case of lubricated core-annular flows has persuaded us that this problem is a nearly perfect one to test ideas about nonlinear effects in problems of two-fluid dynamics and perhaps in dynamical systems generally. There are many features of the dynamics which are not explained by linear theory, or are not completely explained.

---

* For example, in a similar problem it took Blennerhasset [1980] more than two pages just to write down the adjoint system and solvability conditions. The chances for errors are great.

There is a lot of information in these equations which we need to analyze. This effort is all the more worthwhile because we have the coefficients, the linear theory and experiments. So when we work out the predictions of the Ginzburg-Landau equations, they can be tested.

REFERENCES

[1] Bai, R., Chen, K., and Joseph, D. D. (1990) In preparation.

[2] Blennerhassett, P. J. 1980 On the generation of waves by wind. *Proc. R. Soc. Lond.* A**298**, 451–494.

[3] Businger, P. S. and Golub, G. H. 1969 Algorithm 358: Singular value decomposition of a complex matrix. *Comm. Assoc. Comp. Mach.* **12**, 564–565.

[4] Charles, M. E., Govier, G. W. and Hodgson, G. W. 1961 The horizontal pipeline flow of equal density of oil-water mixtures. *Can. J. Chem. Engng.* **39**, 17–36.

[5] Charles, M. E. and Lilleleht, L. U. 1966 Correlation of pressure gradients for the stratified laminar-turbulent flow of two immiscible liquids. *Can. J. Chem. Engng.* **44**, 47–49.

[6] Chen, K., Bai, R. and Joseph, D. D. 1990 Lubricated pipelining III: Stability of core-annular flow in vertical pipes. *J. Fluid Mech.* (to appear).

[7] Chen, K. and Joseph, D. D. 1990 Application of the singular value decomposition to the numerical computation of coefficients of amplitude equations and normal forms. *Appl. Num. Math.* (to appear).

[8] Chen, K. and Joseph, D. D. (1990) Lubricated pipelining V. In preparation.

[9] Frenkel, A. L. 1988 Private communication.

[10] Frenkel, A. L., Babchin, A. J., Levich, B. G., Shlang, T. and Shivashinsky, G. I. 1987 Annular flows can keep unstable films from breakup: Nonlinear saturation of capillary instability. *J. Colloid Interface Sci.* **115**, 225–233.

[11] Gemmel, A. R. and Epstein, N. 1962 Numerical analysis of stratified laminar flow of two immiscible Newtonian liquids in circular pipes. *Can. J. Chem. Engng.* **40**, 215–224.

[12] Hickox, C. E. 1971 Instability due to viscosity and density stratification in axisymmetric pipe flow. *Phys. Fluids* **14**, 251–262.

[13] Hu, H. H. and Joseph, D. D. 1989 Lubricated pipelining: Stability of core-annular flow. Part 2. *J. Fluid Mech.* **205**, 359–396.

[14] Hu, H. H. and Joseph, D. D. 1989 Stability of core-annular flow in a rotating pipe. *Phys. Fluids* A**1**(10), 1677–1689.

[15] Joseph, D. D., Renardy, Y. and Renardy, M. 1984 Instability of the flow of immiscible fliquids with different viscosities in a pipe. *J. Fluid Mech.* **141**, 319–345.

[16] Oliemans, R. V. A. and Ooms, G. 1986 Core-annular flow of oil and water through a pipeline. In *Multiphase Science and Technology,* Vol. 2 (ed. G. F. Hewitt, J. M. Delhaye and N. Zuber). Hemisphere.

[17] Papageorgiou, D. T., Maldarelli, C. and Rumschitzki, D. S. 1990 Nonlinear interfacial stability of core-annular film flows. *Phys. Fluids* (to appear).

[18] Preziosi, L., Chen, K. and Joseph, D. D. 1989 Lubricated pipelining: Stability of core-annular flow. *J. Fluid Mech.* **201**, 323–356.

[19] Russell, T. W. F. and Charles, M. E. 1959 The effect of the less viscous liquid in the laminar flow of two immiscible liquids. *Can. J. Chem. Engng.* **39**, 18–24.

[20] Russell, T. W. F., Hodgson, G. W. and Govier, G. W. 1959 Horizontal pipeline flow of mixtures of oil and water. *Can. J. Chem. Engng.* **37**, 9–17.

Main Features of Modulated Wave Compatible with Quasilinear Dissipative Systems of First Order

N. Manganaro
Dipartimento di Matematica
Universita' di Messina
Contrada Papardo,Salita Sperone 31
98166 Sant'Agata, Messina, Italy.

Abstract

We consider a dissipative quasilinear first order system of partial differential equations which is compatible with different levels of description of the wave processes. For the model in point we study the propagation of a modulated quasi-monochromatic wave into a constant state and we deduce the evolution equation ruling the wave amplitude.

1. Asymptotic analysis and wave modulation.

Several physical theories such as irreversible thermodynamics, channel flows, active media lead to governing mathematical models for which different levels of description of the wave processes are possible.

Within such a context the results obtained in [1], [2] for certain dissipative hyperbolic first order systems of P. D. E. of physical interest point out that, in the far field behaviour, the wave amplitude is ruled by a Burger's-like evolution equation or by a generalized form of it.

In this paper we consider the following system of first order

$$A^{\alpha}(U,V)\,\partial_{\alpha}U + B^{\alpha}(U,V)\,\partial_{\alpha}V = C(U,V) \tag{1}$$

$$M^{\alpha}(U,V)\,\partial_{\alpha}V + N^{\alpha}(U,V)\,\partial_{\alpha}U = P(U,V) \tag{2}$$

where $\partial_{\alpha}=\partial/\partial x^{\alpha}$, $x^0=t$ and $x^1=x$ are, respectively, space and time coordinates, U and V are column vectors of R^n and R^m, respectively, representing the field; A^{α}, B^{α}, M^{α}, N^{α} are suitable matrix coefficients while C and P are column vectors. Moreover a subscript denotes partial derivative with respect to the indicated variable.

The use of a well established approach [3] shows that for the system of equations (1) and (2) the modulation of a quasi-monochromatic wave propagating into a constant state is governed by a Schrödinger-like equation.

However a lower order description of the wave process is possible when the left-hand side of the equation (2) is negligible so that

$$P(U,V) = 0. \tag{3}$$

If $|\nabla_v P| \neq 0$, $\nabla_v=(\partial/\partial V)$, then from the relation (3) $V=V(U)$ and consequently (1) gives rise to the "reduced" governing model

$$\bar{A}^{\alpha}(U)\partial_{\alpha}U = \bar{C}(U) \tag{4}$$

where

$$\bar{A}^{\alpha} = (B^{\alpha}\cdot \nabla_u V + A^{\alpha}); \qquad \bar{C} = C(U,V(U)) \tag{5}$$

and $\nabla_u = (\partial/\partial U)$.

Here we are interested in the situation where the order of magnitude of the left-hand side of (2) is small but not negligible. Namely the following relation holds

$$\epsilon^2(\,M^{\alpha}\partial_{\alpha}V + N^{\alpha}\partial_{\alpha}U\,) = P \tag{6}$$

where $\epsilon \ll 1$ is a real parameter.

In the present case it is interesting to investigate the influence of the higher order terms involved in the left hand side of (6) on the propagation ruled by the reduced system (4).

Searching for solutions describing the modulation of a quasi-monochromatic wave, we assume the following asymptotic developments to hold [3]

$$U = U_0 + \epsilon U_1 + \epsilon^2 U_2 + \epsilon^3 U_3 + 0(\epsilon^4) \tag{7}$$

$$V = V_0 + \epsilon V_1 + \epsilon^2 V_2 + \epsilon^3 V_3 + 0(\epsilon^4) \tag{8}$$

where

$$U_n = \sum_{l=-\infty}^{+\infty} U_n^{(l)}(\xi,\tau) \exp\{il(kx-\omega t)\} \tag{9}$$

$$\xi = \epsilon(x - \lambda t); \quad \tau = \epsilon^2 t \tag{10}$$

λ being a quantity which will be specified later, while U_0 and V_0 are constants satisfying the conditions $C(U_0,V_0)=P(U_0,V_0)=0$.

The stretched variables (10) are introduced to specify the slow variations of the amplitude compared with the phase of the carrier waves [3].

In order to U_n be real we have to require that

$$U_n^{(l)} = U_n^{(-l)*} \tag{11}$$

where * means for complex conjugate.

Substituting (7), (8) into the system (1), (6) and cancelling the coefficients of ϵ, ϵ^2,ϵ^3 we obtain six partial differential equations involving the vectors U_1 , U_2, V_1, V_2 as well as the matrix coefficients of the system (1), (6). The interested reader can recover the explicit expression of these cumbersome equations in [5]. The compatibility of the obtained equations gives:

$$V_1 = F_0 U_1 \tag{12}$$

where $F = -(\nabla_v P)^{-1}(\nabla_u P)$ and $_0$ indicates that a certain quantity has been evaluated at $U = U_0$ and $V = V_0$;

$$W_l U_1^{(l)} = 0 \tag{13}$$

where

$$W_l = -il\omega(A_0^0 + B_0^0 F_0) + ilk(A_0^1 + B_0^1 F_0) - (\nabla_u C)_0 - (\nabla_v C)_0 F_0 \tag{14}$$

For describing the modulation of quasi-monochromatic waves with wavenumber k and frequency ω and for excluding harmonic resonances we have to assume for $|l| \neq 1$

$$\det W_l \neq 0 \tag{15}$$

while for $|l| = 1$

$$\det W_l = 0. \tag{16}$$

For $l = 1$, (16) yields the usual dispersion relation associated with the linearized version of the model (4) which, in general, defines ω as a complex function of k.

Therefore, on account of (15) and (16), from (13) we obtain

$$U_1^{(1)} = \phi d \tag{17}$$

$$U_1^{(l)} = 0 , \qquad |l| \neq 1 \tag{18}$$

where the column vector d satisfies the relation

$$W_1 d = 0 \tag{19}$$

As already shown in [3] it is possible to ascertain that

$$\lambda = \frac{\partial \omega_R}{\partial k} \tag{20}$$

Finally it is not difficult [3,5] to obtain for ϕ the following evolution equation:

$$\frac{\partial \phi}{\partial \tau} + \eta \frac{\partial^2 \phi}{\partial \xi^2} + \mu |\phi|^2 \phi + \nu \phi = 0 \tag{21}$$

where $\eta = -(i/2)(\partial^2 \omega_R / \partial k^2)$, while the complex coefficients μ and ν are given by cumbersome expressions which, for the sake of simplicity, are not written here, although it is not difficult to obtain them.

2. Conclusions and remarks.

A cubically nonlinear diffusion equation like (21) occurs in superconductivity, in fluid mechanics mainly for problems in convection and plane parallel flow as well as in the nonlinear propagation of waves in piezo-electric semi-conductors. A large body of literature on this subject can be foun in [3,4] where also the main features of (21) are presented.

Because of the linear lower order term in (21), the higher order terms involved in the left hand side of (2) cause dissipation in the wave propagation governed by the reduced system (4).

It is noteworthy that,if $\nu_R = 0$, by means of the change of variable

$$\phi = u \exp(-i\nu_I \tau) \tag{22}$$

the equation (21) reduces to the form

$$\frac{\partial u}{\partial \tau} + \eta \frac{\partial^2 u}{\partial \xi^2} + \mu |u|^2 u = 0 \tag{23}$$

If $\mu_R=0$, the equation (23) specializes to the classical Schrödinger equation.

In general the coefficients of the equation under interest are not pure imaginary so that it cannot be integrated through inverse scattering [4]. Therefore the present model deserves further attention as far as the research of exact solutions is concerned.

Finally in [5] the general results obtained for the system of equations (1), (2) are used for studying the propagation of modulated waves in nerve fibers.

References

[1] D. FUSCO, *Meccanica* 17,128 (1982).

[2] D. FUSCO and N. MANGANARO, *J. de Mecanique Theorique et Appliquee* 6, 6, 761 (1987).

[3] A. JEFFREY and T. KAWAHARA, *Asymptotic Methods in Nonlinear Wave Theory*, Pitman, London (1982).

[4] R. K. DODD, J. C. EILBECK, J. D. GIBBON and H. C. MORRIS, *Solitons and Nonlinear Wave Equations*, Academic Press, New York (1982).

[5] D. FUSCO, N. MANGANARO, *On Some Features of the Wave Modulation in Nonlinear Media*, Int. J. of Eng. Sc. 28, 8, 821-827 (1990).

A DOMAIN OF INFLUENCE THEOREM IN FINITE ELASTICITY

Paolo Maremonti and Remigio Russo

Dipartimento di Matematica e Applicazioni

Università "Federico II"

via Mezzocannone, 8 - 80134 Napoli, ITALY

Abstract - *By using a mild growth hypothesis on the acoustic tensor and assuming that the elasticity tensor is uniformly strongly elliptic, it is shown that the disturbances in a nonlinear elastic body travel with a finite speed.*

1. Introduction

The domain of influence theorem is of special interest in the dynamical theory of elasticity since it assures that any perturbation in an elastic medium travels with a finite speed.

Assuming that the elasticity tensor is uniformly strongly elliptic and, if the body is unbounded, the acoustic tensor satisfies the hyperbolicity condition (cf. Section 2) we can state a domain of influence theorem for regular motions of a nonlinear elastic body. From this result, which can be viewed as an extension of a theorem of [1], we deduce a uniqueness theorem for regular solutions to the *displacement problem* of finite elastodynamics. Of course, since, as is well-known (see, e.g., [2-4]), regular motions of a nonlinear elastic body can exist only in a finite (small for arbitrary initial data) time interval $[0,\tau)$, the above results hold only in $[0,\tau)$.

In linear elasticity the above question has been examined in [5,6] under the hypothesis of positive semi-definiteness for the elasticity tensor.

NOTATION - Letters in ordinary form indicate scalars; bold face lower-case letters different from $\mathbf{o}$ and $\mathbf{x}$ indicate vectors (in $\mathbb{R}^3$), while $\mathbf{o}$ and $\mathbf{x}$ stand respectively for the origin of an assigned reference frame and a generic point of $\mathbb{R}^3$; bold face upper-case letters indicate second-order tensors (linear transformations from $\mathbb{R}^3$ into $\mathbb{R}^3$), *Lin* is the space of all second order tensors, $Lin^* = \{\mathbf{L} : \det \mathbf{L} + \mathbf{I} > 0\}$ and $Sym = \{\mathbf{L} : \mathbf{L} = \mathbf{L}^T\}$; $\nabla\mathbf{u}$ is the second-order tensor with components $(\nabla\mathbf{u})_{ij} = u_{i,j}$ $(,j = \partial/\partial x_j)$; $\operatorname{div}\mathbf{S}$ is the vector with components $S_{ij,j}$; a superimposed dot denotes differentiation with respect to time; $S_R(\mathbf{x}_0)$ is the ball of radius R centered at $\mathbf{x}_0$.

2. Basic equations. Hypotheses.

Let *B* be an elastic body. As is customary, we identify *B* with the domain B of $\mathbb{R}^3$ it occupies in an assigned reference configuration and we assume that the boundary ∂B of B is sufficiently smooth to admit application of the divergence theorem. To fix idea, we assume that B is unbounded. However, as will be clear from the context, our results hold for an arbitrary regular domain B.

Let $Q = B\times(0,T)$, with $T>0$. Let us assign:

i) a continuous and positive scalar field ρ on $\bar{B}$ (*mass density*);

ii) a continuous vector field $\mathbf{b}$ on $\bar{Q}$ (*body force per unit volume*);

iii) a regular map $\mathbf{S}^*$ from $Lin^* \times \bar{B}$ into *Lin* such that $\mathbf{S}^*(\mathbf{QL},\mathbf{x}) = \mathbf{Q}\mathbf{S}^*(\mathbf{L},\mathbf{x})$, for any $\mathbf{L} \in Lin^*$ and any orthogonal $\mathbf{Q}$ (*responce function*).

A motion of *B* in the time interval $[0,T)$ is a regular solution to the system

$$\text{(1)}\qquad \begin{array}{l} \rho\ddot{\mathbf{u}} = \operatorname{div}\mathbf{S} + \mathbf{b} \\ \mathbf{S} = \mathbf{S}^*(\nabla\mathbf{u},\mathbf{x}) \end{array} \qquad \text{on } \bar{Q}.$$

In (1) $\mathbf{S}$ denotes the *Piola-Kirchhoff stress tensor*.

Set $\mathbf{S}^*(\mathbf{L}) = \mathbf{S}^*(\mathbf{L},\mathbf{x})$ and $\mathbb{C}(\mathbf{L}) = \partial_{\mathbf{L}}\mathbf{S}^*(\mathbf{L})$ (*elasticity tensor for the deformation* $\mathbf{L}$. By Taylor's theorem we have

$$\text{(2)}\qquad \mathbf{S}^*(\mathbf{M}) = \mathbf{S}^*(\mathbf{L}) + \mathbb{C}(\mathbf{L})\mathbf{H} + \tfrac{1}{2}\Big[\int_0^1 \partial^2_{\mathbf{L}}\mathbf{S}^*(\mathbf{L}+\alpha\mathbf{H})\,d\alpha\Big]\mathbf{H}\otimes\mathbf{H}, \qquad \mathbf{H} = \mathbf{M} - \mathbf{L}\ ,$$

for any $\mathbf{L}, \mathbf{M} \in Lin^*$. Here, the elasticity tensor $\mathbb{C}$ and $\partial^2_{\mathbf{L}}\mathbf{S}^*$ are respectively fourth-order and sixth-order tensors.

We require that $\mathbb{C}$ is *symmetric*, i.e.,

$$\mathbf{M}\cdot\mathbb{C}(\mathbf{L})\mathbf{N} = \mathbf{N}\cdot\mathbb{C}(\mathbf{L})\mathbf{M}, \quad \forall \mathbf{M}, \mathbf{N}, \ \forall \mathbf{L} \in Lin^*, \tag{3}$$

and *uniformly strongly elliptic* in an open set Ω of Lin^*, i.e.,

$$\mathbf{M}\cdot\mathbb{C}(\mathbf{L})\mathbf{M} \geq \mu \mathbf{M}^2, \quad \forall \mathbf{L} \in \Omega, \ \forall \mathbf{M} = \mathbf{a} \otimes \mathbf{b}, \tag{4}$$

for some positive constants μ independent of $\mathbf{L}$.

Assumptions (3) and (4) imply that B is *hyperelastic* and the quasi linear system (1) is *hyperbolic* in Ω. We have to limit ourselves to an open set Ω of Lin^*, since there are some examples in which it is shown that the ellipticity condition (4) fails to hold for large $\mathbf{L}$ (cf. [4]).

From now on we shall consider only solutions $\mathbf{u}$ to System (1) such that $\nabla\mathbf{u}(\mathbf{x},t) \in \Omega$, $\forall (\mathbf{x},t) \in \bar{Q}$.

Let $\mathbf{A}(\mathbf{L},\mathbf{m})$ be the *acoustic tensor for the deformation* $\mathbf{L}$ *and the direction* $\mathbf{m}$, defined by

$$\rho\mathbf{A}(\mathbf{L},\mathbf{m})\mathbf{a} = [\mathbb{C}(\mathbf{L})\mathbf{a} \otimes \mathbf{m}]\mathbf{m}, \quad \forall \mathbf{a},$$

and let

$$\phi(\mathbf{L},\xi) = \sup\{|\mathbf{A}(\mathbf{L},\mathbf{m})|, \ \mathbf{x} \in B \cap S_\xi(\mathbf{o}), \ \mathbf{m}: \ |\mathbf{m}| = 1\},$$

$$p(\mathbf{L},r) = \int_{r^*}^{r} [\phi(\mathbf{L},\xi)]^{-\frac{1}{2}} d\xi \qquad (r^* > 0).$$

Note that $[\phi(\mathbf{L},\xi)]^{\frac{1}{2}}$ represents the maximum speed of propagation in $B \cap S_\xi(\mathbf{o})$ corresponding to $\mathbb{C}(\mathbf{L})$ and ρ, and p is a positive, smooth and increasing function of r $(=|\mathbf{x}-\mathbf{o}|)$.

Definition - The acoustic tensor is said to satisfy the hyperbolicity condition with respect to $\mathbf{L}$ if

$$\lim_{r \to +\infty} p(\mathbf{L},r) = +\infty. \quad \square \tag{5}$$

The above property on $\mathbf{A}$ has been introduced in the framework of linear elasticity [5,6] where $\mathbf{A}$ is independent of $\mathbf{L}$. It has been called hyperbolic, because it completely characterizes the linear elastic material that can support wave propagation. On the other hand, we easily realize that $p(\bar{L},r)$ majorizes the time employed by a signal to reach $B \cap \partial S_r(o)$ by starting from $B \cap \partial S_{r*}(o)$. Thus, roughly speaking, (5) assures that the time needed by a signal to *touch* the infinity can not be finite.

3. The domain of influence theorem. Uniqueness for the displacement problem.

Let $\mathbf{u}_1$, $\mathbf{u}_2$ be two solutions to System (1) corresponding to the body force fields $\mathbf{b}_1$ and $\mathbf{b}_2$ respectively and such that $\mathbf{u}_1 = \mathbf{u}_2$ on $\partial B\times[0,T)$.

Let

$$D_t^* = \{\mathbf{x} \in B \;:\; \mathbf{u}_1(\mathbf{x},0) \neq \mathbf{u}_2(\mathbf{x},0) \;\; or \;\; \dot{\mathbf{u}}_1(\mathbf{x},0) \neq \dot{\mathbf{u}}_2(\mathbf{x},0) \;\; or \;\; \exists\, s \in [0,t] \;:\; \mathbf{b}_1(\mathbf{x},s) \neq \mathbf{b}_2(\mathbf{x},s)\}.$$

Moreover let

$$D_t = \{\mathbf{x}_0 \in B \;:\; D_t^* \cap S_{p^{-1}(p(r_0)+\chi t)-r_0}(\mathbf{x}_0) \neq \emptyset\} \;,$$

where p is the function $p(\nabla\mathbf{u}_1,r)$ and $r_0 = |\mathbf{x}_0 - \mathbf{o}|$, for some $\chi > 1$.

Observe that, if $\mathbf{A}$ satisfies the hyperbolicity condition with respect to $\nabla\mathbf{u}_1$, then p is one-to-one.

The following theorem, whose proof is given in [7], asserts that the perturbation $(\mathbf{u}_1 - \mathbf{u}_2)(\mathbf{x},0)$, $(\dot{\mathbf{u}}_1 - \dot{\mathbf{u}}_2)(\mathbf{x},0)$ and $\mathbf{b}_1 - \mathbf{b}_2$ could have effect, at instant t, only in the points of D_t.

DOMAIN OF INFLUENCE THEOREM - Let $\mathbf{u}_1$ and $\mathbf{u}_2$ be two solutions to System (1) corresponding to the body force fields $\mathbf{b}_1$ and $\mathbf{b}_2$ respectively and such that $\mathbf{u}_1 = \mathbf{u}_2$ on $\partial B\times[0,T)$. If the acoustic tensor satisfies the hyperbolicity condition with respect to $\nabla\mathbf{u}_I$, then

$$\mathbf{u}_1 = \mathbf{u}_2 \quad \text{on } \overline{(B - D_t)} \times [0,t],\ \forall\, t \in [0,T). \qquad \square$$

The *displacement problem of finite elastodynamics* consists in finding a solution $\mathbf{u}$ to System (1) which satisfies the *initial conditions*

(6) $$\mathbf{u} = \mathbf{u}_0 \ , \quad \dot{\mathbf{u}} = \dot{\mathbf{u}}_0 \qquad \text{on } \bar{B} \times \{0\},$$

and the *boundary conditions*

(7) $$\mathbf{u} = \mathbf{s} \qquad \text{on } \partial B \times [0,T),$$

where $(\mathbf{u}_0, \dot{\mathbf{u}}_0)$ and $\mathbf{s}$ are assigned regular vector functions on $\bar{B}$ and $\partial B \times [0,T)$.

Let

$\mathcal{J}$ = {regular motions $\mathbf{v}$ of B : $\mathbf{A}$ satisfies the hyperbolicity condition with respect to $\nabla \mathbf{v}$}.

UNIQUENESS THEOREM FOR THE DISPLACEMENT PROBLEM - The displacement problem of finite elastodynamics has at most one solution in $\mathcal{J}$. $\square$

Proof - Let $\mathbf{u}_1$ and $\mathbf{u}_2$ be two solutions to System (1)-(6)-(7) and let $\mathbf{u} = \mathbf{u}_1 - \mathbf{u}_2$. Since $\mathbf{u}(\mathbf{x},0) = \mathbf{0}$ and $\dot{\mathbf{u}}(\mathbf{x},0) = \mathbf{0}$ on $B \times \{0\}$ we have that $D_t^* = \emptyset$, so that $D_t = \emptyset$, $\forall\, t \in [0,T)$. Then, we conclude that $\mathbf{u}_1 \equiv \mathbf{u}_2$. $\square$

REFERENCES

[1] Nachlinger, R.R. and Wheeler, L.T., On wave propagation property and uniqueness in one-dimensional elastic bodies, J. Elasticity **3**, 141 (1973).

[2] Dafermos, C.M. and Hrusa,W., Energy methods for quasilinear hyperbolic initial-boundary value problem. Applications to Elasto_ dynamics, Arch. Rational Mech. Anal. **87**, 267 (1985).

[3] John, F., Finite amplitude waves in a homogeneous isotropic elastic solid, Comm. Pure Appl. Math. **30**, 421 (1977).

[4] Hughes, T.J.R., Kato T. and Marsden J.E., Well-posed quasi-linear second-order hyperbolic systems with apllication to nonlinear elasto_ dynamics and general relativity, Arch. Rational Mech. Anal. **63**, 273 (1977).

[5] Carbonaro, B. and Russo, R., Energy inequalities and the domain of influence theorem in classical elastodynamics, J. Elasticity **14**, 163 (1984).

[6] Knops, R.J., Rionero, S. and Russo, R., Continuous data dependence and stability in linear elastodynamics, forthcoming.

[7] Maremonti, P. and Russo, R., On wave propagation and uniqueness in finite elasticity, submitted for publication.

HYDRODYNAMICAL STABILITY PROBLEMS IN INFINITELY LONG CYLINDRICAL DOMAINS: EXISTENCE OF DEFECTS IN TRAVELLING WAVES

Gérard Iooss[†] & Alexander Mielke[¥]

[†] Laboratoire de Mathématiques, U.A.CNRS 168, Université de Nice, Parc Valrose, F-06034 Nice.

[¥] Mathematisches Institut A, Universität Stuttgart, Pfaffenwaldring 57, D-7000 Stuttgart 80.

ABSTRACT: We investigate all bifurcating time-periodic solutions, not restricted to spatially periodic ones, when a Hopf bifurcation occurs in an hydrodynamic stability problems in an infinite cylinder, possessing reflection symmetry. We show the existence of spatially quasi-periodic flows, and more particularly, we also show the existence of heteroclinic solutions which connect two symmetric travelling waves staying at each end of the cylinder.

1. INTRODUCTION

Many classical hydrodynamical stability problems deal with flows in a very long domain. This is often theoretically modelled by an infinite domain, which simplifies the linear analysis, and is generally physically justified when one looks not too closely to the ends of the domain. Let us consider flows in cylindrical domains of a one or two dimensional bounded cross-section Ω. Typical examples of such a situation are Taylor-Couette flow between two concentric rotating cylinders, where the section Ω is a 2-dimensional annulus, and Bénard convection problem of a liquid heated from below in a long box, and where the section Ω is a rectangle. In all these problems the system is invariant under translations parallel to the generators of the cylinder and it is also invariant under reflection symmetry through any cross-sectional plane.

In the present work we consider hydrodynamical stability problems where the most unstable mode is *oscillatory at criticality*, leading to a classical "Hopf bifurcation", when one assumes spatial periodicity for the flow (see for instance Iudovich [12], Sattinger [21] and Iooss [7]). In what follows, *we suppress the spatial periodicity assumption*. Contrary to a previous work [9] where we concentrated on steady bifurcating solutions, we cannot do the same here, since the critical modes are oscillatory in time. To be able to avoid the difficulty of continuous spectra for linear operators as in [9], we only study *time-periodic solutions*, the unknown period appearing as an additional parameter. Moreover, we only consider the reflection symmetric case. By the way, we recover the already known bifurcating time and space periodic solutions (travelling waves and standing waves), and we obtain new solutions, no longer spatially periodic.

In previous works K.Kirchgässner [14] [15] and Collet & Eckmann [3][4] studied time dependent solutions for some similar evolution problems in a strip, or on the real line. The Kirchgässner's analysis reduces to the search for steady solutions in a moving frame, as in [9], while the analysis of [3][4] deals with propagating fronts, which have not a form compatible with our job here. In fact, the work which is the closest to ours comes back to M.Renardy [20], who treats reaction-diffusion equations on the real line. He focusses his analysis on two different types of non spatially periodic solutions, the first one approaching a constant at infinity, the second one approaching periodic wave trains at infinity, the direction of propagation being opposite at both infinities.

In this paper, we present results developed in the more detailed paper [11]. We first show that one can use the center manifold theorem to obtain bifurcating time-periodic solutions of Navier-Stokes equations in a cylindrical domain. This tool was not used by Renardy in [20], but his problem was of simpler structure. The idea of using center manifold theory for elliptic problems in cylindrical domains was initiated by Kirchgässner [14] and is now extensively used for water wave problems (see Mielke [18] and Amick & Kirchgässner [1]), and for elasticity problems (long beams) see Mielke [19]. We also used it for steady solutions of Navier-Stokes equations in [9]. Then, using symmetry arguments and normal form theory, as in [9], we derive the reduced ordinary differential system for the amplitudes in the space variable which gives all bifurcating bounded solutions. It appears that one can completely solve at least the truncated system (at any order), and that this gives many unusual non spatialy-periodic solutions. For the full system (untruncated), we are able to prove the persistence of most of these solutions. The physically most interesting solution is a "defect" solution occuring for instance in the Couette-Taylor problem, which appears in the reduced system as an heteroclinic connection between the two symmetric regimes of travelling waves (spiral waves in this example). This regime is effectively observed in experiments with counter-rotating cylinders.

2. THE CENTER MANIFOLD

In this section we consider the Navier-Stokes equations in an infinite cylinder $Q = \mathbb{R}\times\Omega$, where Ω is a smooth bounded domain in $\mathbb{R}^2$. We give the result, proved in [11] on the center manifold theorem for *time-periodic solutions,* the evolution variable is then the unbounded space variable $x \in \mathbb{R}$. This result allows to reduce our problem into an ordinary differential equation in a two or four dimensions space, the evolution variable being x.

The Navier-Stokes equations are as follows:

$$(1)\qquad \begin{cases} \frac{\partial V}{\partial t} + (V.\nabla)V + \nabla p = \nu\Delta V + f(\mu,.), \\ \nabla.V = 0 \qquad\qquad \text{in } Q\ , \\ V = g(\mu,.) \text{ on } \partial Q = \mathbb{R}\times\partial\Omega, \end{cases}$$

where V represents the velocity vector, p the pressure, both functions of $(x,y,t) \in \mathbb{R}\times\Omega\times\mathbb{R}^+$, μ represents the set of parameters (kinematic viscosity ν being included), and f, g are functions of the cross-sectional variable $y \in \Omega$ (resp.$\partial\Omega$) only. We assume the existence of a family of (x,t)-independent solutions $V = V^{(0)}(\mu,.) \in C^1(\bar{\Omega},\mathbb{R}^3)$.

We are interested in the appearence of time-periodic solutions staying close to $V^{(0)}$ for all time t and all values of the axial variable $x \in \mathbb{R}$. The method developed here will of course recover the well-known travelling wave solutions which are also periodic in x. This is the classical result for Hopf bifurcation with an O(2) symmetry group acting non trivially on the critical eigenspace (see for instance [8] in the context of Navier-Stokes equations). However, as the x-behavior is now not prescribed in advance we also find new types of solutions being quasi-periodic, homoclinic or heteroclinic in the spatial direction, always modulated with some time frequency ω.

Looking for solutions with period $T = 2\pi/\omega$, it is convenient to introduce the scaled time $\tau = \omega t \in S^1$; so that the solutions will be 2π-periodic in τ, and ω appears as a parameter in the equation. We decompose the velocity V into a longitudinal component V_x and a transversal component $V_\perp$, and introduce the notations:

$$(2)\qquad U = (U_x,U_\perp) = V-V^{(0)},\ W = (W_x,W_\perp) = (-p, \nu\frac{\partial}{\partial x}U_\perp).$$

Moreover, setting $\mathfrak{V} = (U,W)$, equation (1) takes the form

$$(3)\qquad \frac{d\mathfrak{V}}{dx} = \mathcal{K}_{\mu,\omega}\mathfrak{V} + \mathcal{B}_\mu(\mathfrak{V},\mathfrak{V}),$$

where there are no longer x-derivatives on the right hand side. From now on, we consider $\mathfrak{V} = (U,W)$ as a (vector-valued) function of $(\tau,y) \in S^1\times\Omega$, which will vary, according to (3), along the axial variable $x \in \mathbb{R}$. In particular, we introduce the following Hilbert spaces

$$\mathcal{H}^{\theta,s} = H^\theta[S^1,H^s(\Omega)],$$

where $H^s(\Omega)$ is the classical Sobolev space (see for instance [16]), and $H^\theta[S^1,H^s(\Omega)]$ denotes the space of $H^s(\Omega)$-valued functions $u(\tau,.) = \sum_{n\in\mathbb{Z}} u_n(.)e^{in\tau}$, with $u_n \in H^s(\Omega)$, such that

(4) $\|u\|^2_{\theta,s} = \sum_{n\in\mathbb{Z}} (1 + n^2)^\theta \|u_n\|^2_{H^s(\Omega)} < \infty .$

The phase space $\mathcal{X}$ is set equal to

(5) $\mathcal{X} = \{(U,W) \in (\mathcal{H}^{1,-1}\cap\mathcal{H}^{0,1})\times(\mathcal{H}^{0,1})^2\times(\mathcal{H}^{0,0})^3;\ U = 0 \text{ on } S^1\times\partial\Omega\},$

and the linear operator $\mathcal{K}_{\mu,\omega}$ has domain;

(6) $D(\mathcal{K}) = \{(U,W)\in\mathcal{X};\ U\in(\mathcal{H}^{1,0}\cap\mathcal{H}^{0,2})^3,\ W\in(\mathcal{H}^{0,1})^3,\ \nabla_\perp.U_\perp = W_\perp = 0 \text{ on } S^1\times\partial\Omega\}.$

For the nonlinear mapping $\mathcal{B}_\mu$ we then obtain the following theorem proved in [11] (the interpolation functor $[.,.]_\theta$ being defined as in [16]):

Theorem 1. *For* $\theta > \frac{1+\sqrt{17}}{8} = 0.64...$, *the quadratic mapping* $\mathcal{B}_\mu$ *is an analytic function from* $D(\mathcal{K}^\theta) := [D(\mathcal{K}),\mathcal{X}]_\theta \subset (\mathcal{H}^{\theta,0}\cap\mathcal{H}^{0,1+\theta})^3\times(\mathcal{H}^{0,\theta})^3$ *into the closed subspace* $\mathcal{Y}\subset\mathcal{X}$ *given by* $\mathcal{Y} := \{(U,W)\in\mathcal{X};\ U = 0\}$.

The analysis of the linear operator $\mathcal{K}_{\mu,\omega}$ involves some difficulties which arise from the unsymmetric form of $\mathcal{A}_\mu$, involving the divergence equation in the first component and the pressure in the fourth. In fact, it is not possible to define $\mathcal{K}_{\mu,\omega}$ on a domain $D(\hat{\mathcal{K}})$ being compactly embedded in $\mathcal{X}$, such that the resolvent $(\hat{\mathcal{K}}_{\mu,\omega} - \lambda)^{-1}: \mathcal{X} \to D(\hat{\mathcal{K}})$ exists for some $\lambda\in\mathbb{C}$. One obvious reason is that $\mathcal{K}_{\mu,\omega}$ has an infinite-dimensional kernel spanned by $(U,W) = (0,0,\alpha(\tau),0)$ with arbitrary periodic functions α depending only on τ.

Even cutting out this kernel does not resolve the problem. The problem arises because the pressure at time τ, which is $-W_x$ in our notation, is only a function of the velocity field at time τ. Hence the time-dependence is not smoothed out, which would be necessary to obtain compactness.

To deal with the infinite-dimensional kernel, we define the projections

(7) $Qf = f - [f], \qquad [f](.) = \frac{1}{|\Omega|}\int_\Omega f(.,y)dy,$

(8) $\mathcal{Q}(U_x,U_\perp,W_x,W_\perp) = (QU_x,U_\perp,QW_x,W_\perp),$

and decompose $\mathcal{V}\in\mathcal{X}$ into $\bar{\mathcal{V}} + \tilde{\mathcal{V}}$ where $\bar{\mathcal{V}} = ([U_x],0,[W_x],0)$ and $\tilde{\mathcal{V}} = \mathcal{Q}\mathcal{V}$. We then obtain :

(9) $$\frac{d}{dx}\tilde{\mathcal{V}} = \tilde{\mathcal{K}}_{\mu,\omega}\tilde{\mathcal{V}} + \tilde{\mathcal{B}}_\mu([U_x],\tilde{\mathcal{V}}),$$

where $$\tilde{\mathcal{K}}_{\mu,\omega}: \begin{cases} D(\tilde{\mathcal{K}}) = \mathcal{Q}D(\mathcal{K}) \to \tilde{\mathcal{X}} = \mathcal{Q}\mathcal{X}; \\ \qquad \tilde{\mathcal{V}} \to \mathcal{Q}\mathcal{K}_{\mu,\omega}\tilde{\mathcal{V}}, \end{cases}$$

and $$\tilde{\mathcal{B}}_\mu([U_x],\tilde{\mathcal{V}}) = \mathcal{Q}\mathcal{B}_\mu(\bar{\mathcal{V}} + \tilde{\mathcal{V}}, \bar{\mathcal{V}} + \tilde{\mathcal{V}}) \in \mathcal{Q}\mathcal{Y} =: \tilde{\mathcal{Y}}.$$

Now, $[W_x]$ does not appear in $\widetilde{\mathfrak{B}}_\mu$ and $\beta(\tau) = [U_x(\tau,.)]$ is a x-independent volumic flux, which is a consequence of the incompressibility ($\nabla_\perp . U_\perp + \frac{\partial}{\partial x} U_x = 0$) and the boundary condition $U_\perp = 0$ on $\partial\Omega$.

Hence, having fixed $\beta = \beta(\tau)$, we may first solve (9) and then the mean value of the pressure $-[W_x]$ can be obtained by simple integration. In the following, we will restrict the analysis to the case when $[U_x] \equiv 0$, which is reasonable for instance for the Taylor-Couette problem.

In [11] we prove the following results on the linear operator $\widetilde{\mathfrak{K}}_{\mu,\omega}$:

Theorem 2.

There is a positive δ *such that the resolvent* $(\widetilde{\mathfrak{K}}_{\mu,\omega} - \lambda)^{-1} : \widetilde{\mathfrak{X}} \to D(\widetilde{\mathfrak{K}})$ *is a meromorphic function of* $\lambda \in \mathbb{C}_\delta := \{\lambda \in \mathbb{C};\ |\mathrm{Re}\lambda| < \delta(1 + |\mathrm{Im}\lambda|)\}$ *(see figure 1).*

i) *In* $\mathbb{C}_\delta$ *there are only finitely many eigenvalues of* $\widetilde{\mathfrak{K}}_{\mu,\omega}$ *(i.e. poles of* $(\widetilde{\mathfrak{K}}_{\mu,\omega} - \lambda)^{-1}$*) each having a finite-dimensional generalized eigenspace. Functions in these eigenspaces are finite sums of terms like* $e^{in\tau}\mathfrak{V}_n(y)$.

ii) *Moreover, for any fixed* $\theta \in [0,1]$ *and for* $\lambda \in \mathbb{C}_\delta$ *with* $|\lambda| \to \infty$ *the following estimates hold:*

$$\|(\widetilde{\mathfrak{K}}_{\mu,\omega} - \lambda)^{-1}\|_{\mathfrak{L}(\widetilde{\mathfrak{X}},\widetilde{\mathfrak{X}})} = \mathcal{O}(1), \tag{10}$$

$$\|(\widetilde{\mathfrak{K}}_{\mu,\omega} - \lambda)^{-1}\|_{\mathfrak{L}(\widetilde{\mathfrak{Y}},QD(\mathfrak{K}^\theta))} = \mathcal{O}(1/|\lambda|^{1-\theta}). \tag{11}$$

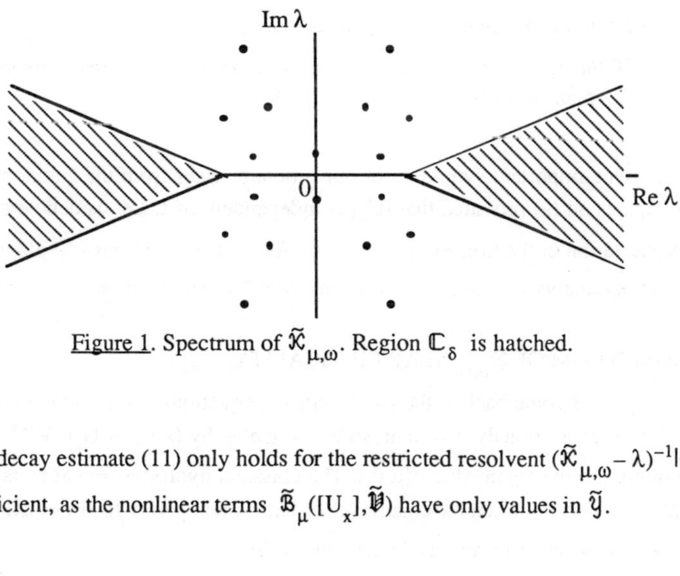

Figure 1. Spectrum of $\widetilde{\mathfrak{K}}_{\mu,\omega}$. Region $\mathbb{C}_\delta$ is hatched.

Remarks:

i) The decay estimate (11) only holds for the restricted resolvent $(\widetilde{\mathfrak{K}}_{\mu,\omega} - \lambda)^{-1}|_{\widetilde{\mathfrak{Y}}}$, but this is sufficient, as the nonlinear terms $\widetilde{\mathfrak{B}}_\mu([U_x],\widetilde{\mathfrak{V}})$ have only values in $\widetilde{\mathfrak{Y}}$.

ii) We cannot expect $(\mathcal{K}_{\mu,\omega}-\lambda)^{-1}$ to be meromorphic in the whole complex plane (see an example in the Appendix of [11]).

Theorem 2 implies that $\tilde{\mathcal{K}}_{\mu,\omega}$ has for any (μ,ω) only a finite-dimensional isolated spectral part on the imaginary axis; let us call it $\mathcal{X}^0$ for $(\mu,\omega) = (0,\omega_0)$ (to be specified later), and the corresponding $\tilde{\mathcal{K}}_{0,\omega_0}$– invariant projection $\mathcal{P}^0$ (defined by a Dunford integral). Hence, $\mathcal{K}^0 = \tilde{\mathcal{K}}_{0,\omega_0}|_{\mathcal{X}_0}$ has only eigenvalues on the imaginary axis. Now, for (μ,ω) close to $(0,\omega_0)$, we know by perturbation theory (see Kato [13]) that one can define a $\tilde{\mathcal{K}}_{\mu,\omega}$– invariant projection $\mathcal{P}^0_{\mu,\omega}$ on a finite-dimensional invariant subspace $\mathcal{X}^0_{\mu,\omega}$, and the restriction $\mathcal{K}^0_{\mu,\omega} = \tilde{\mathcal{K}}_{\mu,\omega}|_{\mathcal{X}^0_{\mu,\omega}}$ has only a finite number of isolated eigenvalues all being close to the imaginary axis. Considering the restriction of $\tilde{\mathcal{K}}_{\mu,\omega}$ to the complementary invariant subspace, we are now able to build the Green's function (for θ in $\left((1+\sqrt{17})/8,1\right)$ following theorem 1), which is of fundamental use in the center manifold theorem (see [17] or [22]). It results the following

Theorem 3.

For each integer k *there are neighborhoods* $\mathcal{O}_k$ *of* 0 *in* $\mathcal{X}^0$ *and* σ_k *of* $(0,\omega_0)$ *in* $\mathbb{R}^2$ *and a* C^k *function* $\Phi: \sigma_k \times \mathcal{O}_k \to \mathcal{X}^1$ *such that the manifold*

$$M_c = \{\tilde{\mathbf{V}} = \tilde{\mathbf{V}}_0 + \Phi(\mu,\omega,\tilde{\mathbf{V}}_0);\ (\mu,\omega,\tilde{\mathbf{V}}_0) \in \sigma_k \times \mathcal{O}_k\}$$

contains every small bounded solution of (9) and (1). Moreover, every solution of the reduced differential equation on M_c *yields through* $\tilde{\mathbf{V}} = \tilde{\mathbf{V}}_0 + \Phi(\mu,\omega,\tilde{\mathbf{V}}_0)$ *a solution of (9).*

If the system (9) is equivariant with respect to some symmetry or if it is reversible, then so is the reduced system on M_c.

One important symmetry of our problem arises from the autonomy of (1) which propagates on (9) provided that $[U_x]$ is independent of τ. Let us denote by $\mathbf{T}_\alpha$ the linear representation of the time-shift $\tau \to \tau + \alpha$. As we see in section 4, it plays a fundamental role in our analysis, contrary to the steady case [9], where this action is trivial.

3. SPECTRUM OF $\tilde{\mathcal{K}}_{\mu,\omega}$ NEAR CRITICALITY

Let us come back to the Navier-Stokes equations (1) linearized around $V^{(0)}$. This system is translationaly invariant, so let us denote by $(x,y) \to \hat{U}_k(y)e^{ikx}$ an eigenvector belonging to the eigenvalue $\sigma(\mu,ik)$. The classical hydrodynamical instability threshold (criticality) occurs when the eigenvalue of largest real part $\sigma_0(\mu,ik)$ satisfies the following properties (where criticality is defined by $\mu = 0$):

$$(12)\qquad \mathrm{Re}\,\sigma_0(0,ik_c) = 0, \qquad \frac{\partial}{\partial k}\mathrm{Re}\,\sigma_0(0,ik_c) = 0, \quad \frac{\partial}{\partial \mu}\mathrm{Re}\,\sigma_0(0,ik_c) > 0,$$

and Re $\sigma_0(\mu,ik)$ is maximum at $k = k_c$, $\mu = 0$ (see figure 2a). We assume in what follows that $k_c \neq 0$. It results from these properties that one can write the Taylor expansion of $\sigma_0(\mu,\lambda)$ for (μ,λ) in $\mathbb{R}\times\mathbb{C}$ near $(0,ik_c)$ under the form (indices r and i mean real and imaginary parts):

$$(13)\qquad \sigma_0(\mu,\lambda) = i\omega_0 + a\mu + e_1(\lambda - ik_c) + e_2(\lambda - ik_c)^2 + e_3\mu^2 + e_4\mu(\lambda - ik_c) + \text{h.o.t.}$$

where $a_r > 0$, $e_1 \in \mathbb{R}$, $e_{2r} > 0$. We check that the neutral stability curve of figure 2b, given by Re $\sigma_0(\mu,ik) = 0$, takes the form:

$$(14)\qquad \mu = \mu_c(k) = \frac{e_{2r}}{a_r}(k - k_c)^2 + \text{h.o.t.},$$

and, on this curve, we have Im $\sigma_0(\mu_c(k),ik) = \omega_c(k)$.

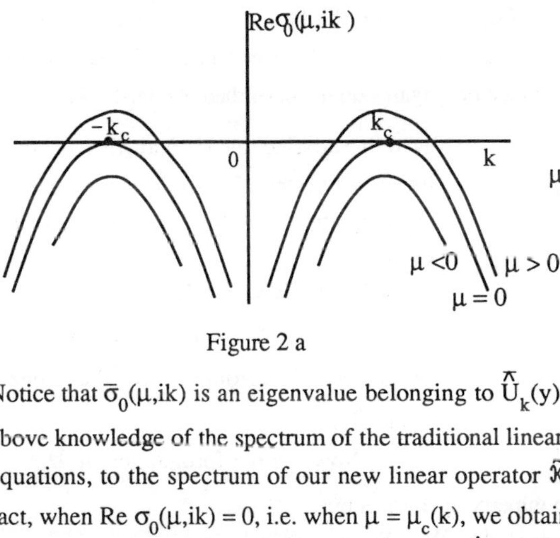

Figure 2 a

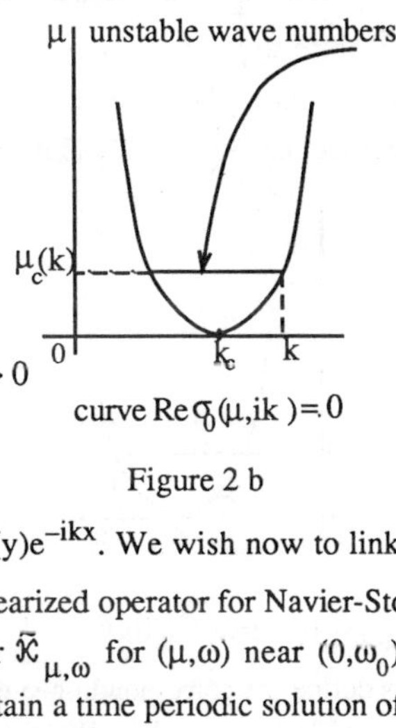

Figure 2 b

Notice that $\bar{\sigma}_0(\mu,ik)$ is an eigenvalue belonging to $\bar{\hat{U}}_k(y)e^{-ikx}$. We wish now to link the above knowledge of the spectrum of the traditional linearized operator for Navier-Stokes equations, to the spectrum of our new linear operator $\tilde{\mathcal{K}}_{\mu,\omega}$ for (μ,ω) near $(0,\omega_0)$. In fact, when Re $\sigma_0(\mu,ik) = 0$, i.e. when $\mu = \mu_c(k)$, we obtain a time periodic solution of the linearized Navier–Stokes equations, of the form $\hat{U}_k(y)e^{ikx}\,e^{i\omega t}$, with $\omega = \omega_c(k)$. With our new formulation in $\mathbb{V} = (U,W)$, this corresponds to the existence of an eigenvalue ik for $\mathcal{K}_{\mu,\omega}$, hence, since $k \neq 0$, to the same eigenvalue for $\tilde{\mathcal{K}}_{\mu,\omega}$, with an eigenvector of the form $(\tau,y) \to \mathbb{V}(y)e^{i\tau}$. In addition, we also observe that $\bar{\mathbb{V}}(y)e^{-i\tau}$ is an eigenvector belonging to the eigenvalue $-ik$ of $\tilde{\mathcal{K}}_{\mu,\omega}$. So, for $\mu = \mu_c(k)$, $\omega = \omega_c(k)$, we know eigenvalues $\pm$ ik of $\tilde{\mathcal{K}}_{\mu,\omega}$ on the imaginary axis, and, by construction, other eigenvalues of this operator do not belong to this axis.

Now, what happens for the spectrum of $\tilde{\mathcal{K}}_{\mu,\omega}$ if (μ,ω) lies in a neighborhood of $(0,\omega_0)$, not necessarily on the neutral curve $\mu = \mu_c(k)$, $\omega = \omega_c(k)$?

We may observe, more generally on (13), that if Re $\sigma_0(\mu,\lambda) = 0$, we have an eigenvalue λ for $\widetilde{\mathcal{K}}_{\mu,\omega}$ where $\omega = \text{Im } \sigma_0(\mu,\lambda)$. Here λ is not necessarily pure imaginary, so that we obtain a new information on the spectrum of $\widetilde{\mathcal{K}}_{\mu,\omega}$. Let us solve with respect to λ the system

$$\text{Re } \sigma_0(\mu,\lambda) = 0, \qquad \omega = \text{Im } \sigma_0(\mu,\lambda).$$

Provided that coefficient e_1 is not zero, we obtain:

$$(15)\quad \lambda(\mu,\omega) = ik_c - \frac{1}{e_1}[a\mu - i(\omega-\omega_0)] - \frac{e_2}{e_1^3}[a\mu - i(\omega-\omega_0)]^2 - \frac{e_3}{e_1}\mu^2 + \frac{e_4}{e_1^2}\mu[a\mu - i(\omega-\omega_0)] + \text{h.o.t.}$$

and Re $\lambda = 0$ gives again the neutral curve, now under the form (see figure 3):

$$(16)\qquad \mu = \mu^c(\omega) = \frac{e_{2r}}{a_r e_1^2}(\omega - \omega_0)^2 + \text{h.o.t.}$$

It results that for (μ,ω) close to $(0,\omega_0)$ the spectrum of $\widetilde{\mathcal{K}}_{\mu,\omega}$ contains two eigenvalues λ and $\bar{\lambda}$ close to the imaginary axis. Let us now use the fact that our original system is invariant under the reflection symmetry $x \to -x$, and assume that $V^{(0)}$ respects this invariance, i.e. $V_x^{(0)} = 0$. This invariance propagates on the perturbed Navier-Stokes system satisfied by U, and also on the linearized system. Let us define the linear representation S of the symmetry $x \to -x$, as follows:

$$S(U_x,U_\perp)(x,y) = (-U_x,U_\perp)(-x,y).$$

Then, if $\hat{U}_k(y)e^{ikx}$ is an eigenvector, so is $S\hat{U}_k(y)e^{-ikx}$ belonging to the same eigenvalue.

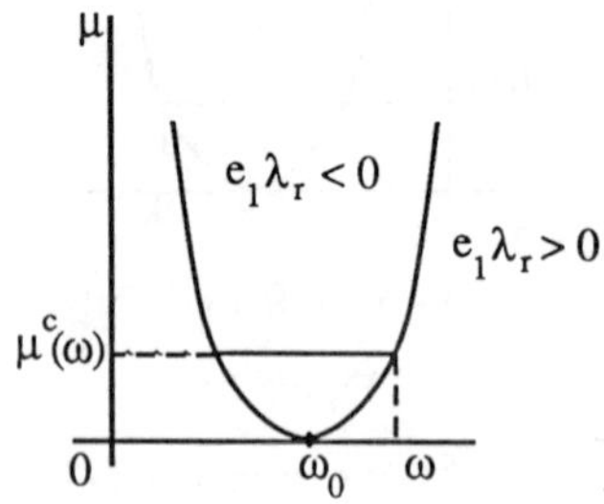

Figure 3: curve $\lambda_r(\mu,\omega) = 0$

Now, for our formulation in $\mathcal{V}$, let us define the corresponding symmetry operator S by:

$$(17)\qquad S(U,W) = (-U_x,U_\perp,W_x,-W_\perp), \qquad S^2 = I,$$

then our system (3) is now *reversible*. This means that we have the anticomutation properties:

$$(18)\qquad S\widetilde{\mathcal{K}}_{\mu,\omega} = -\widetilde{\mathcal{K}}_{\mu,\omega}S, \quad S\widetilde{\mathcal{B}}_\mu([U_x],\widetilde{\mathcal{V}}) = -\widetilde{\mathcal{B}}_\mu(-[U_x],S\widetilde{\mathcal{V}}).$$

so, this system is also reversible whenever $[U_x] = 0$, which is assumed in the following.

As for all reversible systems, if λ is an eigenvalue , then $-\lambda$ is an eigenvalue too, coresponding to the symmetric eigenvector. Finally, for $(\mu,\omega) = (0,\omega_0)$ we have two double semi-simple eigenvalues $\pm ik_c$ with the following eigenvectors:

$$(19)\qquad \begin{cases} \mathbf{V}_0(y)e^{i\tau},\ \mathbf{S}\overline{\mathbf{V}}_0(y)e^{-i\tau} & \in ik_c, \\ \overline{\mathbf{V}}_0(y)e^{-i\tau},\ \mathbf{S}\mathbf{V}_0(y)e^{i\tau} & \in -ik_c. \end{cases}$$

When (μ,ω) is near $(0,\omega_0)$, these two eigenvalues split into 4 simple eigenvalues $\pm\lambda,\pm\overline{\lambda}$, symmetric with respect to both axis in the complex plane, and λ is given by formula (15).

4. NORMAL FORM FOR AMPLITUDES EQUATIONS

Following the results of section 3 and Theorem 3, the bifurcating time-periodic solutions, bounded for $x \in \mathbb{R}$, are solutions of the following system of complex amplitudes equations:

$$(20)\qquad \begin{cases} \dfrac{dA}{dx} = ik_cA + N_0(\mu,\omega,A,B,\overline{A},\overline{B})\ , \\ \dfrac{dB}{dx} = -\ ik_cB + N_1(\mu,\omega,A,B,\overline{A},\overline{B}), \end{cases}$$

where A and B are defined by the coordinates of the 4-dimensional vector $\widetilde{\mathbf{V}}_0$ in $\mathcal{X}^0$ and

$$(21)\qquad \widetilde{\mathbf{V}} = Ae^{i\tau}\mathbf{V}_0(y) + \overline{A}e^{-i\tau}\overline{\mathbf{V}}_0(y) + Be^{i\tau}\mathbf{S}\mathbf{V}_0(y) + \overline{B}e^{-i\tau}\mathbf{S}\overline{\mathbf{V}}_0(y) + \Phi(\mu,\omega,A,\overline{A},B,\overline{B}),$$

the vector field Φ taking values in $\widetilde{\mathcal{X}}$ and giving explicitely the center manifold.

We know from theorem 3, that (20) is equivariant under the group $\mathbf{T}_\alpha$ and is reversible. This first means that, for any α, we have for j = 0,1:

$$(22)\qquad N_j(\mu,\omega,e^{i\alpha}A,e^{i\alpha}B,e^{-i\alpha}\overline{A},e^{-i\alpha}\overline{B}) = e^{i\alpha}N_j(\mu,\omega,A,B,\overline{A},\overline{B}).$$

Secondly, reversibility of (20) has to be understood with the representation of the action of S on (A,B), which is by construction: $(A,B) \rightarrow (B,A)$. For (20) this leads to:

$$(23)\qquad N_0(\mu,\omega,B,A,\overline{B},\overline{A}) = -\ N_1(\mu,\omega,A,B,\overline{A},\overline{B}).$$

Now, we can also arrange the system (20) into "normal form". This means that we can choose suitable coordinates, such that our system, truncated at any fixed arbitrary order, looks the simplest possible. This normalization results directly from the structure of the linear operator in (20), which is diagonal. It is a classical result (see [6] for instance) that one can choose coordinates such that the truncated vector field (P_0,P_1) commutes with the action $(A,B) \rightarrow (e^{ik_cx}A,Be^{-ik_cx})$ for any $x \in \mathbb{R}$.

The normal form of the amplitude equation (20) is then as follows:

$$(24)\qquad \begin{cases} \dfrac{dA}{dx} = ik_cA + AP[\mu,\omega,|A|^2,|B|^2], \\ \dfrac{dB}{dx} = -ik_cB - BP[\mu,\omega,|B|^2,|A|^2], \end{cases}$$

where P is a polynomial in its two last arguments taking complex values, such that $P[0,\omega_0,0,0] = 0$, and where, by construction $ik_c + P(\mu,\omega,0,0) = \lambda(\mu,\omega)$ is given by formula (15). In fact, the full system (20) may be written under the form of a vector field of the form (24) completed by adding a vector field $(R_0,R_1) = O[(|A| + |B|)^N]$ which satisfies invariance properties (22), (23). This remark is fundamental for studying the precise structure of bounded solutions of (20).

Let us study all bounded solutions of the normal form (24). For this, let us introduce polar coordinates:

$$(25) \qquad A = r_0 e^{i(k_c x+\psi_0)}, \qquad B = r_1 e^{-i(k_c x+\psi_1)},$$

we then obtain a system in r_0 and r_1 uncoupled from the phases:

$$(26) \qquad \begin{cases} \dfrac{dr_0}{dx} = r_0 P_r(\mu,\omega,r_0^2,r_1^2) = r_0[\lambda_r(\mu,\omega) + b_r r_0^2 + c_r r_1^2 + \text{h.o.t.}], \\ \dfrac{dr_1}{dx} = -\, r_1 P_r(\mu,\omega,r_1^2,r_0^2) = -\, r_1[\lambda_r(\mu,\omega) + c_r r_0^2 + b_r r_1^2 + \text{h.o.t.}], \end{cases}$$

We are reduced to study the two-dimensional vector field in (r_0,r_1), which appears to be *integrable*. We can of course restrict the analysis to the quadrant $r_0 \geq 0$, $r_1 \geq 0$. In fact, if $b_r \neq 0$, and $c_r - b_r \neq 0$, we have an explicit integral for the system (26) truncated at the cubic order:

$$(27) \qquad H(r_0,r_1) = (r_0 r_1)^{2b_r/(c_r-b_r)}[\lambda_r/b_r + r_0^2 + r_1^2].$$

We have no explicit integral for the higher order system, but it is not hard to show that the trajectories are very similar to the ones obtained for the cubic vector field. In fact, equilibria, other than 0, are $(r_0,0)$, $(0,r_0)$ which exist for $\lambda_r b_r < 0$, and (r_1,r_1) which exists for $\lambda_r(b_r + c_r) < 0$, and where

$$(28) \qquad P_r(\mu,\omega,r_0^2,0) = 0, \qquad P_r(\mu,\omega,r_1^2,r_1^2) = 0.$$

The axis $r_0 = 0$ and $r_1 = 0$ are invariant manifolds, on which the dynamics is easy to determine. Equilibrium points $(r_0,0)$ and $(0,r_0)$ are saddles for $\lambda_r(b_r - c_r) > 0$, while they are nodes for $\lambda_r(b_r - c_r) < 0$, and equilibrium (r_1,r_1) is a saddle for $c_r^2 - b_r^2 < 0$, while it is a center if $c_r^2 - b_r^2 > 0$. Trajectories cut the diagonal orthogonaly, except at the equilibrium points 0 and (r_1,r_1). If $(r_0(x),r_1(x))$ is a solution of (26), then $(r_1(-x),r_0(-x))$ is also a solution (symmetric with respect to the diagonal), then any closed orbit is symmetric with respect to the diagonal. About the trajectories connecting equilibrium points, they are explicit on (27), and the proof of their existence for the higher order system (26) follows from perturbation arguments. We give at figure 4 the phase portraits in the plane (r_0,r_1) in the case $\lambda_r > 0$ (which is not a real restriction), depending on the values of the main non-linear coefficients (b_r and c_r) in P_r.

Equilibrium solution 0 corresponds to the basic solution $V^{(0)}$ of (1). Now, the solution of the form $(r_0,0)$ gives

$$(29) \qquad \psi_0 = \beta_0 x + \varphi_0, \text{ with } k_0 = k_c + \beta_0 = \lambda_i(\mu,\omega) + b_i r_0^2 + \text{h.o.t.},$$

and the corresponding flow $\tilde{\mathbb{V}}_0$ on the center manifold has the form:

$$(30) \qquad \tilde{\mathbb{V}}_0(\tau,x,y) = r_0 e^{i[k_0 x + \tau + \varphi_0]}\mathbb{V}_0(y) + \text{c.c.},$$

which leads to the following invariance properties: let us denote by σ_a the shift $x \rightarrow x + a$, then we have the following invariance property: $\mathbf{T}_{k\alpha}\sigma_{-\alpha}\tilde{\mathbb{V}}_0 = \tilde{\mathbb{V}}_0$, where $k = k_c + \beta_0$. This invariance propagates on the center manifold, due to the commutativity of $\mathbf{T}$ and σ with Φ (see theorem 3), hence the full solution also satisfies:

$$(31) \qquad \mathbf{T}_{k\alpha}\sigma_{-\alpha}\tilde{\mathbb{V}} = \tilde{\mathbb{V}}.$$

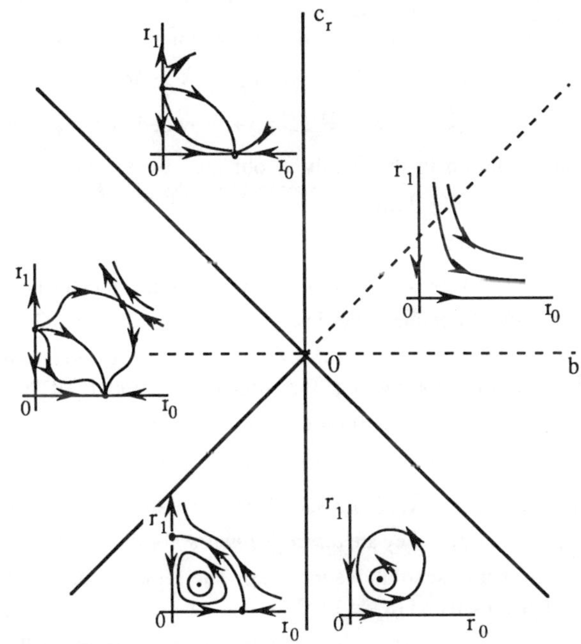

Figure 4: Phase portraits ($\lambda_r > 0$) depending on coefficients of P_r

This solution then corresponds to travelling waves, as well as the symmetric solution $(0,r_0)$ of (26) which travels in the opposite direction. In fact, we just showed, up to now, this result is true on the truncated system (at any order!), and it is possible to show that we have indeed *travelling waves for the full problem*, when taking account of the "flat terms" (R_0, R_1) which are not on normal form (see [11]). We recover, in fact one of the classical solutions obtained when assuming spatial periodicity, giving in such a case Hopf bifurcation in presence of O(2) symmetry. In the case of Couette-Taylor problem between counter-rotating cylinders, the *spiral structure of the travelling waves* results from the structure of $\mathfrak{V}_0(y)$, which is a vector field function of the radial coordinate r multiplied by $\exp(im\theta)$, leading to a velocity vector field only depending on (r, kx + ωt + mθ) (see Chossat & Iooss [2]).

The solution of the form (r_1,r_1) gives

$$(32)\qquad \psi_j = \beta_1 x + \varphi_j,\ j = 0, 1, \quad \text{with} \qquad k_1 = k_c + \beta_1 = \lambda_i(\mu,\omega) + (b_i + c_i)r_1^2 + \text{h.o.t.},$$

and the corresponding flow $\tilde{\mathfrak{V}}_0$ on the center manifold has the form:

$$(33)\qquad \tilde{\mathfrak{V}}_0(\tau,x,y) = r_1 e^{i[\tau + k_1 x + \varphi_0]}\mathfrak{V}_0(y) + r_1 e^{i[\tau - k_1 x - \varphi_1]}S\mathfrak{V}_0(y) + \text{c.c.}.$$

This principal part of the flow satisfies the following invariance properties:

$$(34)\qquad \sigma_{2\pi/k}\tilde{\mathfrak{V}}_0 = \tilde{\mathfrak{V}}_0, \qquad \mathbf{T}_\pi \sigma_{\pi/k}\tilde{\mathfrak{V}}_0 = \tilde{\mathfrak{V}}_0,$$

$$(35)\qquad S\tilde{\mathfrak{V}}_0(\tau,x,y) = \tilde{\mathfrak{V}}_0(\tau,-x,y) \ \text{provided that}\ \varphi_0 = -\varphi_1.$$

Moreover, any solution in the family is obtained by acting successively the linear operators $\mathbf{T}_{(\varphi_1-\varphi_0)/2}$ and $\sigma_{-(\varphi_1+\varphi_0)/2k}$ on the solution given for $\varphi_0 = \varphi_1 = 0$. These properties propagate on the center manifold, due to the commutativity of $\mathbf{T}$ and σ with Φ in (21), hence they also apply to the full solution $\tilde{\mathfrak{V}}$ of (9). It is then clear that we recover "the standing waves" obtained classically when assuming spatial periodicity (see for instance [2]). We show in [11] that *this family of standing waves indeed exists for the full system* (20), even in considering "flat" high order terms (R_0,R_1).

Remark: when $e_1 < 0$, we see on figure 3 that the case of figure 5 ($\lambda_r > 0$) corresponds to $\mu > \mu^c(\omega)$. Then, we observe that the situation $b_r < 0$, $(b_r - c_r) > 0$ which gives *saddle points* for the travelling waves, corresponds in fact to the situation when, in the classical spatially-periodic analysis, they are *attracting nodes* for the usual stability analysis in the class of spatially-periodic solutions (see [2]). The same remark holds for the standing waves when $(b_r + c_r) < 0$ and $(b_r - c_r) < 0$.

Closed orbits in the (r_0,r_1) plane correspond to periodic solutions $(r_0(x),r_1(x))$ with some period denoted by H. Let us show that this leads to spatially quasi-periodic solutions with two basic frequencies. In fact, we can first choose the origin in x such that $r_0(0) = r_1(0)$ (there are two of such points), then $r_0(x) = r_1(-x)$ by the uniqueness of the solution of (26) and we have, after integrating the phases:

$$(36)\qquad \begin{cases} \psi_0 = \beta x + h_0(x) + \varphi_0, \\ \psi_1 = \beta x - h_0(-x) + \varphi_1, \end{cases}$$

where h_0 is H-periodic in x with 0 mean value, and where β is the average of $P_i(\mu,\omega,r_0^2(x),r_0^2(-x))$.

Now the principal part $\tilde{\mathbf{V}}_0$ of the flow has two fundamental spatial periods: H and $2\pi/k$ where $k = k_c + \beta$, and this property clearly propagates on the full solution $\tilde{\mathbf{V}}$. For any "good" fixed value of μ and ω, there is a *4-parameters family* of such solutions: one parameter K corresponds to the level curves in the (r_0,r_1) plane, and the three other parameters correspond to independent shifts on the phases (φ_0,φ_1) and on the origin in x.

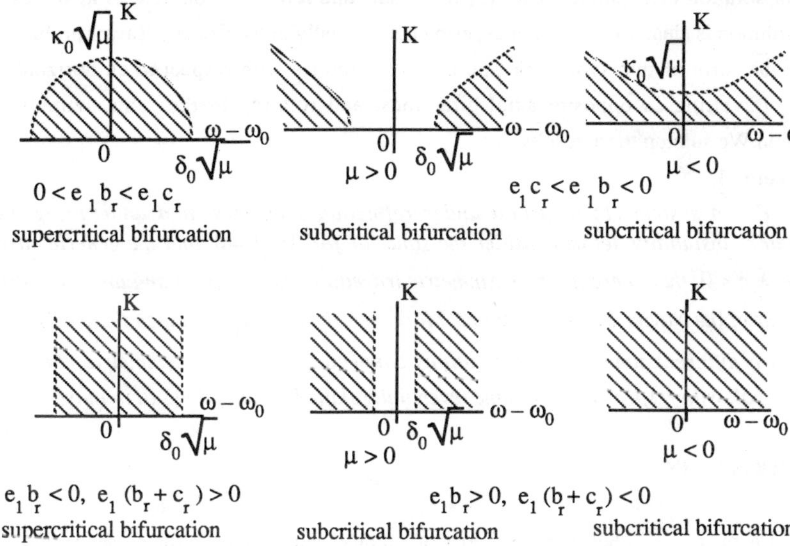

Figure 5 . Regions $\Delta(\mu)$ (hatched) in the parameter space (ω,K) where quasi-periodic solutions are found.

The free shift on the time origin is then included in the phase shifts. We show at figure 5 the regions $\Delta(\mu)$ of the parameter space (ω, K) where quasi-periodic solutions occur for the normal form (24). This shows that we have a large family of solutions both spatially quasi-periodic and time-periodic. We show in [11] that, *most of these solutions persist* when we study the complete vector field (including "flat terms"). The proof is very analogous to the one given in [10].

About the heteroclinic connections between saddles in figure 4, joining the two symmetric travelling waves, we show in [11] that at least two of such solutions persist for the complete system. The technique used rests upon the reduction of the system (20), to a 3-dimensional one, once we use the rotational invariance, and we study the part of the space where |A| ≥ |B|. Then a perturbation technique on the unstable manifold of the travelling waves, and a symmetry argument lead to the result. Other heteroclinic connections do not persist in general. In addition, we show that the separatrices joining the standing waves with the travelling waves and trajectories connecting nodes persist for the complete vector field. All these solutions physically look as juxtapositions of the two limiting regimes connected by a relatively small region of the space. In the case of an heteroclinic connection of the two symmetric travelling wave regimes, this is a "defect" type of solution (see Coullet & al [5] for details and references on defects in waves). This solution is clearly observed in experiments, specially in the Couette-Taylor problem, under the form of two symmetric spiral waves regimes with respect to an horizontal plane, travelling in opposite axial directions, and rotating in the same azimuthal direction.We sum up these results as

Theorem 4.

For a system (1) invariant under reflection symmetry, and undergoing an oscillatory instability, let us consider the situation for $[U_x] = 0$, *and the generic case where* $\lambda_r b_r < 0$, *then there are two symmetric travelling-waves, space and time-periodic solutions of the Navier-Stokes equations. Moreover, if* $\lambda_r(b_r - c_r) > 0$, *then there exists at least two different time-periodic solutions connecting these two symmetric regimes (defects) all with respectively the same (eventually shifted in time) flows at infinity.*

BIBLIOGRAPHY

[1] C.J.Amick, K.Kirchgässner. A theory of solitary water-waves in the presence of surface tension. Arch.Rat.Mech.Anal. 105,1-50, 1989.

[2] P.Chossat, G.Iooss. Primary and secondary bifurcations in the Couette-Taylor problem. Japan J.Appl.Maths 2,1,37-68, 1985.

[3] P.Collet, J.P.Eckmann. The existence of dendritic fronts. Com. Math. Phys. 107, 39-92, 1986.

[4] P.Collet, J.P.Eckmann. The stability of modulated fronts. Helvetica Phys. Acta 60, 969, 1987.

[5] P.Coullet, C.Elphick, L.Gil, J.Lega. Topological defects of wave patterns. Phys. Rev. Let. 59,8,884-887,1987.

[6] C.Elphick, E.Tirapegui, M.E.Brachet, P.Coullet, G.Iooss. A simple global charac--terization for normal forms of singular vector fields. Physica 29D, 95-127,1987.

[7] G.Iooss. Existence et stabilité de la solution périodique secondaire intervenant dans les problèmes d'évolution du type Navier–Stokes. Arch.Rat.Mech.Anal. 47,301-329, 1972.

[8] G.Iooss. Bifurcation and transition to turbulence in hydrodynamics. Bifurcation theory and Applications, CIME 2nd 1983 session, ed.L.Salvadori. Lecture Notes in Maths 1057, 152-201. Springer 1984.

[9] G.Iooss, A.Mielke, Y.Demay. Theory of steady Ginzburg-Landau equation, hydrodynamic stability problems. Europ.J.Mech.B/Fluids,8,3,229-268, 1989.

[10] G.Iooss , J.Los. Bifurcation of spatially quasi-periodic solutions in hydrodynamic stability problems. EEC report series. Maths. Pb. in nonlinear Mech. 26,1989.

[11] G.Iooss, A.Mielke. Bifurcating time-periodic solutions of Navier-Stokes equations in infinite cylinders. Preprint Univ. de Nice 1990.

[12] V.I.Iudovich. Onset of auto-oscillations in a fluid (in russian). Prikl.Mat.Mek. 35, 638-655, 1971.

[13] T.Kato. Perturbation theory for linear operators. Springer-Verlag, Berlin 1966.

[14] K.Kirchgässner. Wave solutions of reversible systems and applications. J.Diff. Equ. 45,113-127, 1982.

[15] K.Kirchgässner. Nonlinear resonant surface waves and homoclinic bifurcation. Adv. Appl. Mech. 26,135-181, 1988.

[16] J.L.Lions, E.Magenes. Problèmes aux limites non homogènes et Applications Vol 1, Dunod 1968.

[17] A.Mielke. A reduction principle for nonautonomous systems in infinite dimensional spaces. J.Diff.Equ. 65,68-88, 1986.

[18] A.Mielke. Steady flows of inviscid fluids under localized perturbations. J.Diff.Equ. 65,89-116, 1986.

[19] A.Mielke. Saint Venant's problem and semi-inverse solutions in nonlinear elasticity. Arch.Rat.Mech.Anal.102,205-229, 1988.

[20] M.Renardy. Bifurcation of singular solutions in reversible systems and applications to reaction - diffusion equations. Adv. in Maths. 3,384-406, 1982.

[21] D.H.Sattinger. Bifurcation of periodic solutions of the Navier-Stokes equations. Arch.Rat.Mech.Anal. 41, 66-80, 1971.

[22] A.Vanderbauwhede, G.Iooss. Center manifold theory in infinite dimensions. Preprint Univ. Nice Juin 1989.

A NOTE ON DIFFERENTIAL MATERIALS

Adriano Montanaro

Dipartimento di Metodi e Modelli Matematici per le Scienze Applicate

Universita` degli studi di Padova via Belzoni 7 35131 Padova

ABSTRACT We consider both classical and Green-Naghdi theories on materials of differential type of complexity one. I define the static internal dissipation and prove that it identically vanishes in the classical case, whereas in the case of Green-Naghdi it equals a term strictly related to the response function for the internal rate of entropy production. In the last case I also prove an analogue of the classical theorem asserting that the gradient of the free-energy function with respect to the deformation tensor yields the static part of the stress. The differences between the two enunciations in the two kinds of theories is directly related to the term which gives the static internal dissipation. In a Green-Naghdi theory in which this last term identically vanishes (i) the enunciations above coincide, and (ii) an indetermination in the response functions, pointed out by Green and Naghdi for general materials, disappears.

N.1 INTRODUCTION (*)

In [5] Green and Naghdi presented a new approach to thermodynamics. Their points of departure were the introduction of (i) *an entropy balance, replacing the Clausius-Duhem inequality*, and (ii) *a response functional* ξ *for the (specific) internal rate of entropy production* ξ, *in addition to those for the stress, heat flux, internal energy and entropy*. Green and Naghdi noticed that (iii) *there is an indetermination in the response functionals of any general material* - see N.4.

(*) This work has been performed within the activity of the Consiglio Nazionale delle ricerche, Group no. 3, in the academic year 1988/89.

In the present work I consider both *classical and Green-Naghdi theories regarding materials of differential type of complexity one* - see Defs.2.2 and 2.2'.(1)

The static internal dissipation δ^*, defined by (3.3), identically vanishes in the classical case, whereas in the case of Green-Naghdi it equals a term related to function $\hat{\xi}$ - see Theors.3.2 and 3.2'; this term does not vanish a priori, and the Euclidean invariance of the response functions does not imply its vanishing - see (5.ϵ), N.5.

The classical theorem asserting that the gradient of the free-energy function with respect to the deformation tensor gives the static part of the stress, has a different enunciation in the Green-Naghdi case; and the difference between them is due to δ^* - see $(3.4)_1$ and $(3.4')_1$. In a Green-Naghdi theory $\mathcal{T}$ in which $\delta^* \equiv 0$ holds,

(1.α) *the aforementioned theorems coincide* - see Theors. 3.2–2' and (4.4); and

(2.α) *the indetermination quoted in* (iii) *disappears in theory* $\mathcal{T}$ - see (4.α).

N. 2 DEFINITIONS AND A PHYSICAL POSSIBILITY AXIOM

Let $\mathcal{B}$ be a body and κ a reference configuration for it. Sometimes we identify a material point X with its position $\mathbf{X}$ $(=X_\kappa)$ in $\mathcal{B}_\kappa = \kappa(\mathcal{B})$, the closed regular region occupied by $\mathcal{B}$ in κ;(2) and $\mathbf{X}$ too is called material point. Spatial [material] Euclidean co-ordinates are denoted by $\{x^a\}$ $[\{X^A\}]$. Only smooth enough (thermokinematic κ-) processes,

$$\hat{p} = (\hat{x}, \hat{\theta}), \quad \hat{x}: \mathcal{B}_\kappa \times R \to R^3, \quad \hat{\theta}: \mathcal{B}_\kappa \times R \to R^3 \tag{2.1}$$

the conceivable and the physically possible ones, will be considered here.(3) The above functions $\hat{x}$ and $\hat{\theta}$ represent the motion and the temperature respectively. As is customary we set

$$\mathbf{F} = \mathrm{GRAD}\,\mathbf{x}, \quad \dot{\mathbf{F}} = \frac{\partial}{\partial t}\mathrm{GRAD}\,\mathbf{x} \quad (\mathbf{x} = \hat{x}(\mathbf{X}, t)) \text{ (deformation tensor and its rate)}, \tag{2.2}$$

$$\theta = \hat{\theta}(\mathbf{X}, t), \quad \mathbf{G} = \mathrm{GRAD}\,\theta \quad \text{(absolute temperature and its material gradient)}. \tag{2.3}$$

Let

$$\mathbf{b} = \hat{b}(\mathbf{X}, t), \qquad r = \hat{r}(\mathbf{X}, t) \tag{2.4}$$

---

(1) Roughly, classical theories on continuous media are those for which the Clausius-Duhem inequality is postulated, whereas the Green-Naghdi ones are those for which the above assertions (i) through (ii) hold.

(2) That is $\mathcal{B}_\kappa$ is the closure of an open and connected subset of the three-dimensional Euclidean space, whose boundary $\partial\mathcal{B}_\kappa$ is piecewise smooth.

(3) As in Bresan [1] we regard the concept of physical possibility as primitive. The physical possibility of a process p is not equivalent to its compatibility with balance laws. Rather, "p is possible" is meant here as "p could be carried out by ideal experimenters".

denote the densities per unit mass in the reference configuration of the body force and heat supply respectively. In Def.2.2 below bodies of differential type of complexity one are defined, in accordance with Day [4, p.81]; we call these *classical* to distinguish them from the *Green-Naghdi* ones – see Def.2.2'. By "differential-type body" we shall mean "body which is of differential type in anyone of the two aforementioned senses".

Def.2.1 [Def.2.1'] *We say that $\mathcal{B}$ is a (κ-regular) classical [Green-Naghdi] body of differential type of complexity one if there is a set of (κ-regular generalized) response functions* $\{\hat{P}_\kappa, \hat{q}_\kappa, \hat{e}_\kappa, \hat{\eta}_\kappa [, \hat{\xi}_\kappa]\}$ *connected with $\mathcal{B}$ and κ,*(4) *in the sense that* (α) *they are Galilean invariant and of class* C^1; (β) *if at time* t *the actual κ-process of $\mathcal{B}$ is* C^2 *and the fields* **b** *and* r *are* C^0, *then (necessarily) the balance laws* (2.6-8) *below, and the entropy inequality* (2.9) [*the entropy equality* (2.9')] *below, hold at any* $\mathbf{X} \in \mathcal{B}$,(5) *where*

$$\chi_\kappa = \chi_\kappa(\mathbf{X}, t) = \hat{\chi}_\kappa(\dot{\mathbf{F}}, \mathbf{F}, \theta, \mathbf{G}, \mathbf{X}) \qquad \text{- see (2.2-3) -} \qquad (2.5)$$

for $\hat{\chi}_\kappa \in \{\hat{P}_\kappa, \hat{q}_\kappa, \hat{e}_\kappa, \hat{\eta}_\kappa [, \hat{\xi}_\kappa]\}$, *and* **v** *is the velocity at* $(\mathbf{X}, t)$.

$$\rho_\kappa \dot{\mathbf{v}} = \rho_\kappa \mathbf{b} + \mathrm{DIV}\,\mathbf{P}_\kappa, \quad \mathbf{P}_\kappa \mathbf{F}^T = (\mathbf{P}_\kappa \mathbf{F}^T)^T, \quad \rho_\kappa \dot{e}_\kappa = \mathbf{P}_\kappa \cdot \dot{\mathbf{F}} + \rho_\kappa r - \mathrm{DIV}\,\mathbf{q}_\kappa, \qquad (2.6\text{-}8)$$

$$\rho_\kappa \dot{\eta}_\kappa \geq \rho_\kappa \theta^{-1} r - \mathrm{DIV}(\theta^{-1}\mathbf{q}_\kappa), \qquad (2.9)$$

$$\rho_\kappa \dot{\eta}_\kappa = \rho_\kappa(\xi_\kappa + \theta^{-1} r) - \mathrm{DIV}(\theta^{-1}\mathbf{q}_\kappa)] \,. \qquad [(2.9')]$$

Notice that, as no dissipation inequality is imposed by Def.2.1', all the theorems proved here hold in any Green-Naghdi theory whatever the dissipation inequality may be.

$\mathbf{P}_\kappa$, $\mathbf{q}_\kappa$, e_κ, η_κ and ξ_κ represent the *first Piola stress*, *heat flux vector*, *(specific) internal energy and entropy*, and *internal rate of entropy production* respectively. The laws (2.8) and (2.9) [(2.9')] imply the *reduced dissipation inequality*

$$\rho_\kappa(\dot{\psi}_\kappa + \dot{\theta}\eta_\kappa) - \mathbf{P}_\kappa \cdot \dot{\mathbf{F}} + \theta^{-1}\mathbf{q}_\kappa \cdot \mathbf{G} \leq 0 \qquad \text{(classical case);} \qquad (2.10)$$

$$\rho_\kappa(\dot{\psi}_\kappa + \dot{\theta}\eta_\kappa) - \mathbf{P}_\kappa \cdot \dot{\mathbf{F}} + \theta^{-1}\mathbf{q}_\kappa \cdot \mathbf{G} + \rho_\kappa \theta \xi_\kappa = 0 \quad \text{(Green-Naghdi case).} \qquad (2.10')$$

Usually a physical possibility axiom similar to Ax.2.1 below, is at least implicitly employed to prove the classical Theor.3.1 below.

Axiom 2.1 *Let $\mathcal{B}$ be a κ-regular body of differential type,*(6) *where κ is a*

---

(4) The notion of generalized stress-function has been introduced in Bressan&Montanaro [2] for the purely mechanical case. In Montanaro [6] generalized sets of response functions are introduced for the thermo-elastic case.

(5) For brevity I write the mentioned laws pointwise.

(6) Namely in anyone of the two types mentioned in Def.3.1.

reference configuration for it, let $\mathbf{X} \in (\mathcal{B}_K)^\circ$ *and* $t \in R$. *For any admissible* $(\dot{\mathbf{F}}, \mathbf{F}, \theta, \mathbf{G})$ *there is an open neighbourhood* $\mathcal{V}(\mathbf{X})$ *of* $\mathbf{O}_{Lin \times R \times R^3}$ *such that for each*

$(\ddot{\mathbf{F}}, \dot{\theta}, \dot{\mathbf{G}}) \in \mathcal{V}(\mathbf{X})$ *there is a physically possible process* $\hat{p} = (\hat{x}, \hat{\theta})$ *which is of class* C^3 *at* $(\mathbf{X}, t)$ *and for which*

$$\frac{\partial}{\partial t}\mathrm{GRAD}\,\hat{x}(\mathbf{X}, t) = \dot{\mathbf{F}}, \quad \mathrm{GRAD}\,\hat{x}(\mathbf{X}, t) = \mathbf{F},$$
$$\hat{\theta}(\mathbf{X}, t) = \theta, \quad \mathrm{GRAD}\,\hat{\theta}(\mathbf{X}, t) = \mathbf{G}, \tag{2.11}$$
$$\frac{\partial^2}{\partial t^2}\mathrm{GRAD}\,\hat{x}(\mathbf{X}, t) = \ddot{\mathbf{F}}, \quad \frac{\partial}{\partial t}\hat{\theta}(\mathbf{X}, t) = \dot{\theta}, \quad \frac{\partial}{\partial t}\mathrm{GRAD}\,\hat{\theta}(\mathbf{X}, t) = \dot{\mathbf{G}}.$$

N. 3 THEOREMS

Theorem 3.1 *Any set of response functions associated with a differential-type body satisfies equalities* $(3.1)_{1-3}$, *thus* $(3.1)_4$ *too.*

$$\frac{\partial \psi_K}{\partial \dot{\mathbf{F}}} \equiv 0 \equiv \frac{\partial \psi_K}{\partial \mathbf{G}}, \quad \eta_K = -\frac{\partial \psi_K}{\partial \theta}, \quad \chi_K = \hat{\chi}_K(\mathbf{F}, \theta, \mathbf{X}) \text{ for } \chi_K \in \{\psi_K, e_K, \eta_K\}. \tag{3.1}$$

For the proof in the Green-Naghdi case one has to proceed (by Ax.2.1) exactly as in the classical one simply by replacing (2.10) with (2.10') – see e.g. Day [4, p. 82].

Recall that $\delta \equiv \theta\dot{\eta}_K - r + \rho_K^{-1}\mathrm{DIV}\mathbf{q}_K$ is the *(specific) internal dissipation* – see Truesdell [8, p.112]. With the help of (2.8) we find

$$\delta \equiv \theta\dot{\eta}_K - (\dot{e}_K - \rho_K^{-1}w) \qquad (w = \mathbf{P}_K \cdot \dot{\mathbf{F}} \text{ stress power}). \tag{3.2}$$

The equilibrium functions $\overset{*}{\hat{\chi}}_K$ associated with $\hat{\chi}_K \in \{\hat{P}_K, \hat{q}_K, \hat{e}_K, \hat{\eta}_K[, \hat{\xi}_K]\}$ are defined by $\overset{*}{\chi}_K = \overset{*}{\hat{\chi}}_K(\mathbf{F}, \theta, \mathbf{X}) = \hat{\chi}_K(\mathbf{O}, \mathbf{F}, \theta, \mathbf{0}, \mathbf{X})$. Now I define the *static internal dissipation* by

$$\delta^* \equiv \theta\dot{\eta}_K - (\dot{e}_K - \rho_K^{-1}w^*) \qquad (w^* = \overset{*}{\mathbf{P}}_K \cdot \dot{\mathbf{F}}). \tag{3.3}$$

The following theorems emphasize the differences between classical and Green-Naghdi theories on differential materials.

Theorem 3.2 [3.2'] *Any set of response functions associated with a classical [Green-Naghdi] body of differential type satisfies equalities* (3.4) [(3.4')] .

$$\overset{*}{\mathbf{P}}_K = \rho_K \frac{\partial \psi_K}{\partial \mathbf{F}}, \qquad \delta^* \equiv 0\,; \tag{3.4}$$

$$\overset{*}{\mathbf{P}}_{\kappa} = \rho_\kappa \left[\frac{\partial \psi_\kappa}{\partial \mathbf{F}} + \theta \frac{\partial \xi_\kappa}{\partial \dot{\mathbf{F}}} \Big|_{\dot{\mathbf{F}}=\mathbf{O}=\mathbf{G}} \right], \quad \delta^* \equiv \left[\theta \frac{\partial \xi_\kappa}{\partial \dot{\mathbf{F}}} \Big|_{\dot{\mathbf{F}}=\mathbf{O}=\mathbf{G}} \right] \cdot \dot{\mathbf{F}},$$

$$\xi_\kappa \Big|_{\dot{\mathbf{F}}=\mathbf{O}=\mathbf{G}} \equiv 0. \tag{3.4'}$$

Proof. Equality $(3.4)_1$ is Coleman's theorem – see [3]. In the Green-Naghdi case, observe that equality (2.10'), and $(3.1)_{3\text{-}4}$ imply

$$[\rho_\kappa \frac{\partial \psi_\kappa}{\partial \mathbf{F}} (\mathbf{F}, \theta, \mathbf{X}) - \hat{P}_\kappa(\dot{\mathbf{F}}, \mathbf{F}, \theta, \mathbf{G}, \mathbf{X})] \cdot \dot{\mathbf{F}} + \rho_\kappa \theta \hat{\xi}_\kappa(\dot{\mathbf{F}}, \mathbf{F}, \theta, \mathbf{G}, \mathbf{X}) +$$

$$+ \theta^{-1} \hat{q}_\kappa(\dot{\mathbf{F}}, \mathbf{F}, \theta, \mathbf{G}, \mathbf{X}) \cdot \mathbf{G} = 0, \tag{3.5}$$

which for $\mathbf{G}=\mathbf{0}$ and $\dot{\mathbf{F}}=\mathbf{O}$ yields $(3.4')_3$ – see Ax.2.1. To obtain $(3.4')_1$ put in (3.5) $\mathbf{G}=\mathbf{O}$, replace $\dot{\mathbf{F}}$ with $\lambda\dot{\mathbf{F}}$ $(\lambda \neq 0)$, divide both sides of the resulting equality by λ, and take the limit as $\lambda \to 0$ – see Ax.2.1; the smoothness of $\hat{\xi}_\kappa$ and $(3.4')_3$ imply

$$[\rho_\kappa \frac{\partial \hat{\psi}_\kappa}{\partial \mathbf{F}} (\mathbf{F}, \theta, \mathbf{X}) - \hat{P}_\kappa(\mathbf{O}, \mathbf{F}, \theta, \mathbf{0}, \mathbf{X}) + \rho_\kappa \theta \frac{\partial \hat{\xi}_\kappa}{\partial \dot{\mathbf{F}}} (\mathbf{O}, \mathbf{F}, \theta, \mathbf{0}, \mathbf{X})] \cdot \dot{\mathbf{F}} = 0,$$

which by the arbitrariness of $\dot{\mathbf{F}}$ gives $(3.4')_1$. To prove $(3.4\text{-}4')_2$ observe that (3.3) and $(3.1)_4$ yield $\rho_\kappa \delta^* = -\rho_\kappa(\frac{\partial \psi_\kappa}{\partial \mathbf{F}} \cdot \dot{\mathbf{F}} + \frac{\partial \psi_\kappa}{\partial \theta} \dot{\theta} + \dot{\theta} \eta_\kappa) + \overset{*}{\mathbf{P}}_\kappa \cdot \dot{\mathbf{F}}$; hence by $(3.1)_3$

$\rho_\kappa \delta^* = (-\rho_\kappa \frac{\partial \psi_\kappa}{\partial \mathbf{F}} + \overset{*}{\mathbf{P}}_\kappa) \cdot \dot{\mathbf{F}}$, which by $(3.4)_1$ $[(3.4')_1]$ yields $(3.4)_2$ $[(3.4')_2]$.(7)

N.4 ON THE GREEN-NAGHDI OBSERVATION OF NON UNIQUENESS FOR THE RESPONSE FUNCTIONALS OF A GENERAL MATERIAL

In Green & Naghdi [5 pp. 258-9] it is noticed that

(4.α) *by replacing the set of response functionals* (4.1) *with the set* (4.2) *below*

$$\hat{\psi}, \hat{\eta}, \hat{\xi}, \hat{T}, \hat{q}, \hat{e}, \tag{4.1}$$

$$\tilde{\psi} = \hat{\psi} - \theta \hat{f}, \quad \tilde{\eta} = \hat{\eta} + \hat{f}, \quad \tilde{\xi} = \hat{\xi} + \dot{\hat{f}}, \quad \tilde{T} = \hat{T}, \quad \tilde{q} = \hat{q}, \quad \tilde{e} = \hat{e} \tag{4.2}$$

– *where* $\hat{f}$ *is an arbitrary functional of the response variables – the dissipation equality* (2.10') *remains unchanged; and if* (4.2) *is substituted into* (2.6-8) *and* (2.10'), *then the same differential equations for the variables* $(\mathbf{x}, \theta)$ *are obtained as those resulting from the set* (4.1). *Moreover "Also the boundary conditions imposed on* $\mathbf{x}$ *or* $\mathbf{T}$ *and*

---

(7) Remark that the same steps of this last proof might be used to show that the result holds for a thermo-elastic Green-Naghdi material too. For a classical one this result is well known.

on Θ or $\mathbf{q}$, *together with the appropriate initial conditions on* $\mathbf{x}$ *and* Θ, *are unaltered*."

This indetermination observation has been reported by Morro in [7] too. To remove this arbitrariness in ξ, η, ψ *Green & Naghdi propose to put*

$$f=\hat{f}(\mathbf{X}), \quad \dot{f}=0 \qquad \text{- see [5, p. 259].}^{(8)} \tag{4.3}$$

We think that assumption (4.3) is not well posed. More thoroughly, by (4.2) they construct a class Γ of physically equivalent sets (of response functional for a given body) ; in [5, p.258] these authors (in effect) suppose that a theorist chooses (or "specifies") an element ($\hat{\psi}$, $\hat{\eta}$, $\hat{\xi}$, $\hat{T}$, $\hat{q}$, $\hat{e}$) of Γ; and their remotion of the (unusual part of the above) indetermination simply consists in suggesting him to consider (or to use) only any other element ($\tilde{\psi}$, $\tilde{\eta}$, $\tilde{\xi}$, $\tilde{T}$, $\tilde{q}$, $\tilde{e}$) of Γ connected with ($\hat{\psi}$, $\hat{\eta}$, $\hat{\xi}$, $\hat{T}$, $\hat{q}$, $\hat{e}$) by means of (4.2) for some function $\hat{f}$ satisfying (4.3).

This suggestion really removes unusual indeterminations of the element of Γ used by any theorist who accepts the suggestion. However any two such theorists may use two arbitrary elements of Γ – related to one another by (4.2) for some $\hat{f}$ with $\dot{\hat{f}} \neq 0$.

Since the suggestion above is extra-physical, in accord with [7] I propose to replace it with some physical condition. More in detail, in order to reduce the indetermination of $\hat{f}$, *we regard both the sets* (4.1) *and* (4.2) – *deprived of the free-energy functions – as (generalized) sets of response functions - in the sense of* Def.2.2'; in this way *function* $\hat{f}$ [$\dot{\hat{f}}$] *is the difference between any two response functions* $\tilde{\eta}$ *and* $\hat{\eta}$ [$\tilde{\xi}$ *and* $\hat{\xi}$]; then I prove that

(4.α) $\hat{f}$ *reduces to a function of the material point, namely it satisfies* (4.3), *provided any response function* $\hat{\xi}$ *satisfies*

$$\left.\frac{\partial \xi_K}{\partial \dot{\mathbf{F}}}\right|_{\dot{\mathbf{F}}=\mathbf{O}=\mathbf{G}} \equiv 0 \,. \tag{4.4}$$

(8) Observe that Green&Naghdi in [5] do not prove that this indetermination is the largest possible. They only ascertain the existence of a (physically unobservable) class of sets of response functionals which are related to a given one through formulae (4.1-2). That such a class is or is not the largest possible must be proved within the theory of the particular material being considered. Here I show that, in the case of materials of differential type of complexity one, by assuming (4.4) this indetermination disappears; however other indeterminations can exist (regarding e.g. the stress or heat flux); in Montanaro [6] indeterminations of the response functions of thermo-elastic bodies are excluded on the basis of uniqueness theorems.

Proof. Put $\mathcal{V} = \{\dot{\mathbf{F}}, \mathbf{F}, \Theta, \mathbf{G}\}$, and assume

$$f = \hat{f}(\mathcal{W}, \mathbf{X}), \quad \mathcal{W} \subseteq \mathcal{V}. \tag{4.5}$$

Then $\dot{\hat{f}}$ is a function of $\mathcal{W}, \dot{\mathcal{W}}, \mathbf{X}$: $\dot{f} = \hat{g}(\mathcal{W}, \dot{\mathcal{W}}, \mathbf{X})$. As $\dot{\hat{f}} = \tilde{\xi} - \hat{\xi}$ – see (4.2) –, $\dot{\mathcal{V}} = \{\ddot{\mathbf{F}}, \dot{\mathbf{F}}, \dot{\Theta}, \dot{\mathbf{G}}\}$; thus $(\mathcal{W} \cup \dot{\mathcal{W}}) \subseteq \mathcal{V}$ *if and only if* $\mathcal{W} \subseteq \{\mathbf{F}\}$;(9) that is

$$f = \hat{f}(\mathbf{F}, \mathbf{X}). \tag{4.6}$$

Now $(4.2)_3$ and (4.6) yield

$$\frac{\partial \tilde{\xi}_K}{\partial \dot{\mathbf{F}}}\bigg|_{\dot{\mathbf{F}}=\mathbf{O}=\mathbf{G}} = \frac{\partial \hat{\xi}_K}{\partial \dot{\mathbf{F}}}\bigg|_{\dot{\mathbf{F}}=\mathbf{O}=\mathbf{G}} + \frac{\partial \dot{\hat{f}}}{\partial \dot{\mathbf{F}}}\bigg|_{\dot{\mathbf{F}}=\mathbf{O}=\mathbf{G}} = \frac{\partial \hat{\xi}_K}{\partial \dot{\mathbf{F}}}\bigg|_{\dot{\mathbf{F}}=\mathbf{O}=\mathbf{G}} + \frac{\partial f}{\partial \mathbf{F}} \tag{4.7}$$

Therefore (4.α) is proved. Lastly remark that, by (3.4-4'), (4.4) implies (1.α).

N. 5 JUSTIFICATION OF ASSUMPTION (4.4)

Now I show that *if* (4.4) *is not assumed the absurdum conclusion that elastic bodies dissipate can be deduced* – see (5.φ) below. First I prove some assertions.

(5.α) *There are (smooth) functions* $f = \hat{f}(\mathbf{F}, \mathbf{X})$ *which are Euclidean invariant together with* $\dot{f} = \dot{\hat{f}}(\dot{\mathbf{F}}, \mathbf{F}, \mathbf{X})$ $\left(= \frac{\partial \hat{f}}{\partial \mathbf{F}}(\mathbf{F}, \mathbf{X}) \cdot \dot{\mathbf{F}}\right)$ *and that satisfy* $\frac{\partial \dot{\hat{f}}}{\partial \dot{\mathbf{F}}} \left(= \frac{\partial \hat{f}}{\partial \mathbf{F}}\right) \not\equiv 0$.

In fact choose $\hat{\alpha} \in C^1(\mathcal{B}_K, R)$, $\hat{h} \in C^1(R, R)$ with $\hat{h}' \not\equiv 0$, and set $f = \hat{f}(\mathbf{F}, \mathbf{X}) = \hat{h}(\alpha J)$ $[= \hat{h}(\alpha K)]$, with $J = \det \mathbf{F}$ $[K = \mathbf{F} \cdot \mathbf{F}]$ $(\alpha = \hat{\alpha}(\mathbf{X}))$; one can easily verify that $\hat{f}$ satisfies all conditions in (5.α).

Now let (4.1) be a set of response functions; by (5.α) one can choose an Euclidean invariant smooth function $f = \hat{f}(\mathbf{F}, \mathbf{X})$ such that

$$\frac{\partial \hat{f}}{\partial \mathbf{F}} \not\equiv 0. \tag{5.1}$$

Consider the set of response functions (4.2) associated with (4.1) through $\hat{f}$; (5.1) and (4.7) imply that *or* $\frac{\partial \tilde{\xi}_K}{\partial \dot{\mathbf{F}}}\bigg|_{\dot{\mathbf{F}}=\mathbf{O}=\mathbf{G}} \not\equiv 0$ *or* $\frac{\partial \hat{\xi}_K}{\partial \dot{\mathbf{F}}}\bigg|_{\dot{\mathbf{F}}=\mathbf{O}=\mathbf{G}} \not\equiv 0$. Thus

---

(9) Incidentally remark that in the elastic case we have $\mathcal{V} = \{\mathbf{F}, \Theta, \mathbf{G}\}$, $\dot{\mathcal{V}} = \{\dot{\mathbf{F}}, \mathbf{F}, \Theta, \mathbf{G}\}$, and $(\mathcal{W} \cup \dot{\mathcal{W}}) \subseteq \mathcal{V}$ if and only if $\mathcal{W} = \emptyset$; thus (4.3) holds, agreeing with an uniqueness theorem on entropy that one would expect - see Montanaro [6, Theor.6.1].

(5.β) *either the set* (4.2) *or the set* (4.1) *defines a set of response functions which does not fulfill* (4.4).

Following e.g. Day [4] and Green&Naghdi [5], we assume that

(5.γ) *to perform an experiment corresponds to posing an initial and boundary-value problem for the (unknown) deformation and temperature fields; the outcome of an experiment is given by the deformation and temperature fields which are the solution of the connected initial and boundary-value problem* ;

(5.δ) *no experiment can distinguish two materials defined through sets of response functionals which differ only by the entropy functionals, in the sense that the differential equations, the boundary and initial conditions, and thus the solution fields too, are not influenced by any entropy indetermination.*

Thus by (4.α) and (5.α–δ) we deduce that

(5.ϵ) *either the set* (4.2) *or the set* (4.1) *defines an ideal material which does not fulfill* (4.4) (*hence by* (5.α) *Euclidean invariance does not imply* (4.4)).

Roughly, the axiom below asserts that any Green-Naghdi differential-type body statically behaves as an elastic body.

Axiom 5.1 *Let* $\mathcal{B}_d$ *be a Green-Naghdi body of differential type. There are an elastic body* $\mathcal{B}_e$ – *said to be elastic body associated with* $\mathcal{B}_d$ –, *and reference configurations* κ_d *and* κ_e *for* $\mathcal{B}_d$ *and* $\mathcal{B}_e$ *respectively, with* $\kappa_d(\mathcal{B}_d)=\kappa_e(\mathcal{B}_e)$, *such that the condition below holds.*

If both $\{\hat{P}_d, \hat{q}_d, \hat{e}_d, \hat{\eta}_d, \hat{\xi}_d\}$ *and* $\{\hat{P}_e, \hat{q}_e, \hat{e}_e, \hat{\eta}_e, \hat{\xi}_e\}$ *denote sets of response functions for* $(\mathcal{B}_d, \kappa_d)$ *and* $(\mathcal{B}_e, \kappa_e)$ *respectively, then*

$$\overset{*}{\hat{P}_d} = \hat{P}_e|_{\mathbf{G}=\mathbf{O}}, \; (\overset{*}{\hat{q}_d} = \hat{q}_e|_{\mathbf{G}=\mathbf{O}} \equiv \mathbf{O}), \quad \hat{e}_d = \hat{e}_e, \quad \hat{\eta}_d = \hat{e}_e, \; (\overset{*}{\hat{\xi}_d} = \hat{\xi}_e|_{\mathbf{G}=\mathbf{O}} (\equiv 0)).$$

By $(3.4')_2$, (3.3), the axiom above, and as $\chi = \hat{\chi}(\mathbf{F}, \theta, \mathbf{X})$ for $\hat{\chi} \in \{\overset{*}{\hat{P}_z}, \hat{e}_z, \hat{\eta}_z\}$, $z \in \{e, d\}$ – see $(3.1)_4$ –, we deduce the equalities

$$\rho\theta \left[\frac{\partial \xi_d}{\partial \dot{\mathbf{F}}} \Big|_{\dot{\mathbf{F}}=\mathbf{O}=\mathbf{G}} \right] \cdot \dot{\mathbf{F}} = \rho \overset{*}{\delta_d} =$$

$$= \rho\theta\left(\frac{\partial \eta_d}{\partial \mathbf{F}} \cdot \dot{\mathbf{F}} + \frac{\partial \eta_d}{\partial \theta}\dot{\theta}\right) - \rho\left(\frac{\partial e_d}{\partial \mathbf{F}} \cdot \dot{\mathbf{F}} + \frac{\partial e_d}{\partial \theta}\dot{\theta} - \overset{*}{\mathbf{P}_d} \cdot \dot{\mathbf{F}}\right) = \tag{5.2}$$

$$= \rho\theta(\frac{\partial\eta_e}{\partial \mathbf{F}}\cdot\dot{\mathbf{F}} + \frac{\partial\eta_e}{\partial\theta}\dot{\theta}) - \rho(\frac{\partial e_e}{\partial \mathbf{F}}\cdot\dot{\mathbf{F}} + \frac{\partial e_d}{\partial\theta}\dot{\theta} - \overset{*}{\mathbf{P}}_e\cdot\dot{\mathbf{F}}) = \rho\delta_e\big|_{\mathbf{G=O}}$$

Now, by (5.β or ϵ) one can choose the function $\hat{\xi}$ such that (4.4) does not hold for it. Thus (3.4')$_2$ and (5.2) imply $0 \neq \delta_d{}^* \equiv \delta_e\big|_{\mathbf{G=O}}$; that is:

(5.φ) *given a Green-Naghdi differential-type body $\mathcal{B}_d$ and a material point X for it, if (4.4) is not postulated, then any elastic body $\mathcal{B}_e$ associated with $\mathcal{B}_d$ dissipates at X (along processes in which* **G=O** *at X).*

This is an absurdum because it contradicts the fact that for an elastic body the internal dissipation identically vanishes along every process – see ftn.7.

REFERENCES

[1] BRESSAN A., *Metodo di assiomatizzazione in senso stretto della Meccanica classica*. Rend. Sem. Mat. Univ. Padova, vol. 32, pp. 55-212, 1962.

[2] BRESSAN A. & MONTANARO A., *On the uniqueness of response stress-functionals for purely mechanical continuous media, from the Mach-Painleve' point of view. Part 1. Hyper-elastic case. Part 2. Case of materials with memory.* Being printed in Atti Accademia Lincei.

[3] COLEMAN B.D. & MIZEL V.J., *Existence of Caloric Equations of state in Thermodynamics.* J. Chem. Phys., **40**, pp. 1116-1125, 1964.

[4] DAY W.A., *The Thermodynamics of Simple Materials with Fading Memory.* Springer-Verlag, Berlin-Heidelberg-New York, 1972.

[5] GREEN A.E. & NAGHDI P.M., *On thermodynamics and the nature of the second law.* Proc. R. Soc. Lond. A., 357, pp. 253-270, 1977.

[6] MONTANARO A., *On the response functions of a thermo-elastic body from the Mach-Painleve' point of view.* Being printed in Atti Accademia Lincei.

[7] MORRO A. & RUGGERI T., *Propagazione del calore ed equazioni costitutive.* Quaderni del C.N.R. G.N.F.M., Bologna, 1984.

[8] TRUESDELL C., *Rational Thermodynamics.* Springer-Verlag, Berlin-Heidelberg-New York, 1984.

ON THE STABILITY OF PLANE PARALLEL CONVECTIVE FLOW

G. MULONE

Dipartimento di Matematica

Viale A.Doria 6

95125 Catania - Italy

ABSTRACT

This paper concerns the stability of parallel convective shear flows in the stress-free boundary case. The Lyapunov direct method has been applied to show that plane Couette and plane Poiseuille flows, with linear temperature profile, are linearly and nonlinearly (conditional) stable whenever the Rayleigh number $\mathcal{R}$ is less than 657.511, *independent* of the Reynolds numbers R.

1. INTRODUCTION

As is well known the problem of stability of parallel convective shear flows is an important extension of the Rayleigh - Bénard problem and has both theoretical and practical significance. In a very simple manner it permits the study of the combined action of shear flows and thermal instabilities. This has practical importance for the understanding of convective processes including those in earth's atmosphere, oceans and seas.

Here I use the Lyapunov direct method to study the nonlinear stability of the basic motion (see [1] in the non-convective case). I consider the stress-free boundary case and I show that the basic motion is linearly and nonlinearly (conditional) stable with respect to the Lyapunov function V(t) (see § 3) whenever the Rayleigh number $\mathcal{R}$ is less than 657.511, *independent* of the Reynolds numbers R.

2. PERTURBATION EQUATION AND STABILITY RESULTS IN THE RIGID CASE.

The non-dimensional form of the equations of a perturbation ($\mathbf{u}$, ϑ, p) to the basic motion $m_o = (\mathbf{U},T,p_1)$ are[2] :

$$\begin{cases} \mathbf{u}_t+RU(z)\mathbf{u}_x+RwU'(z)\mathbf{i}+\mathbf{u}\cdot\nabla\mathbf{u}=-\nabla p+\mathcal{R}\vartheta\mathbf{k}+\Delta\mathbf{u} \\ \nabla\cdot\mathbf{u}=0 \qquad\qquad \text{in } \Omega\times(0,\infty) \\ P_r(\vartheta_t+RU(z)\vartheta_x+\mathbf{u}\cdot\nabla\vartheta) = \mathcal{R}w+\Delta\vartheta \end{cases} \tag{2.1}$$

where $\Omega=\mathbb{R}^2\times(-1/2,1/2)$, with initial condition

$$\mathbf{u}(\mathbf{x},0)=\mathbf{u}_o(\mathbf{x}) \ , \ \vartheta(\mathbf{x},0)=\vartheta_o(\mathbf{x}) \qquad \mathbf{x}\in \mathbb{R}^2\times(-1/2,1/2) \tag{2.2}$$

and boundary conditions

$$\mathbf{u}(x,y,-1/2,t)=\mathbf{u}(x,y,1/2,t)=0, \ \vartheta(x,y,-1/2,t)=\vartheta(x,y,1/2,t)=0, \ t\geq 0 \ , \tag{2.3}$$

in the rigid-rigid case, and

$$w(x,y,\pm 1/2,t) = u_z(x,y,\pm 1/2,t)= v_z(x,y,\pm 1/2,t)= \vartheta(x,y,\pm 1/2,t)=0, \ t\geq 0 \tag{2.4}$$

in the stress-free boundary case.

In (2.1) $U(z)=2z$ for *Couette* flow and $U(z) = 1-4z^2$ for *Poiseuille* flow, $\mathbf{u}$, ϑ,p are the velocity, temperature and pressure perturbation fields, $\mathbf{i}$ and $\mathbf{k}$ are the unit vectors of the x - axis and z - axis, and P_r, $\mathcal{R}^2$, R are the Prandtl, Rayleigh and Reynolds numbers, respectively.

We assume that the perturbations are periodic functions of x and y of periods $\frac{2\pi}{a_x}$, $\frac{2\pi}{a_y}$, respectively, and denote by Ω_1 the periodicity cell

$$\Omega_1= [0,\frac{2\pi}{a_x}]\times[0,\frac{2\pi}{a_y}]\times[-1/2,1/2].$$

As concerns the stability results of linear theory in the rigid case see [3-8]. In particular, Joseph [2),6)] studied the nonlinear stability of the rigid-rigid Couette convective case. He used the "classical energy" norm

$$E(t) = \frac{\|\mathbf{u}\|^2+P_r\|\vartheta\|^2}{2} \quad , \tag{2.5}$$

with $\|\cdot\|$ the $L_2(\Omega_1)$ - norm. Defining $\mathbb{R}_T = (R^2+\mathcal{R}^2)^{1/2}$ he showed that the basic motion is globally and monotonically stable if

$$\mathbb{R}_T^2 < 1708 \ . \tag{2.6}$$

He also showed that "disturbances of plane Couette flow heated from below which are x *independent* decay under the inequality

$$E_1(t) \leq E_1(0) \exp \{-2\pi^2(1-\frac{\mathcal{R}}{\sqrt{1708}})t \} , \tag{2.7}$$

whenever $\mathcal{R} <\sqrt{1708}$, independent of the value of R, with

$$E_1(t) = \frac{1}{2}\int_{\Omega_1} (v^2+ w^2+P_r\vartheta^2)d\Omega_1 . \tag{2.8}$$

Moreover the criterion (2.7) is necessary for instability as well as sufficient for stability to x-independent disturbances".

3. THE STRESS-FREE BOUNDARY CASE

Applying the classical energy method of § 2 to the stress-free boundary case, we obtain globally and monotonically stability conditions *which depend* on the Reynolds number.

In order to obtain stability conditions *independent* of R, we study linear and nonlinear (conditional) stability with the Lyapunov direct method.

In order to choose properly a Lyapunov function we use the idea given in [9-12]. The Lyapunov function V is the sum of two terms $V_o(t)$ and $V_1(t)$. V_o is a Lyapunov function for the linear stability problem and depends on the essential variables, [11-12], while V_1 must dominate the nonlinear terms.

The evolution equations of w and ζ are

$$\zeta_t+RU(z)\zeta_x-RU'(z)w_y+\mathbf{k}\cdot\nabla\times(\mathbf{u}\cdot\nabla\mathbf{u}) = \Delta\zeta \quad , \qquad (3.1)_1$$

$$\Delta w_t+RU(z)\Delta w_x-RU''(z)w_x-\mathbf{k}\cdot\nabla\times(\nabla\times(\mathbf{u}\cdot\nabla\mathbf{u}))=\Delta\Delta w+\mathcal{R}\Delta_1\vartheta \ . \qquad (3.1)_2$$

To (3.1) and $(2.1)_3$ we must add the further boundary conditions

$$w = w_{zz}=w_{zzzz} = \vartheta = \vartheta_{zz} = \vartheta_{zzzz} = \zeta_z = 0 \quad \text{on } z= 1/2,\ z=-1/2,\ \forall\ t\geq 0.$$

First we study the linear stability. For this, we consider the Lyapunov function

$$V_o(t) = \frac{1}{2}\left[\ \|\zeta\|^2 + \beta\ \|\Delta w\|^2 + \gamma P_r\|\vartheta\|^2\right] , \qquad (3.2)$$

where β and γ are positive constants. Now we write the equation of $V_o(t)$

$$\dot{V}_o-R\int_{\Omega_1} U'(z)w_y\zeta d\Omega_1+\mathcal{R}\int_{\Omega_1}[\beta\Delta_1\vartheta\Delta w+\gamma\vartheta w]d\Omega_1- \|\nabla\zeta\|^2- \beta\|\nabla\Delta w\|^2- \gamma\|\nabla\vartheta\|^2. \qquad (3.3)$$

By using the Schwarz inequality, from (3.3) we obtain:

$$\dot{V}_o \leq Rm\|w_y\|\ \|\zeta\| -\|\nabla\zeta\|^2+\mathcal{R}\beta[(\Delta_1\vartheta,\Delta w)+ \sigma(\vartheta,w)]- \beta[\|\nabla\Delta w\|^2 + \sigma\|\nabla\vartheta\|^2], \qquad (3.4)$$

with $m = \max\limits_{[-1/2,1/2]} |U'(z)|$, $\sigma = \dfrac{\gamma}{\beta}$. From (3.4) it follows

$$\dot{V}_o \leq Rm\|w_y\|\ \|\zeta\| -\|\nabla\zeta\|^2-\mathcal{R}\beta(\|\nabla\Delta w\|^2 + \sigma\ \|\nabla\vartheta\|^2)\left[\frac{1}{\mathcal{R}} - \frac{\mathfrak{I}}{\mathfrak{D}}\right], \qquad (3.5)$$

$$\text{with} \qquad \mathfrak{I} = (\Delta_1\vartheta,\Delta w)+ \sigma(\vartheta,w),\ \mathfrak{D} = \|\nabla\Delta w\|^2 + \sigma\ \|\nabla\vartheta\|^2. \qquad (3.6)$$

(3.5) and (3.6) imply

$$\dot{V}_o\leq Rm\|w_y\|\ \|\zeta\| -\|\nabla\zeta\|^2-\mathcal{R}\beta\mathfrak{D}\left[\frac{1}{\mathcal{R}} - \frac{1}{\mathcal{R}_c}\right] , \qquad (3.7)$$

where

$$\frac{1}{\mathcal{R}_c} = \max_{\mathcal{G}} \frac{(\Delta_1\vartheta, \Delta w)+ \sigma(\vartheta, w)}{\|\nabla\Delta w\|^2 + \sigma\ \|\nabla\vartheta\|^2} \quad , \qquad (3.8)$$

with

$$\mathcal{G} = \{ \, w,\vartheta \text{ regular fields, periodic in x and y, } 0<\|\nabla\Delta w\|^2 + \sigma\|\nabla\vartheta\|^2 < \infty,$$
$$w = w_{zz} = w_{zzzz} = \vartheta = \vartheta_{zz} = \vartheta_{zzzz} = 0 \quad \text{on } z=1/2,\ z=-1/2\}. \tag{3.9}$$

By solving the Euler-Lagrange equations associated to (3.8)-(3.9), we have:

$$\mathcal{R}_c^2 = \frac{27}{4}\pi^4 \cong 657.511\ . \tag{3.10}$$

Now we assume

$$\mathcal{R}^2 < \mathcal{R}_c^2\ , \tag{3.11}$$

then, from (3.7), using the Poincaré inequality, we obtain

$$\dot{V}_o \le Rm\|\nabla w\|\ \|\zeta\| - \|\nabla\zeta\|^2 - (1 - \frac{\mathcal{R}}{\mathcal{R}_c})[\beta\|\nabla\Delta w\|^2 + \gamma\|\nabla\vartheta\|^2] \le \frac{mR}{\pi^3}\|\nabla\Delta w\|\ \|\nabla\zeta\|$$

$$-(1-\frac{\mathcal{R}}{\mathcal{R}_c})[\beta\|\nabla\Delta w\|^2 + \gamma\|\nabla\vartheta\|^2] \le \left[\frac{m^2R^2}{2\varepsilon\pi^6} - \beta(1-\frac{\mathcal{R}}{\mathcal{R}_c})\right]\|\nabla\Delta w\|^2 + (\frac{\varepsilon}{2} - 1)\|\nabla\zeta\|^2 - \gamma(1-\frac{\mathcal{R}}{\mathcal{R}_c})\|\nabla\vartheta\|^2.$$

Choosing

$$\varepsilon = 1 \quad , \quad \beta = \frac{R^2m^2}{\pi^6(1 - \frac{\mathcal{R}}{\mathcal{R}_c})}, \quad \text{we obtain}$$

$$\dot{V}_o \le -\frac{\beta}{2}(1 - \frac{\mathcal{R}}{\mathcal{R}_c})\|\nabla\Delta w\|^2 - \frac{1}{2}\|\nabla\zeta\|^2 - \gamma(1 - \frac{\mathcal{R}}{\mathcal{R}_c})\|\nabla\vartheta\|^2$$

$$\le -\pi^2(1 - \frac{\mathcal{R}}{\mathcal{R}_c})\{\frac{\beta}{2}\|\Delta w\|^2 - \frac{1}{2}\|\zeta\|^2 - \frac{\gamma}{2}\|\vartheta\|^2\} \le -\frac{\pi^2}{c_o}(1 - \frac{\mathcal{R}}{\mathcal{R}_c})\, V_o(t), \tag{3.12}$$

with $c_o = \max\,(1,\ P_r)\ .$ (3.13)

Integrating this last inequality, it follows

$$V_o(t) \le V_o(0)\,\exp\{-\frac{\pi^2}{c_o}(1 - \frac{\mathcal{R}}{\mathcal{R}_c})\,t\}\ . \tag{3.14}$$

So we have proved the

Theorem 3.1 *If* $\mathcal{R}^2<\mathcal{R}_c^2$, *with* $\mathcal{R}_c^2$ *given by (3.10), then the basic motion* m_o *is linearly stable in the norm* $V_o(t)$ *independent of the Reynolds numbers.*

We now study the nonlinear stability. We introduce the Lyapunov function

$$V(t) = V_o(t) + bV_1(t)\quad , \tag{3.15}$$

(b is a positive constant), with

$$V_1(t) = \frac{1}{2}\,(\|\nabla\mathbf{u}\|^2 + \|\nabla(\nabla\times\mathbf{u})\|^2 \tag{3.16}$$

and write the evolution equation of V(t):

$$\dot{V}(t) = I_o - D_o + B_o + N_o + bI_1 - bD_1 + bN_1 + bB_1\quad , \tag{3.17}$$

with $\qquad I_o = R(U'(z)w_y,\zeta)+\mathcal{R}(\beta(\Delta_1\vartheta,\Delta w)+ \gamma(\vartheta,w))$,

$$D_o=\|\nabla\zeta\|^2+\beta[\|\nabla\Delta w\|^2+\gamma\|\nabla\vartheta\|^2], \qquad N_o=\int_{\Omega_1}[-\mathbf{k}\cdot\nabla\times(\mathbf{u}\cdot\nabla\mathbf{u})\zeta+\beta\mathbf{k}\cdot\nabla\times\nabla\times(\mathbf{u}\cdot\nabla\mathbf{u})\Delta w]d\Omega_1,$$

$$I_1= -\int_{\Omega_1}\Big[R\{U'(z)[\mathbf{u}_z\cdot\mathbf{u}_x+\nabla u\cdot\nabla w+w_{xyz}(w_y-v_z)+v_{zz}w_{xy}+u_{zz}w_{xx}+ w_{xxz}(w_x-u_z) - \zeta_x\zeta_z$$

$$- w_{zzz}u_z-v_{zz}\zeta_z-\mathbf{u}_x\cdot\nabla(\Delta w)+\Delta\zeta v_x-\Delta(\nabla\times\mathbf{u}\cdot\mathbf{j})w_z+(w_z\mathbf{j}-w_y\mathbf{k})\cdot\Delta(\nabla\times\mathbf{u})]+ U''(z) [u_z w$$

$$+ \Delta w w_x + w\mathbf{j}\cdot\Delta(\nabla\times\mathbf{u})]\}+ \mathcal{R}[\vartheta\Delta w+ (\vartheta_x \mathbf{j}-\vartheta_y\mathbf{i})\Delta(\nabla\times\mathbf{u})]\Big]d\Omega_1,$$

$$D_1=\|\Delta\mathbf{u}\|^2+\|\Delta(\nabla\times\mathbf{u})\|^2] , \quad N_1=\int_{\Omega_1}\{\mathbf{u}\cdot\nabla\mathbf{u}\cdot\Delta\mathbf{u}+[\nabla\times\mathbf{u}\cdot\nabla\mathbf{u} - \mathbf{u}\cdot\nabla(\nabla\times\mathbf{u})]\cdot\Delta(\nabla\times\mathbf{u}) \} d\Omega_1,$$

finally B_o and B_1 are the boundary integrals that vanish because of the boundary conditions and the periodicity. From (3.3) and (3.17) we have

$$\dot{V}(t) \le - \frac{1-M}{2} D_o + N_o + bI_1 - bD_1 + bN_1, \tag{3.18}$$

with

$$M = \frac{\mathcal{R}}{\mathcal{R}_c} . \tag{3.19}$$

Now we have the following estimate of the term I_1:

$$bI_1 \le D_2 , \tag{3.20}$$

where

$$D_2 = \frac{1-M}{4} D_o + \frac{b}{2} D_1 . \tag{3.21}$$

In a similar way it is possible to estimate the non-linear terms:

$$N_o + bN_1 \le \frac{\sqrt{2}\,C}{b}\left[\frac{2\sqrt{2}}{\sqrt{1-M}}(1+2\sqrt{\beta}) + \frac{5}{\sqrt{b}}\right]D_2V^{1/2} , \tag{3.22}$$

where C is a computable constant, (cf.[10]) and β has been given above.

Defining
$$A = \frac{\sqrt{2}\,C}{b}\left[\frac{2\sqrt{2}}{\sqrt{1-M}}(1+2\sqrt{\beta}) + \frac{5}{\sqrt{b}}\right] , \tag{3.23}$$

then we have

$$\dot{V} \le - \frac{1-M}{2} D_o - bD_1 + \frac{1-M}{4} D_o + \frac{b}{2} D_1 + AD_2V^{1/2} \le - D_2 (1 -AV^{1/2}) . \tag{3.24}$$

This last inequality implies the following nonlinear stability theorem.

THEOREM 3.2. *Let* $V(0) < A^{-2}$, *with* A *given by (3.23),* $\mathcal{R}^2 < \mathcal{R}_c^2$, *then the non-*

linear asymptotical stability holds

$$V(t) \le V(0) \exp\{-\pi^2 k^*[1 - AV(0)^{1/2} t]\}, \tag{3.25}$$

independent of the Reynolds number; $k^* = \max\left(1, \frac{2}{c_o(1-M)}\right)$ *and* c_o *is given by (3.13).*

The proof follows from (3.24) by a recursive argument (cf.[12]). Moreover, by classical imbedding theorems, we have the following corollary.

COROLLARY. *In the hypothesis of theorem 3.2, we have the following pointwise (exponential) decay of the perturbation to the velocity field* **u**:

$$\sup_{\mathbb{R}^2\times[-1/2,1/2]} |\mathbf{u}(\mathbf{x},t)| \to 0 \quad \text{as } t \to \infty .$$

Acknowledgement

This research has been partially supported by the italian Ministry for Education (M.P.I.) under 40% and 60% contracts, and by C.N.R.-S.P.A.I.M.

REFERENCES

1. Rionero,S. and Mulone,G., On the nonlinear stability of parallel shear flows, (*to appear*).
2. Joseph,D.D., *Stability of fluid motions* (2 vols.) Springer Tracts in Natural Philosophy, vols. 27 and 28., Berlin, Springer-Verlag, (1976).
3. Jeffreys,H., *Proc. Roy. Soc. A* 118, 33–49 (1928).
4. Brunt,D., Compendium of Meterology, *American Meterological Soc.*, p.1255, Boston (1957).
5. Gallacher,A.P. and Mercer,A. McD., *Proc. Roy. Soc. A* 286, 117–128 (1965).
6. Joseph,D.D., *Arc. Rational Mech. Anal.* 22, 163–184 (1966).
7. Orszag,S.A., *J. Fluid Mech.* 50, 689–703 (1971).
8. Gage,K.S., and Reid,W.H., *J. Fluid Mech.* 33, 21–32 (1968).
9. Galdi,G.P., *Arch. Rational Mech. Anal.* 87, 167–186 (1985).
10. Galdi,G.P. and Straughan,B., *Proc.R. Soc. Lond. A* 402, 257–283 (1985).
11. Rionero,S. and Mulone,G., *Arch. Rational Mech. Anal.* 103, 347–368 (1988).
12. Mulone,G. and Rionero,S., *J.Mat.Anal.App.* 144, 109–127 (1989).

ON THE STABILITY OF A BIAXIALLY-LOADED RUBBER SHEET

by
Ingo Müller
Physikalische Ingenieurwissenschaft
TU Berlin

ABSTRACT

In a recent publication, E. A. Kearsly has shown that a symmetrically-loaded, square rubber sheet does not deform symmetrically beyond a critical load. While the symmetric deformation is, of course, possible, it is unstable. This paper will investigate this phenomenon more closely than did the Kearsly paper and will present the results of this investigation.

This problem provides an elegant example of how a well-behaved--albeit non-linear--constitutive function can give rise to unstable behavior, hysteresis, and catastrophic breakthrough in the suitably symmetric geometrical setting. We will also see how symmetrybreaking prevents bifurcation.

1. THE PROBLEM.

We consider a square rubber sheet which has side lengths L and thickness T in its natural configuration, see Figure 1. The rubber is considered a Mooney-Rivlin material in that the stress-strain relation has the form

$$\underset{\sim}{t} = s_{-1}\underset{\sim}{B} + P1 + s_1\underset{\sim}{B} \,, \tag{1.1}$$

while the stored energy density is given by

$$\rho W = \frac{1}{2} s_1(I - 3) - \frac{1}{2} s_{-1}(II - 3) . \tag{1.2}$$

$\underset{\sim}{B}$ represents the left Cauchy-Green tensor with principal invariants I and II. P is the hydrostatic pressure, and s_1, s_{-1} are material constants. For rubber we have

$$s_1 = 3\cdot 10^5 N/m^2 \qquad s_{-1} = -0.2s_1. \tag{1.3}$$

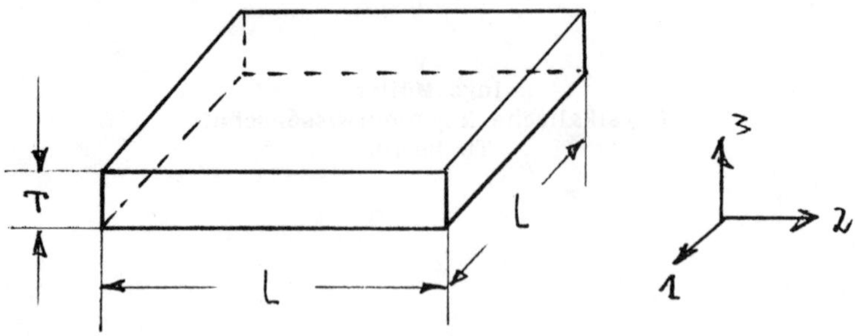

Figure 1: A square membrane in its natural state.

Throughout this paper we consider the top and bottom areas of the sheet as unloaded, whereas we do apply loads in the 1- and 2-directions. The resulting deformation gradient will be denoted by

$$\underline{F} = \begin{bmatrix} \lambda & & 0 \\ & \mu & \\ 0 & & 1/\lambda\mu \end{bmatrix} \tag{1.4}$$

and the loads are $K_\lambda = t_{11}\mu L(1/\lambda\mu)T$ and $K_\mu = t_{22}\lambda L(1/\lambda\mu)T$ respectively. Dividing the loads F_λ, F_μ by $-s_{-1}LT$, we obtain the non-dimensional loads f_λ, f_μ. These may be calculated from (1.1) by elimination of P, and we obtain

$$f_\lambda = (\lambda - \frac{1}{\lambda^3\mu^2})(K + \mu^2), \qquad f_\mu = (\mu - \frac{1}{\lambda^2\mu^3})(K + \lambda^2), \tag{1.5}$$

where K is defined as $-s_1/s_{-1}$. In all calculations we take K to be equal to 5.

For the deformation gradient (1.4), the energy density ρw, or its non-dimensional form $W = \rho w/(-s_{-1}TL^2)$ may be calculated from (1.2). We obtain

$$W = \frac{1}{2}[K(\lambda^2 + \mu^2 + \frac{1}{\lambda^2\mu^2} - 3) + (\lambda^2\mu^2 + \frac{1}{\mu^2} + \frac{1}{\lambda^2} - 3)]. \tag{1.6}$$

In this paper we shall

i.) Calculate the possible strains λ, μ when $f_\lambda = f_\mu$ is given,

ii.) Calculate the possible strains λ, μ when $f_\lambda \neq f_\mu$ is given,

iii.) Calculate $f_\mu(\mu)$ and $f_\mu(\lambda)$ for fixed f_λ.

Close inspection of these problems will reveal a surprising wealth of phenomena, which, in turn, makes the consideration of this simple case of non-linear elasticity interesting.

2. THE CASE OF THE SYMMETRICALLY LOADED SHEET.

Setting $f_\lambda = f_\mu$ in (1.5), we obtain a relation between the strains, viz.

$$(\lambda - \frac{1}{\lambda^3\mu^2})(K + \mu^2) = (\mu - \frac{1}{\lambda^2\mu^3})(K + \lambda^2). \qquad (2.1)$$

After some rearrangement this relation assumes the form

$$(\lambda - \mu)\{(\lambda^3\mu^3+1)K + (\lambda^2+\lambda\mu+\mu^2-\lambda^4\mu^4)\} = 0. \qquad (2.2)$$

One possible solution is obviously $\lambda=\mu$, the symmetric solution in which the sheet remains a square, but there is a second solution defined by the vanishing of the curly bracket term. The evaluation of this solution requires the determination of the roots of an algebraic equation of the fourth order, which is most easily done numerically. Figure 2 shows both solutions, and we see that the second solution corresponds to non-symmetric strains even though the loads are symmetric. The two curves intersect at $\lambda=\mu=2.27$, which, by (1.5) corresponds to a load $f_\lambda=f_\mu=22.9$. We shall refer to this number as the critical load.

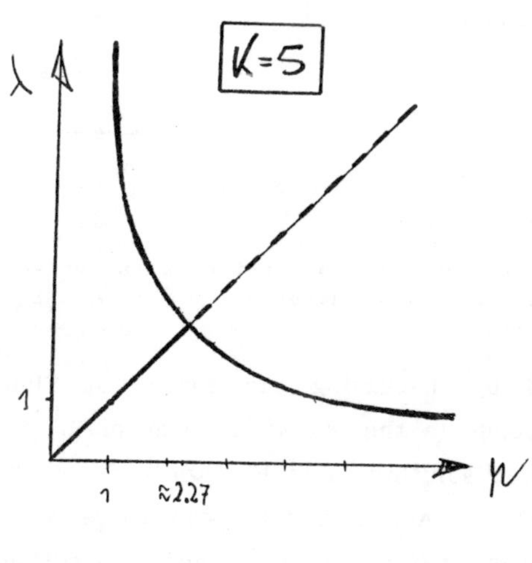

Figure 2: The two possible functions $\lambda=\lambda(\mu)$ for the symmetrically-loaded membrane.

We conclude that for large loads there is a bifurcation. Upon loading beyond $f_\lambda=f_\mu=22.9$, the membrane may continue to deform symmetrically--as it began doing at $\lambda=\mu=1$--or it may become non-symmetric. The question now arises which one of the two solutions has stability.

In order to answer this question we consider the free energy P of the loaded sheet. This is the sum of the stored energy (1.6) and of potential energy of the loads, i.e. with $f_\lambda=f_\mu=f$

$$P = \frac{1}{2}[K(\lambda^2 + \mu^2 + \frac{1}{\lambda^2\mu^2} - 3) + (\lambda^2\mu^2 + \frac{1}{\mu^2} + \frac{1}{\lambda^2} - 3)] - f(\lambda+\mu). \quad (2.3)$$

Figure 3 represents this function for two cases: On the left we have P for the symmetric solution $\lambda=\mu$, and on the right we have P for the non-symmetric solution $\lambda=\lambda(\mu)$ given by the hyperbola-shaped curve of Figure 2. The load f is the curve parameter in both cases.

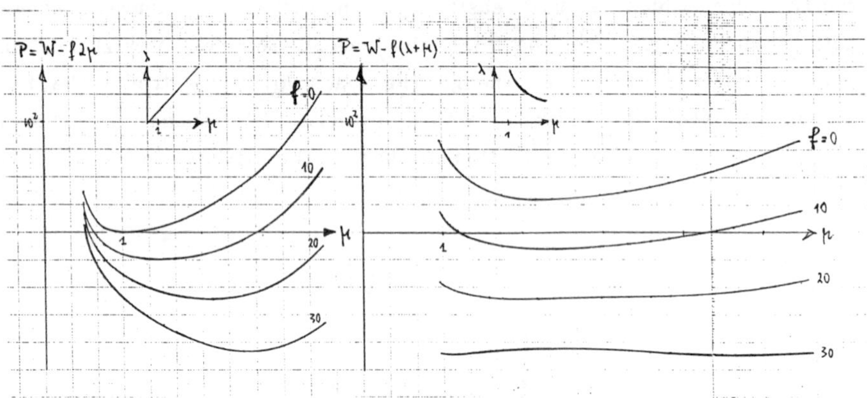

Figure 3: Free energy as a function of μ for various equal loads f. Symmetric deformation (left). Unsymmetric deformation (right).

We proceed by discussing the curves of Figure 3. The stable solution corresponds to the lowest minimum of P. Thus, for P=0,10, and 20, the symmetric solution is stable, while for P=30 the non-symmetric solution is stable. Actually, for P=30 there are two minima in the non-symmetric case, namely at $\mu\approx1.2$ and $\mu\approx4.4$. These values correspond to $\lambda\approx4.4$ and $\lambda\approx1.2$ respectively, and it is clear why this is so. Indeed, when the symmetric sheet becomes unstable, either $\lambda > \mu$ or $\mu > \lambda$ will result; a random fluctuation will determine which.

Another aspect of the same phenomenon has been represented in Figure 4. This figure shows lines of constant P as functions of λ and μ for two values of f (see 12.3). For f=12.5, there is only one minimum for $\lambda=\mu$, while for f=25, two minima have appeared with pairs (λ,μ) that lie symmetric to the bisector. From the previous considerations, we conclude that the splitting of the minimum into two minima occurs at the critical load f=22.9. Beyond this value, the symmetric deformation is unstable.

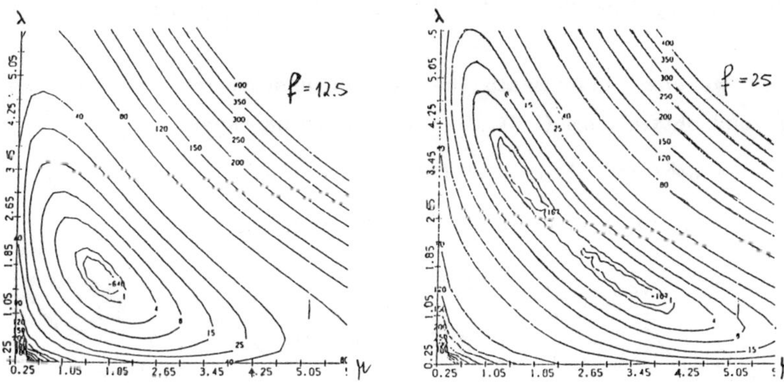

Figure 4: Contour lines of free energy at two loads, one sub-critical, the other supercritical.

We may also consider stability for the clamped sheet when the clamps are located at values of λ and μ that satisfy equation (2.2). In this case, the sheet will assume the smallest value of W as μ is increased. Figure 5 shows that for increasing values up to $\mu \approx 2.27$, the symmetric case $\lambda=\mu$ provides a smaller value of W than the non-symmetric case, whereas for greater values of μ, the situation is reversed. Thus, we see again that the symmetric and the unsymmetric branches exchange stability at the critical deformation $\lambda=\mu=2.27$.

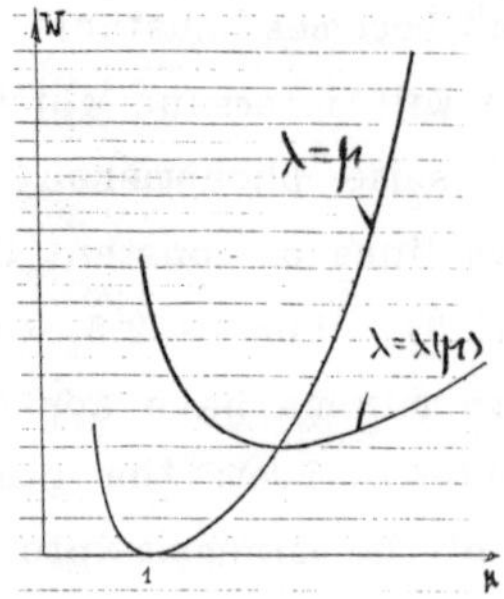

Figure 5: Free energy of the symmetric and unsymmetric case.

3. THE CASE OF THE NON-SYMMETRICALLY LOADED SHEET.

Now we shall investigate the case of unsymmetric loading. We set $f_\lambda=\varphi f_\mu$ and obtain from (1.5) after some rearrangement

$$\lambda^5-{}^1/_\varphi(\mu+{}^K/_\mu)\lambda^4-({}^1/_{\mu^4}-K)\lambda^3-({}^K/_{\mu^4})\lambda+{}^1/_\varphi({}^1/_\mu+{}^K/_{\mu^3}) = 0. \quad (3.1)$$

This is an algebraic equation of the fifth order as was (2.2). Yet

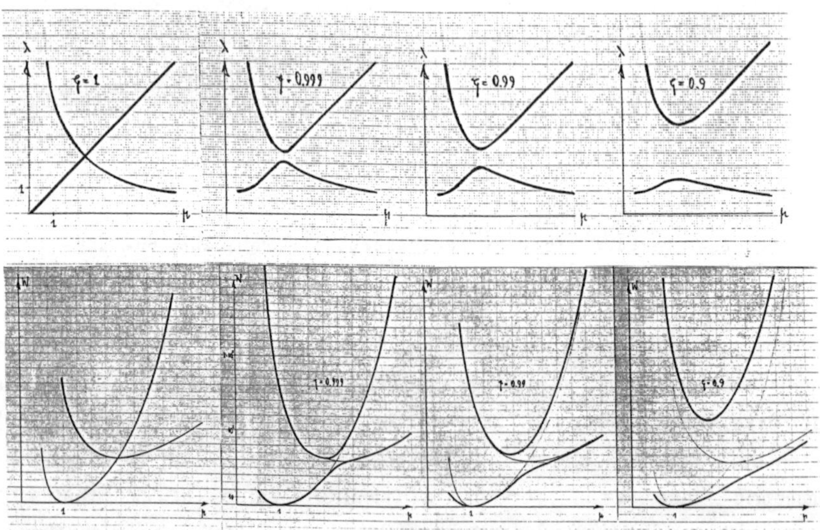

Figure 6: (λ,μ)–relation and stored energy for increasingly non-symmetric loading.

unlike (2.2), the left-hand side of (3.1) does not factorize, and we must solve the entire equation numerically in order to obtain $\lambda=\lambda(\mu)$. The upper row of Figure 6 shows the results for several values $\varphi<1$. For easy comparison, the previously treated case $\varphi=1$ is also shown. We conclude that there are two values of λ remaining for each μ, although the bifurcation of the symmetrically-loaded case is now absent, and the upper branch cannot be reached continuously from $\lambda=\mu=1$.

The lower part of Figure 6 shows the stored energy as a function of μ. These curves no longer intersect, and there is a smooth transition from the initial branch where $\lambda\approx\mu$ to the branch where μ becomes much larger than λ, as f_μ is larger than f_λ. In other words there is no point at which the sheet can choose between two configurations.

Actually, this phenomenon is quite common in physics: symmetry-breaking promotes uniqueness, whereas symmetry provides non-uniqueness, as well as a random choice between several possibilities. The best-known example is the story of "Buridan's Ass".

Here again it is instructive to consider the contour lines of the free energy in the (λ,μ)-diagram. Since the loads are now different, the free energy will read $P=W-f_\lambda\lambda-f_\mu\mu$. Figure 7 shows a case with small f's and only one minimum. Obviously this occurs for unequal values of λ and μ.

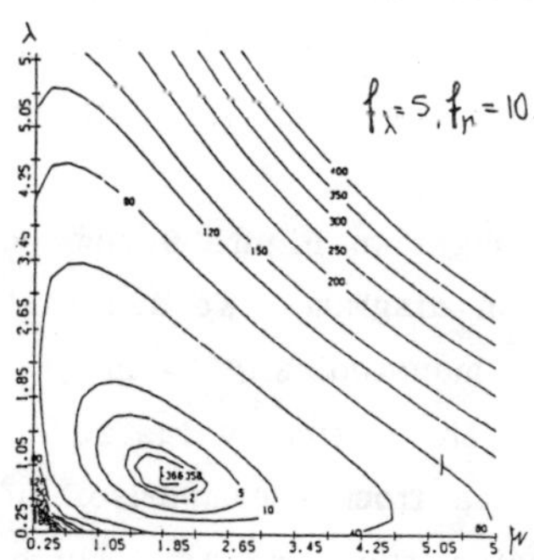

Figure 7: Contour lines of the free energy for small unequal loads.

More interesting is the case of larger and "more equal" loads. Figure 8 shows a sequence of cases in which $f_\mu=30$ is held constant, while f_λ varies from 34 to 26.

The observation we may make in this sequence is that as f_λ falls from $f_\lambda=34$--with f_μ always at 30--the original single minimum for large λ, and small μ is joined by another minimum at large μ and small λ. This new minimum is at first shallower than the other minimum, and then progressively deeper, yet there always remains a barrier between the minimum as long as they are there. Hence, a sheet that began with the left minimum is eliminated when finally $F_\lambda=26$. At that point the sheet will flip over to the right minimum. Subsequently, as F_λ is again increased, the sheet will remain in the right minimum until the value $F_\lambda=34$ is reached, whereupon it will flip back. In other words, we may observe a hysteresis in the loading-unloading diagram $f_\lambda(\lambda)$. We may now proceed to investigate this hysteresis more closely.

4. HYSTERESIS.

We now recall the equations (1.5), and we assign f_λ the value $\overset{\circ}{f}_\lambda$. Equation $(1.5)_1$ then provides a bi-quadratic equation for $\mu=\mu(\lambda;\overset{\circ}{f}_\lambda)$, which may easily be solved to give

$$\mu^2 = \left(\frac{1}{2}\frac{\overset{\circ}{f}_\lambda}{\lambda} + \frac{1}{\lambda^4} - K\right) + \sqrt{\frac{1}{4}\left(\frac{\overset{\circ}{f}_\lambda}{\lambda} + \frac{1}{\lambda^4} - K\right)^2 + \frac{K}{\lambda^4}} \quad . \tag{4.1}$$

If we eliminate μ between $(1.5)_2$ and (4.1), we obtain an equation for $f_\mu(\lambda;\overset{\circ}{f}_\lambda)$ or, making use of (4.1), if λ is replaced by μ, we may numerically obtain values for $f_\mu(\mu;\overset{\circ}{f}_\lambda)$. Both sets of curves are represented in Figure 9.

The curves on the right in Figure 9 come closest to representing a uniaxial load-deformation diagram. We see that for small values of $\overset{\circ}{f}_\lambda$, the load f_μ increases monotonically with μ. However, when $\overset{\circ}{f}_\lambda$ is assigned a value beyond the critical value 22.9, the resulting curve $f_\mu(\mu)$ is non-monotone. We see from a comparison of both plots in Figure 9 that this branch contains the case $\lambda=\mu$ which is unstable for supercritical loads.

Non-monotone curves imply hysteresis. Indeed, if f_μ is increased for $\overset{\circ}{f}_\lambda=25$, the deformation μ will increase smoothly until f_μ reaches the maximum of the $f_\mu(\mu;\overset{\circ}{f}_\lambda)$-curve. Any further increase of f_μ will result in a catastrophic breakthrough until μ has reached the right branch of the

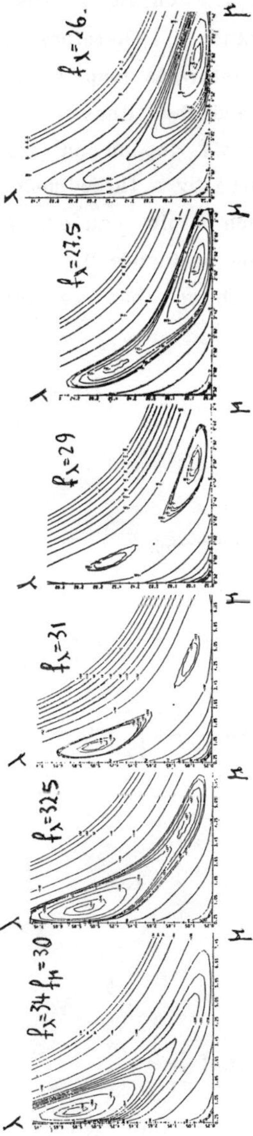

Figure 8: Contour lines of $P=W-f_\lambda\lambda-f_\mu\mu$ for $f_\mu=30$ and varying f_λ.

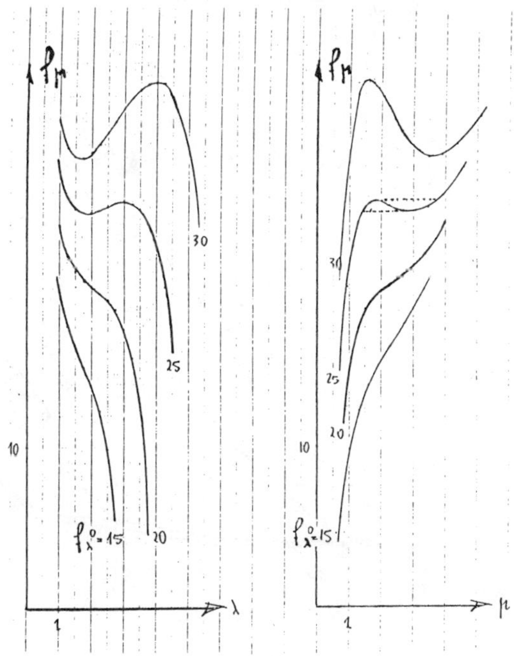

Figure 9: f_μ as a function of λ and μ for various values of f^{o}_{λ}.

curve. Once on that curve, the load may be further increased or lowered. If it is lowered below the minimum of the curve, another breakthrough will occur back to the original branch. Therefore a loading-unloading process will progress through a hysteresis loop whose horizontal composition is indicated in Figure 9 by the dashed lines.

The observation that breakthrough and hysteresis depend on the chosen value of f_λ can most interestingly be illustrated by a threedimensional plot where λ is a function of f_λ and f_μ. Such a plot can easily be constructed from the curves on the left-hand side of Figure 9, while Figure 10 gives a qualitative view. Inspection shows that we are dealing with a <u>fold singularity</u>.

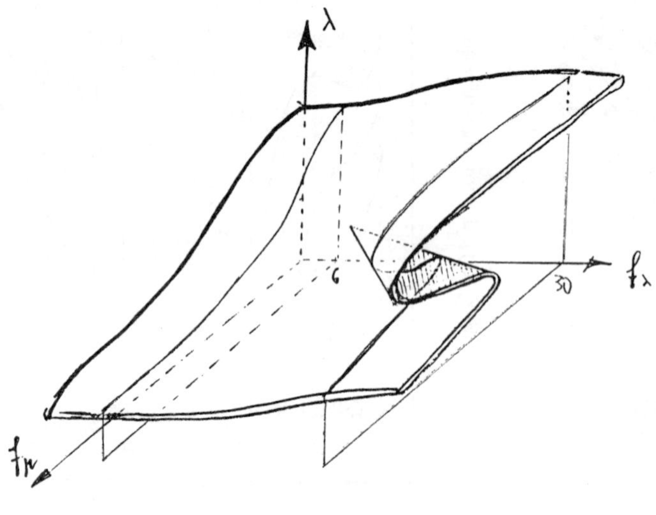

Figure 10: Fold singularity in the function $\lambda=\lambda(f_\mu,f_\lambda)$.

REFERENCE:

Kearsley, E.A., Asymmetric Stretching of a Symmetrically-loaded Elastic Sheet, Int. J. of Solids and Structures (1985).

Shock waves formation in gravitational gases

Augusto Muracchini
Dipartimento di Matematica e CIRAM, Università di Bologna,
Via Saragozza 8, 40123 Bologna, Italy

1 General Remarks

The problem of the non linear wave propagation has been studied in the last years by many techniques. In particular we wish to quote

I) The method of asymptotic waves [1].

II) The method of discontinuity (or acceleration) waves [2], [3], [4], [5].

To study the formation of shock waves in a gravitational atmosphere the first method is applied, for example, in [6] and the second in [7]. In both cases the dissipative effects are neglected and so the field equations of hyperbolic type are taken into account.

Obviously it is more interesting, in view of the physical applications, to consider the problem of the non linear wave propagation when the viscosity and the heat flux are present. However in order to obtain a satisfying model for studying the shock wave formation, it is necessary that the system of balance laws be hyperbolic: in fact, if the field equations are parabolic, as it happens if we use the Fourier-Navier Stokes constitutive equations, we find the well known *paradox of infinite velocities of propagation* for the acceleration waves.

The Extended Thermodynamics for a classical dissipative ideal gas represents a possible approach to this problem since it is based on an hyperbolic system of thirteen balance laws containing, as field variables, in addition to mass density, momentum and energy also the stress deviator and the heat flux.

In this framework the general non-linear theory of acceleration waves has been applied, recently, in the case of a thermoviscous ideal gas without gravity [8].

Because the gravity force behaves in a different way for the upwards and downwards traveling waves it seems interesting to consider the role which play the gravity and the dissipation effects together in the problem of shock wave formation.

In [9] is studied the propagation of acceleration waves in a thermoviscous ideal gas with a gravitational force.

2 Discontinuity waves in a gravitational gas

Consider a quasi-linear first order hyperbolic system of the type

$$\mathbf{u}_t + \mathbf{A}(\mathbf{u})\mathbf{u}_x = \mathbf{f}(\mathbf{u}) \tag{2.1}$$

$\mathbf{u}(x,t)$: an R^N unknown column vector of the system, $\mathbf{A}$: a $N \times N$ matrix, $\mathbf{f}$: the R^N production vector.

For a generic system of the type (2.1) it is possible to consider a particular class of solutions characterizing the so called *weak discontinuity waves.*

In [2] $\div$ [5] are obtained many results concerning the theory of discontinuity waves: we refer the reader to these quoted papers for the comprehension of what follows and the meaning of symbols.

The one dimensional field equations of classical Extended Thermodynamics [10] for a monoatomic ideal gas in presence of the gravity may be written in the form (the x axis is vertical and we ignore the inertial forces)

$$\begin{cases} \dot{\rho} + \rho v_x = 0 \\ \rho\dot{v} + (p-\sigma)_x = -\rho g \\ \rho\dot{e} + q_x + (p-\sigma)v_x = 0 \\ \dot{\sigma} - \frac{8}{15}q_x + \frac{7}{3}\sigma v_x - \frac{4}{3}p v_x = \tau_o \sigma \\ \dot{q} + \frac{16}{5}q v_x + \frac{1}{2}\left[\Gamma\left(\frac{5}{7}p - \sigma\right)\right]_x - \frac{1}{\rho}(\sigma\sigma_x - \sigma p_x - \frac{5}{2}p\sigma_x + \frac{5}{2}p p_x) = \tau_1 q \end{cases} \tag{2.2}$$

ρ, v, σ, e, q, g are respectively: the density, the velocity, the viscous stress, the internal energy, the heat flux and the gravity acceleration. The dot indicates the material derivative and the index x represents the partial derivative $\partial/\partial x$.

The pressure p and the other constitutive functions Γ, τ_o, τ_1 are given, for a non degenerate gas, by

$$p = \frac{2}{3}\rho e = \mathcal{R}\rho\theta, \quad \Gamma = 7\mathcal{R}\theta, \quad \tau_o = -\frac{p}{\mu}, \quad \tau_1 = -\frac{5}{2}\frac{\mathcal{R}p}{\chi} \tag{2.3}$$

$\mathcal{R}$: the gas constant for unit molecular weight, θ: the absolute temperature, μ: the viscosity coefficient, χ: the heat conductivity.

Notice that the reciprocal of $|\tau_o|$ and $|\tau_1|$ are the relaxation times of the model

$$\tau_\sigma = \frac{1}{|\tau_o|} = \frac{\mu}{p}, \qquad \tau_q = \frac{1}{|\tau_1|} = \frac{2}{5}\frac{\chi}{\mathcal{R}p}. \tag{2.4}$$

The system (2.2) is a particular case of (2.1) with $N = 5$. In the present case, choosing as field

$$\mathbf{u} \equiv (\rho, v, \theta, \sigma, q)^T, \tag{2.5}$$

we have immediately

$$\mathbf{A} \equiv \begin{pmatrix} v & \rho & 0 & 0 & 0 \\ \frac{\mathcal{R}\theta}{\rho} & v & \mathcal{R} & -\frac{1}{\rho} & 0 \\ 0 & \frac{2}{3}(\theta - \frac{\sigma}{\mathcal{R}\rho}) & v & 0 & \frac{2}{3\mathcal{R}\rho} \\ 0 & \frac{7}{3}\sigma - \frac{4}{3}p & 0 & v & -\frac{8}{15} \\ \frac{\sigma\mathcal{R}\theta}{\rho} & \frac{16}{5}q & \frac{5}{2}\mathcal{R}(p-\sigma) & -(\mathcal{R}\theta + \frac{\sigma}{\rho}) & v \end{pmatrix}; \qquad \mathbf{f} \equiv \begin{pmatrix} 0 \\ -g \\ 0 \\ \tau_o\sigma \\ \tau_1 q \end{pmatrix} \tag{2.6}$$

where the superscript T indicates the transpose of a matrix.

By considering the propagation problem of discontinuity waves associated to the system (2.1) ÷ (2.6) we have that the characteristic velocities λ in the equilibrium case ($q = \sigma = 0$) and $v = 0$ are

$$\lambda_o = 0, \qquad \lambda_o \simeq \pm 0.63c, \qquad \lambda_o \simeq \pm 1.65c, \tag{2.7}$$

where c is the sound velocity: $c^2 = p_\rho = (\partial p/\partial \rho)_S$ (S is the entropy for unit mass).

Observe that:

i) $\lambda_o = 0$ is an exceptional wave and so there is no critical time for it;

ii) the calculations which follow are performed for the fastest wave $\lambda_o \simeq +1.65c$ only: in fact, the wave with velocity $\lambda_o \simeq -1.65c$ can be dealt in the same way without difficulty by changing c into $-c$;

iii) we consider as unperturbed field the barotropic equilibrium static isothermal solution with zero velocity of (2.2)

$$v_o = \sigma_o = q_o = 0, \qquad \rho_o = \tilde{\rho} exp\left(-\frac{gx}{\mathcal{R}\theta}\right), \qquad \theta_o = \tilde{\theta} \tag{2.8}$$

which is the only static equilibrium solution of the system (2.2) in presence of gravity. In (2.8) $\tilde{\rho}$ and $\tilde{\theta}$ are positive constants (henceforth $\tilde{\psi}$ indicates the value of a generic quantity ψ at $x = 0$ and $\psi(0)$ is the value of ψ at time $t = 0$ along the characteristic line starting at $x = x_o$).

By many lengthy and cumbersome calculations (see for the details in [9]) we find the expressions of the coefficients (a and b) of the Bernoulli equation [2],[4]

$$a(t) = 3.749\frac{\alpha\lambda}{\rho} = 3.749\frac{\alpha_1\lambda}{\rho(0)^2}e^{\frac{2g\lambda t}{\mathcal{R}\theta}} \tag{2.9}$$

$$b(t) = \text{const.} = -1.737\frac{g}{c} + 0.203\mid \tau_o \mid + 0.198 \mid \tau_1 \mid . \tag{2.10}$$

where $\alpha = \alpha_1/\rho$; $\alpha_1 = 1/4\mathcal{R}\theta\rho^2(\frac{13}{3}\lambda^2 - 5\mathcal{R}\theta)$ and $\rho = \rho(0)exp(-g\lambda t/\mathcal{R}\theta)$.

Then the amplitude (see [2]) is

$$\Pi(t) = \frac{\Pi(0)e^{-bt}}{1 - \Pi(0)a_1\alpha_1\frac{\lambda}{\rho(o)^2\nu^*}(e^{-\nu^* t} - 1)}, \tag{2.11}$$

where

$$\nu^* = b - \nu = -5.070\frac{g}{c} + 0.203 \mid \tau_o \mid +0.198 \mid \tau_1 \mid \tag{2.12}$$

with $\nu = \frac{2gc}{R\theta} = \frac{2g\gamma}{c}$, $a_1 = 3.749$.

Putting equal to zero the denominator of (2.11) and remembering that the amplitude Π represents the jump of the normal derivative of the density and this is related with G through the relation $G = -\Pi\alpha_o\lambda_o^2/\rho_o$ we get the possible critical time in terms of the initial acceleration jump $G(0)$

$$t_{cr} = -\frac{1}{\nu^*}\log\left\{1 - \frac{G_{th}}{G(0)}\right\} \tag{2.13}$$

where

$$G_{th} = \nu^*\frac{\lambda}{a_1} = -2.231g + c(0.089 \mid \tau_o \mid +0.087 \mid \tau_1 \mid). \tag{2.14}$$

For the existence of shock waves two cases need to be considered.

$\boxed{\text{Case } \lambda > 0}$: Because the critical time must be a positive quantity it is easily verified that, the following conditions for the initial acceleration jump $G(0)$ must hold

$$\text{if} \quad \nu^* < 0 \quad \rightarrow \quad G(0) > 0 \tag{2.15}$$

$$\text{if} \quad \nu^* > 0 \quad \rightarrow \quad G(0) > G_{th} > 0 \tag{2.16}$$

$\boxed{\text{Case } \lambda < 0}$: Actually, changing c into -c, ν^* becomes

$$\nu_-^* = 5.070\frac{g}{c} + 0.203 \mid \tau_o \mid +0.198 \mid \tau_1 \mid > 0, \tag{2.17}$$

and the critical time is a positive quantity if and only if

$$G(0) < G_{th}^- < 0 \tag{2.18}$$

where

$$G_{th}^- = \nu_-^*\frac{\lambda}{a_1} = -2.231g - c(0.089 \mid \tau_o \mid +0.087 \mid \tau_1 \mid). \tag{2.19}$$

The previous conditions are easily explained:

If (2.15), is verified the critical time (i.e. a shock wave) exists for any initial *compressive* amplitude of the acceleration initial jump $G(0)$.

If (2.16), (2.18) hold the critical time (i.e. a shock wave) exists only for a sufficiently large initial amplitude of $G(0)$.

In other words:

When $\nu^* > 0$ the critical time of the wave traveling upwards and downward exists only for a sufficiently large initial amplitude of the acceleration jump $\mid G(0) \mid$ (see (2.16), (2.18)). In this case the critical time and the critical amplitude are given by (2.13),(2.14),(2.19) respectively.

The case of $\nu^* < 0$ is compatible with the wave traveling upwards only and the critical time exists for any initial compressive acceleration jump (see (2.15)).

3 Stability of the isothermal solution

About the stability of the solution (2.8) we obtain, according to results proven in [5], that in the case of $\nu^* > 0$ we have stability.

On the contrary, if $\nu^* < 0$ the solution is unstable: this is an obvious result by remembering that if $\nu^* < 0$ the shock wave exists for any initial acceleration jump.

By summarizing:

The isothermal solution (2.8), *when* $\nu^* > 0$, *is asymptotically stable with respect to the upwards and downward wave (if the initial acceleration* $G(0)$ *satisfies the condition* $|G(0)| < |G_{th}|$).

If $\nu^* < 0$ *the solution for the upwards wave is unstable.*

4 Conclusive remarks

The formation of shock waves as degeneration of discontinuity waves was studied, at first, in [11] considering a perfect non gravitating fluid: in this case the shock is always formed and the solution is unstable.

In [8] is considered the case of a thermoviscous ideal gas without gravity, in the framework of Extended Thermodynamics, proving that:

I) There exists always a very large critical amplitude $G_{cr.}$, depending on the dissipative effects, such that the shock formation is possible only if the initial acceleration jump $G(O)$ satisfies the condition $G(O) > G_{cr.}$.

II) The isothermal solution is always stable.

In [7] the shock formation in a gravitating ideal atmosphere has been examined finding that:

I) In the isothermal case there exists always a critical time for a compression ($G(O) > 0$) upwards acoustic wave [1] whereas the critical time of the acoustic wave traveling downward exists only for a sufficient large initial compressive amplitude of the acceleration jump $G(O)$.

II) The isothermal solution is unstable in both cases ($c > 0, c < 0$).

By considering now the physical aspects of the present problem, it should be pointed out that in our case, unlike what happens in absence of gravity where necessarily $\nu^* > 0$ and only the case (2.16) is possible, the situation (2.15) also could occur.

We prove that in the normal physical situations it happens that $\nu^* \gg 0$ having then, in practice, only one case even now.

[1] In this case the fastest discontinuity wave propagates with the characteristic velocity $\lambda = v + c$: therefore this wave is very different from thermo-acceleration waves which arise in dissipative case.

Consider to clear up this question the following example.

In the Helium at $300^o K$, $p = 1$ atm. one finds from $(2.3)_{3,4}$

$$|\tau_o|, |\tau_1| \sim 10^9 s^{-1} \tag{4.1}$$

having $\mu = 5.023\theta^{0.647}$ micropoise and $\tau_1 = \frac{2}{3}\tau_o$ (2/3 is the Prandtl number for a monoatomic gas). The gravity contribution in ν^* is very small compared to $|\tau_o|, |\tau_1|$ ($g/(c|\tau_o|) \sim 1/10^{10}$ in the terrestrial atmosphere) and, besides, from (2.14),(2.19) $|G_{th}| \sim 10^{10} g$.

In any case, if $|G_{th}|$ is very large and it is surpassed, then t_{cr} is very small or, in other words, the shock wave forms immediately (it is a trivial consequence of (2.13)).

Observe also that having G_{th} an enormous value, it happens in the physical world that $|G(0)| \ll |G_{th}|$ and so the amplitude $G(t)$ practically vanishes since (see (2.11))

$$G(t) \sim G(0)e^{-bt} \tag{4.2}$$

The acceleration wave damps almost immediately: in practice, owing to dissipation a discontinuity wave disappears and can never degenerate into a shock wave.

In order to obtain a formation of shock waves as degeneration of acceleration waves we must have a small G_{th} i.e.: or an enormous value of g, as perhaps it occurs in some astrophysical phenomena, or very high relaxation times.

Let us summarize these results in the following statements :

In the usual physical situations no shock formation is possible for acceleration waves.

The isothermal solution is asymptotically stable with respect to the upwards and downward wave since in the ordinary physical cases the initial acceleration $G(0)$ satisfies the condition $|G(0)| < |G_{th}|$.

Note finally that the absorption is of the same order of magnitude as that of plane waves in the range of high frequencies (for the dispersion law of Extended Thermodynamics without gravity see [12] p. 141). This coincidence is due to the fact that it is possible to prove, in the case of constant unperturbed field (i.e. in absence of gravity),[2] that the value of b appearing in the Bernoulli equation coincides, for a generic hyperbolic system and in the limit of high frequencies, with the attenuation coefficient K_i times $-\lambda$ of harmonic waves of small amplitude

$$b = -\lambda \lim_{\omega \to \infty} K_i. \tag{4.3}$$

More details on these matters will be found in a forthcoming paper.

---

[2] Our problem approximates this case because the gravity acceleration g is negligible with respect to dissipative terms.

References

[1] **Y. Choquet-Bruhat** *Ondes Asymptotiques et Approchées pour des Systémes d' Équation aux Derivées Partielles non Linéaires.* J. Math. Pures et Appl. **48**, 117 (1969).

[2] **G. Boillat** *La Propagation des Ondes.* Gauthier-Villars, Paris (1965).

[3] **P. Chen** *Growth and Decay of Waves in Solids* in Mechanics of Solids III, Handbuch der Physik, **6A/3**, 303 Springer-Verlag (1973).

[4] **G. Boillat-T. Ruggeri** *On the Evolution Law of Weak Discontinuities for Hyperbolic Quasi-linear Systems.* Wave Motion **1**, 149 (1979).

[5] **T. Ruggeri** *Stability and Discontinuity Waves for Symmetric Hyperbolic Systems.* In: *Nonlinear Wave Motion.* Edited by A. Jeffrey, Longman (1989).

[6] **A.M. Anile-G. Mulone-S. Pluchino** *Critical Time for Asymptotic Acoustic Waves in a Gravitational Atmosphere.*Wave Motion **2**, 267 (1980).

[7] **A. Muracchini-T. Ruggeri** *Acceleration Waves, Shock Formation and Stability in a Gravitating Atmosphere.* Astrophysics and Space Science. **153**, 127 (1989).

[8] **T. Ruggeri-L. Seccia** *Hyperbolicity and Wave Propagation in Extended Thermodynamics.* Meccanica. **24**, 127 (1989).

[9] **A. Muracchini-L. Seccia** *Thermo-acceleration waves and shock formation in Extended Thermodynamics of gravitational gases.* Continuum Mechanics and Thermodynamics **1**, 227 (1989).

[10] **Liu I. Shih-I. Müller** *Extended Thermodynamics of Classical and Degenerate Ideal Gases.* Arch. Rat. Mech. Anal. **83**, 285 (1983).

[11] **T. Ruggeri** *Interaction between a discontinuity wave and a shock wave: critical time for the fastest transmitted wave, example of the polytropic fluid.* Applicable Analysis **11**, 103 (1980).

[12] **I. Müller** *Thermodynamics.* Pitman, Boston (1985).

NONLINEAR STABILITY IN NON-NEWTONIAN FLOWS

John A. Nohel
Center for the Mathematical Sciences
University of Wisconsin-Madison
Madison, Wisconsin 53705 U.S.A.

Abstract: Based on joint work with R. L. Pego and A. E. Tzavaras, we discuss nonlinear stability of co-existing phases for a model initial - boundary value problem in one space dimension for isothermal shear flow of a non-Newtonian fluid between parallel plates, driven by a constant pressure gradient. The key feature of the model is a non-monotone relation between the total steady shear stress and shear strain-rate; the non- Newtonian contribution to the shear stress is assumed to satisfy a simple differential constitutive law. We show that the initial-boundary value problem governing the flow is globally well posed for smooth initial data of arbitrary size and every solution tends to a unique steady state as $t \to \infty$; moreover, discontinuous steady states that take their values on the strictly increasing parts of the steady stress - strain rate curve are nonlinearly asymptotically stable.

Reference:

J. A. Nohel, R. L. Pego & A. E. Tzavaras, Stability of Discontinuous Steady States in Shearing Motion of a Non-Newtonian Fluid, Proc. Royal Soc. Edinburgh (Series A), to appear.

INSTABILITY OF MARANGONI AND COMBINED FREE CONVECTION

Luigi G. Napolitano & Antonio Viviani

Istituto "Umberto Nobile", Università di Napoli
P.le V.Tecchio 80, 80125 Napoli (Italy)
also
MARS-Microgravity Advanced Research and Support center
via Diocleziano 328, 80125 Napoli (Italy)

ABSTRACT

The paper deals with the linear stability analysis of Marangoni, natural and combined free convection in rectangular cavities of small aspect ratio, open at the top and with differentially heated end walls. The class of flow considered is that of parallel flows, away from the end walls, for arbitrary values of the Marangoni, Grashof and Prandtl numbers. Stability analyses of this class of flows have been published by Smith (1982) and Hart (1983), who have dealt with the cases of Marangoni and natural convection, respectively.

In this work we consider combined convection and give the critical Reynolds number for the dynamic Bond number ranging from zero (Marangoni convection) to infinite (natural convection), for a wide range of the Prandtl number, considering instability with respect to tridimensional disturbances. In the particular case of Marangoni convection, we confirm the results of Smith and, in the case of natural convection, we extend the results of Hart, whose analysis is limited to low Prandtl numbers and two-dimensional disturbances. The limiting value of the critical Reynolds number, for $Pr \rightarrow \infty$ is also calculated, and asymptotic laws for high Prandtl numbers are proposed, in the whole range of definition of the dynamic Bond number.

1. INTRODUCTION

Convection in the absence of imposed velocities and/or pressure gradients in fluid systems with interfaces can be classified (Napolitano 1982) as natural, Marangoni or combined free convection, according to whether the motion is caused by buoyant forces, Marangoni stresses (surface tension gradients) or both.

The subject matter of this paper is free convection in shallow rectangular cavity with differentially heated end walls and an upper free surface.

The bottom boundary is rigid and insulated. The top liquid-air interface is assumed to be flat and undeformable (this places constraints on the values of

the contact angle and of the capillary number) and there is not dynamic and thermal coupling between the liquid and the surrounding air.
The surface tension along the interface is assumed to be a linear decreasing or increasing function of the temperature. In most papers dealing with surface tension effects, the surface tension is assumed to be a linearly decreasing function of the temperature. This behaviour is typical of a large class of fluids, but one may find systems like some alloys, molten salts or liquid crystals with a surface tension growing linearly with temperature (Guyon and Pantaloni 1980). This last behaviour has been accounted for in various papers, particularly in papers on Marangoni instability of liquid layers heated from below (see Lebon 1988). However, from the theoretical point of view, it must be pointed out that, if the surface tension depends only on the temperature, a model of surface tension linearly increasing with the temperature implies that the surface phase is thermodynamically unstable, as proved by the senior author (Napolitano 1989).
The imposed horizontal temperature gradient induces a circulating convection flow into the cavity. The motion is generated by density differences in the gravitational field and by surface tension gradients along the free surface.
For sufficiently small temperature differences between the side walls, the flow is laminar and steady, and for a sufficient small aspect ratio and far away from the end walls it is parallel to the horizontal boundaries, with parabolic (resp. cubic) velocity profile for Marangoni (resp. natural) convection. In the general case of combined free convection we have different profiles (linear combination of the previous ones) depending on the value of the dynamic Bond number. When the surface tension increases with the temperature, the resulting Marangoni convection is from cold to hot region near the free surface, in opposition to the motion due to the gravity forces (we do not consider the case of a negative thermal expansion coefficient), so that in this case two superimposed counter-rotation cells can be observed, and some particular profiles (e.g. zero velocity at the free surface) can form.
By increasing the imposed horizontal temperature gradients, the "base flow", just described, will become unstable and a new state, marked by "secondary flows", emerges.
Aims of this study are: evaluate, by means of the classical linear stability analysis, the critical conditions for the onset of instability, in terms of appropriate non-dimensional characteristic numbers; assess the nature (steady or oscillatory in time) of the new state and the type of the secondary flows (two-dimensional: transverse or longitudinal; tridimensional).
The reason for work on this problem arises from tasks such as the growing of semiconductor crystals (Hurle, Jakeman and Johnson 1974; Gill 1974), tracking the dispersion of pollutants in estuaries (Cormack, Stone and Leal 1975) or from meteorology (Hart 1983).
Our interest comes from the growth of crystals from melt by the horizontal zone-melting and horizontal normal-freeze techniques (Hurle and Jakeman 1981) both on earth and in microgravity environments. One of the problems associated with these techniques is the development of striations or segregations bands parallel to the melt-crystal interface. The cause of this is recognized to be a time-dependent crystal-growth speed due to temperature oscillations in the melt.
The only theoretical works on the subject are those of Smith (1982), see also

Smith and Davis (1983), and Hart (1983).
Smith studies Marangoni convection (zero Bond number) and finds an oscillatory instability with respect to tridimensional disturbances for all the values of the Prandtl number.
Hart analyzes the case of natural convection; he finds an oscillatory instability but his analysis is limited to two-dimensional longitudinal disturbances and low Prandtl numbers.
Instability due to both gravity forces and surface tension gradients are widely studied in the literature when the temperature gradient is vertical (the base flow is null), see the monograph of Chandrasekar (1961) and, for later developments, the review paper of Lebon (1988), which account for recent papers where a negative Marangoni number (surface tension increasing with the temperature) has been considered.
In the above case of a vertical temperature gradient both linear and energy stability analyses are available.
Here we present the first linear stability analysis for combined convection with Bond numbers ranging from zero to infinity by assuming tridimensional disturbances, and the energy stability analysis will be presented in a forthcoming paper (Napolitano and Viviani 1990, see also Viviani 1988). In the particular case of Marangoni convection we confirm the results obtained by Smith (1982). In the case of natural convection we extend the results obtained by Hart (1983) because of our analysis with respect to tridimensional disturbances, and we present new results for higher Prandtl numbers.
The paper runs as follows. After establishing the set of relevant equations and the solution for the base flow (sections 2 and 3), we study in section 4 the instability of the base flow by using the normal mode technique; the resulting differential complex eigenvalue problem is solved by a spectral Galerkin method. Results and discussion are reported in section 5 and final comments in section 6.

2. THE MODEL

Consider the cavity shown in Fig.1, of length L, height H and small aspect ratio H/L, with the left wall at temperature T_C and the right wall at temperature $T_H=T_C+\Delta T$. The liquid filling the cavity is bounded from above by a liquid-gas interface, along which acts the surface tension σ. The gravity vector $\underline{g}$ acts parallel to the side walls and is directed downward.
The state equations for the liquid density ρ and the surface tension σ are assumed to be, respectively:

$$\rho=\rho_r\,[1-\beta_T(T-T_r)] \qquad (1)$$

$$\sigma=\sigma_r-\sigma_T(T-T_r) \qquad (2)$$

where T_r is the reference temperature, ρ_r and σ_r are density and surface tension for $T=T_r$; β_T is the coefficient of thermal expansion; σ_T is the negative rate of change of the surface tension with temperature ($\sigma_T=-d\sigma/dT$), that we assume to be a positive or negative constant. We shall not consider the case of

a negative coefficient of thermal expansion, so that for $\sigma_T > 0$ the Marangoni convection is superimposed to the natural convection.
We assume that: the liquid is Newtonian, non-elastic, non micro-polar, homogeneous, with constant diffusion coefficients; the Boussinesq approximation applies; viscous dissipation and radiation effects are negligible.
With these assumptions the flow in the cavity is governed by the Boussinesq form of the continuity, Navier-Stokes and energy equations, that in non-dimensional form are:

$$\underline{\nabla} \cdot \underline{v} = 0 \tag{3a}$$

$$Re\left(\frac{\partial \underline{v}}{\partial t} + \underline{v} \cdot \underline{\nabla}\, \underline{v}\right) + \underline{\nabla}\, p = \nabla^2 \underline{v} + G\, T\, \underline{i}_z \tag{3b}$$

$$Pe\left(\frac{\partial T}{\partial t} + \underline{v} \cdot \underline{\nabla}\, T\right) = \nabla^2 T \tag{3c}$$

where $\underline{v}$, p, T denote non-dimensional velocity, pressure and temperature, $\underline{i}_z$ is the unit vector of the z-axis.
The non-dimensional form results from scaling the coordinates x, y, z by the height H of the cavity and the velocity $\underline{v}(u,v,w)$ by the reference velocity V_r to be defined below. The non-dimensional temperature is measured above the mean temperature of the side walls and has as its scale $(\Delta T)_r = \Delta T \;\; H/L$. The scales for time and pressure, the latter measured above the hydrostatic pressure in the reference state at the temperature T_r, are H/V_r and $\mu V_r/H$, with μ the dynamic viscosity of the liquid.
The three parameters that appear in Eqs. (3) are the "conditional" Reynolds (Re), Peclet (Pe) and Grashof (G) numbers (defined in terms of a still undetermined reference velocity, Napolitano 1982):
$Re = V_r\, H/\nu$, with ν the kinematic viscosity;
$Pe = Re\;Pr$ with $Pr = \nu/\alpha_T$ the Prandtl number in which α_T is the thermal diffusivity;
$G = [\,g\beta_T\, \Delta T\, H^2]/(\nu\, V_r)\; H/L$.
The Reynolds and Peclet numbers are "response" numbers and can be expressed as the ratio of the reference velocity to the diffusivity velocities: $Re = V_r/V_\nu$ and $Pe = V_r/V_\alpha$ with $V_\nu = \nu/H$ and $V_\alpha = \alpha_T/H$ the viscous and thermal diffusivity velocities.
The Grashof number is a "driving" number, defined as the ratio of the gravitational velocity to the reference velocity: $G = V_g/V_r$ with $V_g = [\,g\beta_T\, \Delta T\, H^2]/\nu\;\; H/L$.
For a systematic study of response, driving and conditional characteristic numbers, see Napolitano (1978, 1982).
The relevant boundary conditions at the free surface are deduced from the general balance equations for surface phases (Napolitano 1979), under the following assumptions:
i) the surface phase is a stream-surface devoid of any thermodynamic

property apart from the surface tension;
ii) the surface phase is in thermal and mechanical equilibrium with the adjacent volume phases;
iii) momentum and energy coupling between the adjacent fluid phases, radiation effects and surface irreversibility are negligible;
iv) the capillary number V_r/V_c with $V_c=\sigma_r/\mu$ the capillary velocity, is much less than one so that the interface preserves its hydrostatic shape, that we assume to be planar (contact angle equal to $\pi/2$).
The boundary conditions at the interface then read:

$$w=0 \qquad z=1 \qquad (4\ a)$$

$$\frac{\partial u}{\partial z}+M\frac{\partial T}{\partial x}=0 \qquad z=1 \qquad (4\ b)$$

$$\frac{\partial v}{\partial z}+M\frac{\partial T}{\partial y}=0 \qquad z=1 \qquad (4\ c)$$

$$\frac{\partial T}{\partial z}=0 \qquad z=1 \qquad (4\ d)$$

where the parameter that appears in Eqs. (4 b,c) is the conditional Marangoni number $M=V_m/V_r$, with $V_m=(\sigma_T\Delta T/\mu)H/L$ the Marangoni velocity and it is a driving number. The boundary conditions at the lower rigid wall are:

$$u=v=w=0 \qquad z=0 \qquad (5\ a,b,c)$$

$$\frac{\partial T}{\partial z}=0 \qquad z=0 \qquad (5\ d)$$

We do not consider boundary conditions at the side walls because the present analysis is devoted to the parallel core-flow solution.
The reference velocity V_r appropriate to the problem under study, according to Napolitano (1982), must be of the order of magnitude of the maximum characteristic driving velocity, i.e.

$$V_r = O(\max[|V_m|, V_g]) \qquad (6)$$

where the absolute value is necessary for σ_T can be either positive or negative. We assume:

$$V_r = |V_m| + V_g \qquad (7)$$

In this way we satisfy the criterion (6) in the complete range $0 \le Bo < \infty$, with $Bo = V_g/|V_m|$ $[=g\Delta\rho H^2/\Delta\sigma;\ \Delta\rho=\rho_r\beta_T\Delta T,\ \Delta\sigma=|\sigma_T|\Delta T]$ the dynamic Bond number, that measures the relative importance of the gravitational and surface tension effects.
Due to the choice (7), the Marangoni and Grashof numbers are at most

unitary and not independent, verifying the relation:

$$|M| + G = 1 \tag{8}$$

By introducing the parameter

$$\lambda = \frac{M}{G} = \frac{V_m}{V_g} = 1/Bo \tag{9}$$

M and G can be eliminated in terms of λ by means of (8) and (9), which give:

$$M = \frac{\lambda}{1+|\lambda|} \quad ; \quad G = \frac{1}{1+|\lambda|} \tag{10 a,b}$$

The choice of λ as parameter of the problem, instead of the more useful Bond number $Bo=1/\lambda$, has been made in order to distinguish the two cases G=0 with M=-1 and G=0 with M=1.

The particular cases of natural and Marangoni convection are characterized by:

$$\lambda = 0 \qquad (M = 0 ; G = 1) \tag{11 a}$$

$$\lambda = +\infty \quad (M = +1; G = 0) \qquad \sigma_T > 0 \tag{11 b}$$

$$\lambda = -\infty \quad (M = -1; G = 0) \qquad \sigma_T < 0 \tag{11 c}$$

Marangoni and natural convection are superimposed in the range $0 < \lambda < \infty$ and in opposition in the range $-\infty < \lambda < 0$.

3. THE BASE FLOW

If the aspect ratio of the cavity is sufficiently small, then a simple exact parallel-flow solution will exist away from thin turning region near the end walls. The parallel flow has been considered in detail by Cormack, Stone and Leal (1975) in the case of natural convection, whereas Sen and Davis (1982) and Strani, Piva and Graziani (1983) studied Marangoni convection.

The solution for combined convection is herein reported for the first time:

$$\bar{u} = a\left[-\frac{M}{4}(3\ z^2 - 2\ z) + \frac{G}{48}(8\ z^3 - 15\ z^2 + 6\ z)\right] \tag{12a}$$

$$\bar{T} = a\,x + a^2\,Pe\left[-\frac{M}{48}(3z^4 - 4z^3) + \frac{G}{48}\left(2\frac{z^5}{5} - 5\frac{z^4}{4} + z^3\right)\right] + b \tag{12b}$$

$$\frac{\partial \bar{p}}{\partial x} = a\left[-\frac{3}{2}M + G\left(z - \frac{5}{8}\right)\right] \tag{12c}$$

The parallel flow arises in response to the horizontal temperature gradient, the lateral walls induce a pressure gradient that sustains the return flow; convection of internal energy by this end-to-end counterflow is then

balanced by vertical diffusion.
The solution (12) has been derived from the application of the singular perturbations method, when the field variables are expanded in an asymptotic series in terms of the small parameter H/L. The constants (a) and (b) must be determined by matching the solution with the flow near the lateral walls. Because only the gradient of $\overline{T}$ appears in the stability problem, the constant (b) is inessential; the constant (a) is at a first approximation unitary, and we shall assume just a=1: this is equivalent to take the leading term of the asymptotic expansion. With these approximations, the base flow for the stability analysis, expressed in terms of the parameter λ instead of M and G, is:

$$\overline{u} = \frac{1}{1+|\lambda|}\left[-\lambda\left(\frac{3}{4}z^2 - \frac{1}{2}z\right) + \frac{z^3}{6} - \frac{5}{16}z^2 + \frac{z}{8}\right] \tag{13a}$$

$$\overline{T} = x + \frac{Pe}{1+|\lambda|}\left[-\lambda\left(\frac{z^4}{16} - \frac{z^3}{12}\right) + \frac{z^5}{120} - \frac{5}{192}z^4 + \frac{z^3}{48}\right] \tag{13b}$$

$$\frac{\partial \overline{p}}{\partial x} = \frac{1}{1+|\lambda|}\left(-\frac{3}{2}\lambda + z - \frac{5}{8}\right) \tag{13c}$$

We point out that the (13) is an exact solution of the problem (3), (4), (5) if one assumes a linear temperature distribution on the bottom wall of the cavity (Napolitano et al. 1990). In this case, the higher approximations of the asymptotic expansion are identically zero, and the lateral walls effect is felt in the core region only through the generation of a pressure gradient in the x-direction, that maintains zero max flux through any vertical plane.
The base flow velocity profiles u(z) for λ=0 (G=1; M=0); λ= +∞ (M=1; G=0); λ=-∞ (M=-1; G=0) are reported in Fig.2a. In Fig.2b the corresponding temperature profiles ($\overline{T}$-x)/Pe are shown. The turning points of the temperature profiles coincide with the points of zero velocity.
In Fig. 3a the velocity profiles are shown for different values of λ and the corresponding temperature profiles are reported in Fig. 3b; for λ<0 we can see two counter rotating cells.
In Fig. 4a and 4b we illustrated three particular profiles, corresponding to the values λ=-1/12, λ=-0.15, λ=-0.25 [profiles (a), (b) and (c), respectively].
For λ=-1/12 the surface velocity is zerp and the corresponding profile is symmetrical with respect to z=1/2. For λ=-0.15 the surface temperature is the same of that of the lower rigid surface and both are less than the temperature in the liquid. For values of λ of about -0.15 there are regions with positive and other ones with negative stratification: this can strongly influence the stability. For λ=-0.25, the lower cell due to the gravity disappears; this case is characterized by d $\overline{u}$/dz = 0 for z=0.

4. THE STABILITY PROBLEM

To investigate for what values of the parameters Re, Pr, M, G (or Re, Pr, λ) the solutions (13) are stable, small disturbances are introduced into the flow. The equations for the evolution of the disturbances are obtained by writing each flow variable as a sum of a base flow quantity and a corresponding small disturbance:

$$(u,v,w,p,T) = (\bar{u}, 0, 0, \bar{p}, \bar{T}) + (u',v',w',p',T'); \qquad (14)$$

where the overbar denotes the base flow and the apex the disturbances. The equations governing the initial growth or decay of the disturbance are obtained by introducing (14) in the Boussinesq system (3) and boundary conditions (4), (5), and dropping terms of second order in small quantity. The disturbances may be expanded in the form of normal modes:

$$(u',v',w',p',T') = \left(\hat{u}(z),\hat{v}(z),\hat{w}(z),\hat{p}(z),\hat{\vartheta}(z)\right) e^{\,i(\alpha x+\beta y)+\gamma t} \qquad (15)$$

and the amplitude functions $\left(\hat{u}(z),\hat{v}(z),\hat{w}(z),\hat{p}(z),\hat{\vartheta}(z)\right)$ are then governed by equations which take the form, by eliminating the symbol $\wedge$ in order to simplify the notation:

$$\left(L - i\alpha Re\ \bar{u} - \gamma\ Re\right) u(z) = i\,\alpha\, p + Re\, w\, D\, \bar{u} \qquad (16\,a)$$

$$\left(L - i\alpha Re\ \bar{u} - \gamma\ Re\right) v(z) = i\,\beta\, p \qquad (16\,b)$$

$$\left(L - i\alpha Re\ \bar{u} - \gamma\ Re\right) w(z) = D\, p - G\,\vartheta \qquad (16\,c)$$

$$\left(L - i\alpha Pe\ \bar{u} - \gamma\ Pe\right) \vartheta(z) = Pe\left(u\ \bar{T}_x + w\, D\ \bar{T}\right) \qquad (16\,d)$$

$$D\,w + i\,\alpha\, u + i\,\beta\, v = 0 \qquad (16\,e)$$

$$u = v = w = D\,\vartheta = 0 \qquad z = 0 \qquad (16\,f)$$

$$w = D\,u + M\,i\,\alpha\,\vartheta = D\,v + M\,i\,\beta\,\vartheta = D\,\vartheta = 0 \qquad z = 1 \qquad (16\,g)$$

where $L = D^2 - k^2$; $D = \dfrac{d}{dz}$; $\qquad k^2 = \alpha^2 + \beta^2$

$\bar{T}_x$ is the x-derivative of the base-flow temperature field; in the present study this derivative is equal to one but, for clearness sake, we retain it in the equations.

The parameters α and β are real wave numbers in the x and y directions and $\gamma = \gamma_r + i\ \gamma_i$ is the complex frequency of the disturbance, the real part of which is an amplification rate and the imaginary part an oscillation frequency. The problem (16) is a dispersion relation between α, β and γ, with Re, Pr, M, G as parameters; for fixed (α, β) the problem is an eigenvalue differential problem for γ. According to the linear theory, the flow is stable if, for given Re, Pr, M, G, the non-dimensional amplification rate γ_r is negative for all values of α and β. Conversely, the flow is unstable if γ_r is positive for some values of α and β.

The states for which $\gamma_r = 0$ are called states of neutral stability and are characterized by

$$\gamma_r\left(\alpha,\beta,Re,Pr,\lambda\right) = 0 \qquad (17)$$

If at the neutral state is $\gamma_i=0$ the instability arises with respect to stationary disturbances and the neutral state is called stationary; if results $\gamma_i \neq 0$ we say that the neutral state is oscillatory.
Eq. (17) defines the neutral stability Reynolds number

$$R_N = R_N(\alpha,\beta,Pr,\lambda) \tag{18}$$

Lastly, that neutral state for which the neutral Reynolds number has a minimum as a function of α and β is called critical state, and the corresponding Reynolds value is the critical Reynolds number:

$$R_C = \min_{\alpha,\beta} R_N(\alpha,\beta,Pr,\lambda) \tag{19}$$

4.1 Transverse and Longitudinal Modes

The system (16) simplifies for strictly two-dimensional transverse (β=0) and longitudinal (α=0) modes, the former being independent of the y coordinate, the latter independent of x.
For transverse modes the v-equation decouples from the set (16); its solution shows the v-component of the velocity to decay independently. The longitudinal modes depend only on y and z, but because of the base flow in the x-direction, the u-component of the disturbance velocity is not zero and the motion is tridimensional.
In stability analyses attention is often limited to transverse and longitudinal modes, and one does not consider stability with respect to tridimensional disturbances, as it is necessary when the Squire theorem is not valid, as in the present case. The above limitation can lead to erroneous results, as our analysis has clearly shown. We shall underline this point in the discussion of the results.

4.2 The Solution Method

We solved the eigenvalue problem posed in equations (16) by the Galerkin method (Mikhlin 1964, Fletcher 1984), after eliminating the pressure and the v-component of the velocity by using the continuity equation. The system results:

$$\left[L^2 - i\alpha Re\left(\bar{u}L - D^2\bar{u}\right) - \gamma\, Re L\right]w = k^2 G\,\vartheta \tag{20 a}$$

$$\left[L - i\alpha Re\,\bar{u} - \gamma\, Re\right]\left(u - \frac{i\alpha}{k^2}Dw\right) = \frac{\beta^2}{k^2} Re\, D\,\bar{u}\,w \tag{20 b}$$

$$\left[L - i\alpha Pe\,\bar{u} - \gamma\, Pe\right]\vartheta = Pe\left(u\,\bar{T}_x + w\,D\,\bar{T}\right) \tag{20 c}$$

$$u = w = Dw = D\,\vartheta = 0 \qquad z=0 \tag{20 d}$$

$$w = D^2w + M k^2\,\vartheta = Du + Mi\alpha\vartheta = D\,\vartheta = 0 \qquad z=1 \tag{20 e}$$

The variables w, u, ϑ, are expanded as

$$w = a_i\, w_i(z) \qquad i=1,\ldots,N \tag{21a}$$

$$u = b_i\, u_i(z) \qquad i=1,\ldots,N \tag{21b}$$

$$\vartheta = c_i\, \vartheta_i(z) \qquad i=1,\ldots,N \tag{21c}$$

where the constant coefficients a_i, b_i, c_i, in general complex, are unknown quantities, while the w_i, u_i, ϑ_i are a-priori given functions; in (21) the summation Einstein convention is utilized.
We assumed for w_i, u_i, ϑ_i the expressions:

$$w_i = z^2(z-1)\, C^{\bullet}_{i-1}(z) \tag{22a}$$

$$u_i = z\ C^{\bullet}_{i-1}(z) \tag{22b}$$

$$\vartheta_i = C^{\bullet}_{i-1}(z) \tag{22c}$$

where $C^{\bullet}_i$ is the i-th modified Chebychev polynomial of first kind defined in the range $0 \le z \le 1$.
The constants a_i, b_i, c_i are determined by inserting the (21) in the equations (20) and then taking the inner product with each trial function over the range $0 \le z \le 1$; use of integration by part allows to account for the natural conditions that are not satisfied by the single trial function.
The result is a generalized complex algebraic eigenvalue problem:

$$\underline{\underline{A}}\,\underline{x} - \gamma\,\underline{\underline{B}}\,\underline{x} = 0 \tag{23}$$

where $\underline{x}^T \equiv \left(a_1,\ldots,a_N,b_1,\ldots,b_N,c_1,\ldots,c_N \right) \equiv \left(\underline{a}^T,\underline{b}^T,\underline{c}^T \right)$ is the transposed of the column vector $\underline{x}$ of the Galerkin coefficient $\underline{a}$, $\underline{b}$, $\underline{c}$. The matrices $\underline{\underline{A}}$ and $\underline{\underline{B}}$ (see Viviani 1988), are of dimension 3N x 3N, and they are real in the case of longitudinal disturbances (α = 0). For β = O the problem can be reduced to dimension 2N x 2N.
The eigenvalue problem (23) has been solved by using the EIGZC subroutine from the IMSL library (Cowell 1984).
For fixed Pr and λ, the neutral state (R_N) is calculated for each α and β by using the secant iteration method to meet the condition $\gamma_r = 0$, where γ_r is the real part of the principal eigenvalue, i.e. the eigenvalue with the greatest real part; then a minimization on α and β determines the critical state (R_c).

5. RESULTS AND DISCUSSION

The linear stability boundary has been defined calculating the critical Reynolds number $R_c = R_c(Pr, \lambda)$, for values of the parameter λ varying in the whole of its range of definition ($-\infty$, $+\infty$) and then covering all the cases of free convection (Marangoni, natural, combined).
For the Prandtl number, we analyzed the range (10^{-4}, 10^3), thus covering both the cases of liquid metals [$Pr \ll 1$] and of silicon oils [$Pr \ge O(1)$], often used

in experiments performed in microgravity conditions, as well as the limit cases $Pr \to 0$ and $Pr \to \infty$.
In all the cases analyzed, the instability occurs always with respect to oscillatory disturbances.

5.1 Marangoni Convection ($\lambda = \pm\infty$)

Fig.5 shows the critical Reynolds number versus Prandtl, for the case of Marangoni Convection ($\lambda = \pm\infty$). The instability occurs with respect to tridimensional disturbances, for any value of the Prandtl number. The critical curves for the two cases $\lambda = \pm\infty$ coincide. The two cases $\lambda = \pm\infty$ correspond to an inversion of velocity and vertical temperature fields (see Fig.2), but for $\lambda = -\infty$ (surface tension increasing with the temperature) the flow is in direction of the horizontal thermal gradient and for $\lambda = +\infty$ it is in opposite direction. Nevertheless one must take into account that the Marangoni stress of the perturbation field is also inverted, so that the physical effect that gives rise to the onset of instability is unchanged. From a mathematical point of view this result follows from the property of invariance of the system (20) under the following transformation:

$$(M,u,w,\vartheta,\alpha, \bar{u},D \bar{T}) \text{ ------> } (-M,-u,w,-\vartheta,-\alpha,- \bar{u},-D \bar{T}) \qquad (24)$$

The case $\lambda = +\infty$ (M=+1) has been analysed by Smith (1982), who solved the eigenvalue problem by using a method of direct integration of the equations (this method does not guaranteee the convergence to the principal eigenvalue and a numerical investigation of the spectrum must be performed, see Lee and Reynolds 1967). Next table shows that the critical Reynolds number, for different values of Prandtl, obtained with the two different methods, are in excellent accordance.

Pr	R_c	
	Present study	Smith (1982)
0.1	223.653	223.667
1	116.028	116.022
5	45.2801	45.2794

We have also evaluated the critical value R_c for Pr=0, by assuming that R_c remains finite for $Pr \to 0$. Then, the value $R_c = R_c(Pr=0, \lambda = \pm\infty)$ can be calculated by setting Pe=0 in the system (20). The computed value is $R_c = 28208$. We consider now the limit $Pr \to \infty$. As Fig.5 shows, for $Pr \to \infty$ $PrR_c \to B$ (constant) and then $R_c \to 0$ as B/Pr.
The asymptotic form of the system (20), for $Pr \to \infty$, is obtained by setting PrRe=B and Re=0. The equations (20 a-c) become

$$L^2 w = k^2 G\vartheta \qquad (25a)$$

$$L\left(u - \frac{i\alpha}{k^2} Dw\right) = 0 \qquad (25b)$$

$$(L - i\alpha B \bar{u} - \gamma B)\vartheta = B(u \bar{T}_x + w D \bar{T}) \qquad (25c)$$

and the corresponding algebraic problem can be formally expressed as:

$$\underline{\underline{U}}\,\underline{c} = \gamma B \underline{\underline{V}}\,\underline{c} \qquad (26)$$

where $\underline{c}^T = (c_1, \ldots\ldots\ldots\ldots\ldots\ldots c_N)$ are the Galerkin coefficients for the temperature ϑ.
The expression for $\underline{\underline{U}}$ is given by

$$\underline{\underline{U}} = \underline{\underline{L}} - B\,\underline{\underline{G}}\,\underline{\underline{C}}^{-1}\,\underline{\underline{D}} - i\alpha B\,(\underline{\underline{Q}} + \underline{\underline{H}}\,\underline{\underline{F}}^{-1}\,\underline{\underline{N}}\,\underline{\underline{C}}^{-1}\,\underline{\underline{D}}) \qquad (27)$$

The different matrices that appear in (27) and the matrix $\underline{\underline{V}}$ in (26) are defined (see Viviani 1988) as submatrices of the matrices $\underline{\underline{A}}$ and $\underline{\underline{B}}$ of the full problem (23).
The asymptotic value for $B=PrR_c$, calculated using the (26), is equal to 397.8, whereas for $Pr=10^3$ the product $Pr\ R_c$ is 395.3. The asymptotic behaviour of R_c for $Pr \rightarrow \infty$ is then $R_c \rightarrow 397.8/Pr$.

5.2 Natural Convection ($\lambda = 0$)

The case of natural convection has been studied by Hart (1983) who considered only longitudinal disturbances. A comparison between the results of the present study and those obtained by Hart is shown in Fig.6, where the critical Reynolds number versus Prandtl is reported. The solid line corresponds to Hart results, found here as particular case of the analysis performed with respect to tridimensional disturbances. For very low values of the Prandtl number ($Pr<10^{-3}$) the less stable "modes" are transverse (see Fig.6). Increasing the Prandtl number the transition towards longitudinal modes occurs by means of tridimensional disturbances. For a Prandtl number close to 0.5 the longitudinal disturbances become very stable, so that the analysis performed by Hart does not provide for cases over this value. In fact, close to the value of $Pr=0.5$ tridimensional modes again rise, joining the curve of Hart with that of the transverse modes, typical of the instability for high values of the Prandtl number. An analysis of the various contributes to the disturbance energy (Viviani 1988) and of their dependence on the Prandtl number for transverse, longitudinal and tridimensional disturbances, should be useful to explain the type of disturbance with respect to which instability sets in.
Then, not only we corrected the stability curve of Hart, but also we pointed out that an analysis with respect to tridimensional disturbances is essential if one is interested in the correct definition of the stability boundary. An instability with respect to two-dimensional disturbances must rise from the general analysis, not characterizing a-priori the kind of the disturbance, but investigating the wave number value at the critical point. For instance, for the case, previously discussed, of Marangoni convection, since we found that the instability always occurs with respect to tridimensional disturbances, the assumption of two-dimensional disturbance should give rise to an over-estimation of the critical Reynolds number, for any value of the Prandtl number. As Fig.6 shows, R_c approaches its limit value, for $Pr \rightarrow 0$, much more quickly than in the case of Marangoni convection. The calculated value of R_c,

for Pr=0 is equal to 7690.
For $Pr \to \infty$, $R_c \to 0$. However the behaviour is different from the case, previously discussed, of Marangoni convection. The numerical results show that in this case R_c decreases as $Pr^{-1/2}$. Therefore, in this case of natural convection, the proper limit for $Pr \to \infty$ is $Pr^{1/2} R_c \to const.$
The expression (25) for the eigenvalue problem, for $Pr \to \infty$, is not more valid, because $Pr\ R_c \to \infty$. The law of variation of $R_c \to 0$ as $Pr^{-1/2}$, for $Pr \to \infty$ has been also found by Gill and Davey (1969) who studied instability of natural convection along a vertical plate.

5.3 Combined Convection (λ finite, different from zero)

Fig.7 shows the critical curves for three different values of the parameter $\lambda > 0$. As λ decreases the critical curves take the shape peculiar to the case of natural convection ($\lambda=0$) (Fig.6). However, we do not see yet the appearance of transverse disturbances for very low Prandtl numbers. They become less stable than the longitudinal disturbances for lower values of the parameter λ. This appears clearly in Fig.8, where the critical curve for $\lambda=-0.04$ and the stability boundary with respect to longitudinal modes are shown. From this figure we can note as the instability to longitudinal disturbances, typical of $\lambda=0$ for $Pr<1$ (Fig.6), is replaced by an instability to tridimensional disturbances. Figs.9 and 10 show the evolution of the stability boundary towards the typical case of Marangoni convection.
The values $R_c(Pr=0, \lambda)$ are reported in Table 1.
Consider now the behaviour of the critical Reynolds number for $Pr \to \infty$. As in the cases of Marangoni and natural convection, for any value of λ, for $Pr \to \infty$ $R_c \to 0$. Except for the case $\lambda=0$, where we found that $R_c \to 0$ as $Pr^{-1/2}$, in all the other cases investigated $[\,|\lambda| \in (10^{-2}, \infty)\,]$, for $Pr \to \infty$, $R_c \to B(\lambda)$. This behaviour is evident from the analysis of Table 2, where the values of R_c and PrR_c, for $Pr=10^2$ and $Pr=10^3$, are reported for different values of the parameter λ, together with the limit for $Pr \to \infty$, i.e. the $B(\lambda)$ values calculated by (26). Then the asymptotic formula for R_c is:

$$Rc \to \frac{B(\lambda)}{Pr} \qquad [\,|\lambda| \in (10^{-2}, \infty)\,] \tag{28}$$

Since for $\lambda = \pm 10^{-2}$ R_c still decreases towards zero as Pr^{-1}, the case $\lambda=0$ seems to have very particular properties. We believe that this case should be investigated thoroughly, studying, with suitable analytical methods, the asymptotic characteristics for $Pr \to \infty$, for small values of the parameter λ.
A theoretical approach appears indispensable because, for high values of the product $Pr\ R_c$, one cannot be certain of the validity of the numerical results. In fact the spectrum for the temperature perturbation (coefficient c_i) could be so wide that, from the numerical point of view, the problem should be solved by means of direct integration of the equations (if the spectrum is very wide there are numerical difficulties, for the solution of the algebraic problem (23), because one obtains matrices of great dimensions).
The results of the three cases previously analysed can be illustrated in the plane (Re, λ), for different values of the Prandtl number. They are shown in

Figs. 11 and 12.
We can see the presence of a peak of stability for negative values of λ, near the value $\lambda=-0.15$ (remind what we said about this particular value of λ when we discussed the base flow in section 3). For very low Prandtl, we can see a lack of stability, near the point $\lambda=0$. This occurrence can be explained if one remember that, for very low values of Prandtl, the instability is due mainly to the Reynolds stress, so that the stability strongly depends upon the velocity profile (see Figs.3 and 4).
For Pr=0.1 the lack of stability disappears, because the thermal convection prevails and then the influence of the velocity profile shape on the onset of instability becomes less important.
Increasing the Prandtl number a strong peak of stability for $\lambda=0$ occurs. Concerning this, two remarks are needed. First, for Prandtl numbers greater than or equal to one, the Marangoni effect considerably contributes to the disturbance energy also for values of λ very close to zero, so that when this effect vanishes, for $\lambda=0$, a strong increase of the stability boundary can occur. On the other hand, as already remarked, the case of natural convection presents particular behaviours and it needs of further investigations; in this regard an analysis of the dependence on λ of the disturbance energy for high Prandtl number will be helpful.

6. CONCLUSIONS

A linear stability analysis of Marangoni, natural and combined free convection in shallow cavity has been performed.
The "base flow", the stability of which has been analyzed, is a close form solution depending on the non-dimensional parameters Marangoni, Grashof, Reynolds and Prandtl numbers, under the hypotheses of planar and undeformable free surface, and of surface tension linearly dcreasing or increasing with the temperature.
The stability analysis has been performed with respect to tridimensional disturbances, and the resulting eigenvalue problem, for the evaluation of the stability boundary, has been solved with the spectral Galerkin method.
The critical Reynolds number has been calculated in the whole range of definition of the dynamic Bond number, and for Pr number ranging from 10^{-4} to 10^{3} as well as for the limiting cases $Pr\rightarrow 0$ and $Pr\rightarrow\infty$. For all values of Prandtl and Bond numbers we always found an oscillatory instability.
The stability boundary, individuated by the critical Reynolds number, has been shown in the Reynolds-Prandtl plane, where the critical curves are reported in terms of a parameter λ, having the meaning of the inverse of the dynamic Bond number.
In the case of Marangoni convection, we have shown that the linear stability boundary is the same for both the cases of surface tension linearly decreasing and increasing with the temperature; our results are in accordance with those ones reported in a previous study of Smith (1982), who considered only the more common case of a linear decreasing dependence of the surface tension on the temperature.
For natural convection, our analysis with respect to tridimensional disturbances allowed to correct and to extend the results of Hart (1983) whose study was limited to two-dimensional longitudinal modes. Thus, we have pointed out that an analysis with respect to tridimensional disturbances is

essential if one is interested in the correct evaluation of the stability boundary.
Asymptotic laws for the critical Reynolds number, for high value of the Prandtl number, has been established. In the case of natural convection we found that R_c decreases as $Pr^{-1/2}$; for the other cases of Marangoni and combined convection R_c decreases as $B(\lambda)/Pr$, where the function $B(\lambda)$ has been numerically evaluated by solving the asymptotic form of the eigenvalue problem for $Pr \rightarrow \infty$.
The critical curves have been also reported in the plane (Re, λ) for various values of the Prandtl number, and in this representation is evident a strong peak of stability, for high values of the Prandtl number, at the point $\lambda = 0$ (natural convection). In view of this result, and of the asymptotic behaviour of the critical Reynolds number for $Pr \rightarrow \infty$ in the case $\lambda=0$, further studies on this subject are needed.
An investigation of the dependence of the disturbance energy on Pr and λ should help to assess the leading mechanism of instability for the various types of disturbances (transverse, longitudinal, tridimensional) and to understand the "preferred mode" of instability. This will be done in a future work, where the authors intend to performe a careful energetic analysis of the disturbance and draw in the plane (Pr, λ) the regions for the various preferred modes. Future works will be also devoted to assess the range of validity of the present results in terms of the cavity aspect ratio (H/L). In fact, the existence of the base flow herein considered depends on the order of magnitude of the Reynolds and Peclet numbers and on the cavity aspect ratio; the next study will be just devoted to analyze in the plane (Re, H/L) or (Pe, H/L) the physical existence of the parallel flow solution when the linear stability boundary is attained.

7. ACKNOWLEDGEMENTS

Part of this work has been developed in the Ph.D. Dissertation of A.Viviani.

8. REFERENCES

CHANDRASEKHAR, S. 1961 *Hydrodynamics and Hydromagnetic stability*, Oxford University Press
CORMACK, D.E., STONE, G.P. & LEAL, L.G. 1975 The effect of upper surface conditions on convection in a shallow cavity with differentially heated end-walls. *Int. J. Heat Mass Transfer* **18**, 635
COWELL, W.R. (Ed.)1984 *Sources and Development of Mathematical Software*, Prentice-Hall
FLETCHER, L.A.J. 1984 *Computational Galerkin methods*, Springer-Verlag
GILL, A.E. & DAVEY, A. 1969 Instabilities of a buoyancy driven system. *J. Fluid Mech.*. **35**, 775
GUYON, E. & PANTALONI, J. 1980 Effect de tension superficielle sur la convection de Rayleigh-Benard. *C.R. Acad. Sci. Paris* **290B**, 301
HART, J. 1983 A note on the stability of low-Prandtl-number Hadley circulations. *J. Fluid Mech.* **132** , 271

HURLE, T.D.J., JAKEMAN, E. & JOHNSON, C.P. 1974 Convective temperature oscillations in molten gallium. *J. Fluid Mech.* **64** , 565
HURLE, D.T.J. & JAKEMAN, E. 1981 Introduction to the techniques of crystals growth. *PhysicoChem. Hydrodynam.* **2**, 237
LEBON, G. 1988 Problems in Marangoni instabilities: A review. *NATO ASI Summer school on physico-chemical hydrodynamics,* La Rabida-Huelva (Spain)
LEE, L.H. & REYNOLDS, N.C. 1967 On the approximate and numerical solution of Orr-Sommerfield problems. *Quart. J. Mech. Applied Math.* **20** , 1
MIKHLIN, S.G. 1964 *Variational Methods of mathematical physics,* Mc Millan
NAPOLITANO, L.G. 1978 Microgravitational fluid Dynamics. *2nd Levitch Conference,* Washington
NAPOLITANO, L.G. 1979 Thermodynamics and dynamics of surface phases. *Acta Astronautica* **6**, 1093
NAPOLITANO, L.G. 1982 Surface and buoyancy driven free convection. *Acta Astronautica* **9**, 199
NAPOLITANO, L.G. 1989 Theoretical modelling of Micro-gravitational Thermofluidynamics.*Workshop on Space Physics: Materials in Microgravity,* International Center for Theoretical Physics, Trieste (Italy)
NAPOLITANO, L.G. & VIVIANI A.1990. *To appear*
NAPOLITANO, L.G., VIVIANI, A. & SAVINO R. 1990. *To appear*
SEN, A.K. & DAVIS, S.H. 1982 Steady thermocapillary flows in two-dimensional slots. *J. Fluid Mech* **121**, 163
SMITH, M.K. 1982 The instabilities of thermocapillary shear layers. *Ph.D. Dissertation* , Northwestern University, Evanston (Illinois)
SMITH, M.K. & DAVIS, S.H. 1983 Instabilities of dynamic thermocapillary liquid layers. *J. Fluid Mech.* **132** , 119
STRANI, M., PIVA, R. & GRAZIANI, G. 1983 Thermocapillary convection in a rectangular cavity: Asymptotic theory and numerical simulation. *J. Fluid Mech.* **130** , 347
VIVIANI, A. 1988 *Ph.D. Dissertation* in Aerospace Engineering, University of Naples, Naples (Italy). in Italian

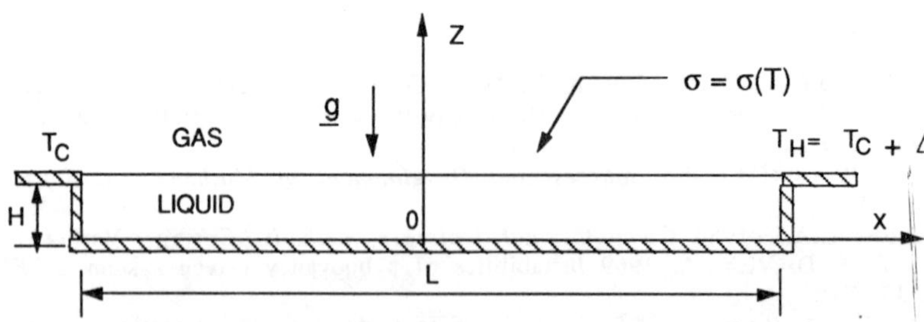

Fig. 1 Geometry of the problem.

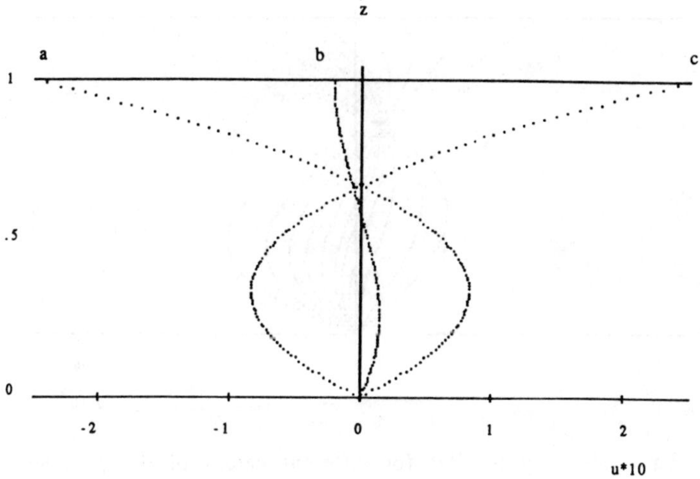

Fig. 2a Velocity profiles for Marangoni convection, (a) : M = 1 and $\lambda = +\infty$; (c) : M = - 1 and $\lambda = -\infty$; and for natural convection, (b) : G = 1 and $\lambda = 0$.

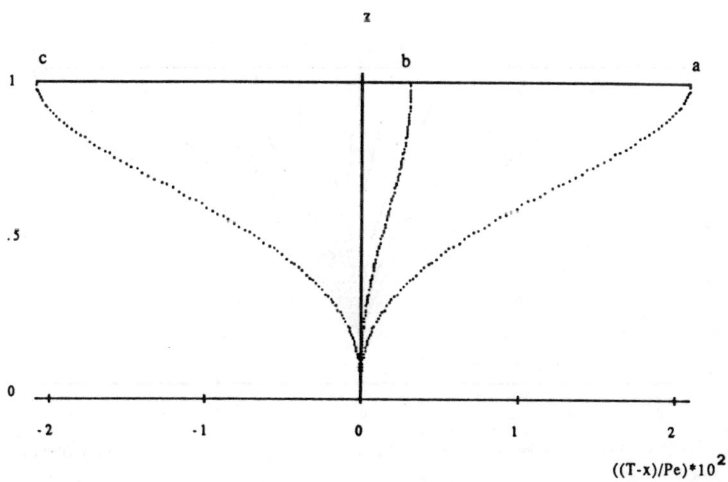

Fig. 2b Temperature difference profiles between free surface and bottom wall corresponding to the velocity profiles of Fig. 2a.

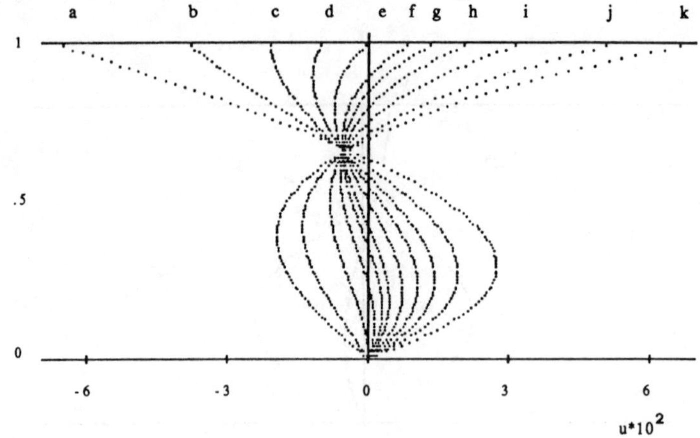

Fig. 3 a Velocity profiles for different values of the parameter λ:
(a : λ = 0.25) (b : λ = 1/12) (c : λ = 0)
(d : λ = -1/24) (e : λ = -1/12) (f : λ = - 0.125)
(g : λ = - 0.15) (h : λ = - 0.1875) (i : λ = - 0.25)
(j : λ = - 0.375) (k : λ = - 0.5).

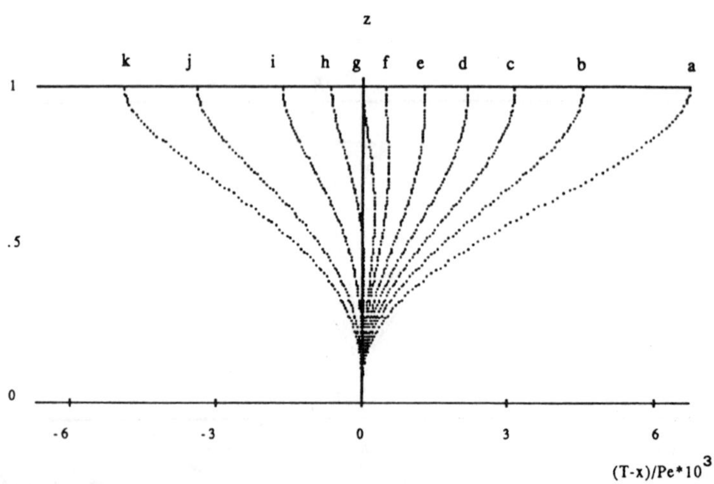

Fig. 3 b Temperature difference profiles between free surface and bottom wall corresponding to the velocity profiles of Fig. 3a.

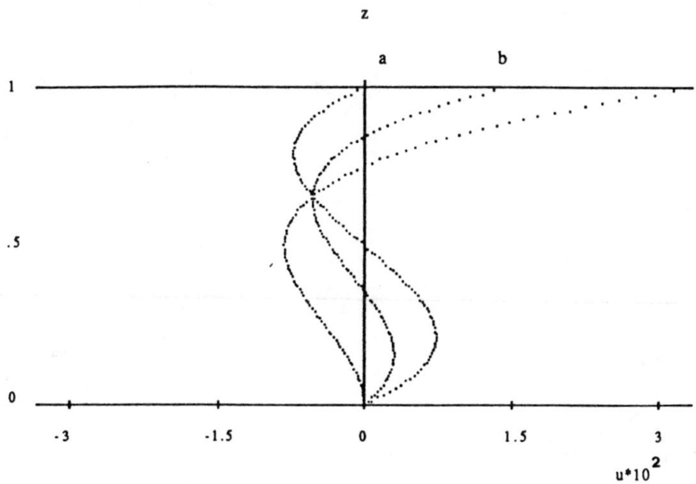

Fig. 4 a Velocity profiles for λ = -1/12 (a); λ = - 0.15 (b); λ = - 0.25 (c).

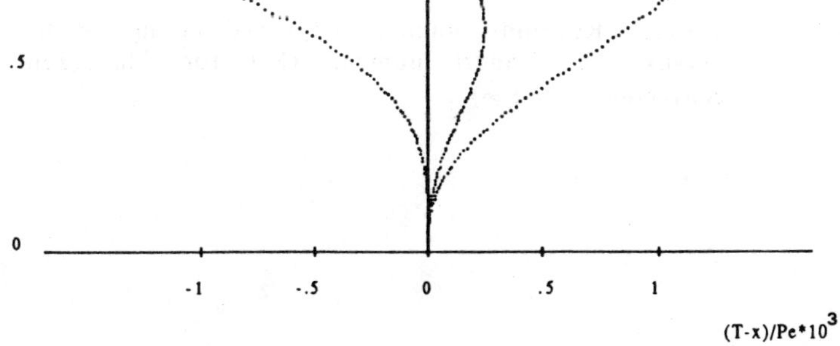

Fig. 4 b Temperature difference profiles between free surface and bottom wall for λ = -1/12 (a); λ = - 0.15 (b); λ = - 0.25 (c).

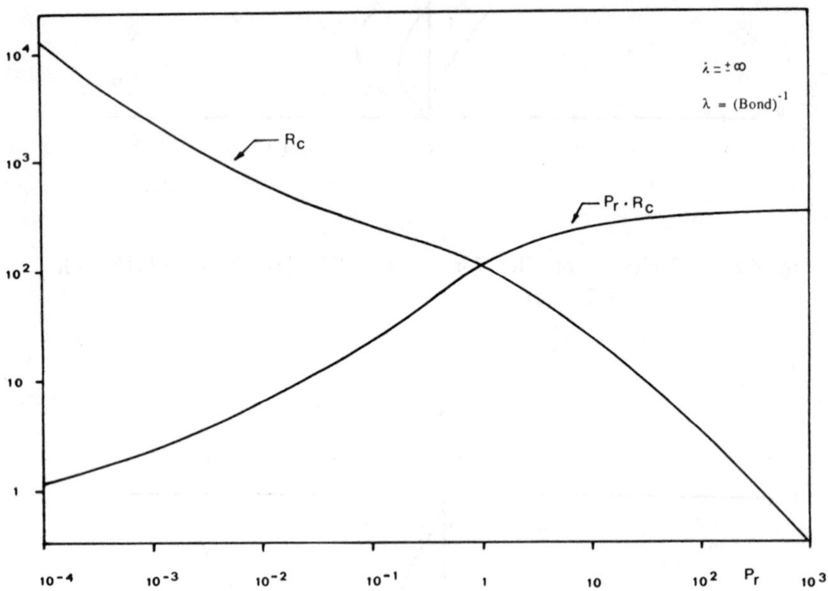

Fig. 5 Critical Reynolds number (R_c) and product $Pr{\bullet}R_c$ versus the Prandtl number (Pr) for Marangoni convection ($\lambda = \pm\infty$).

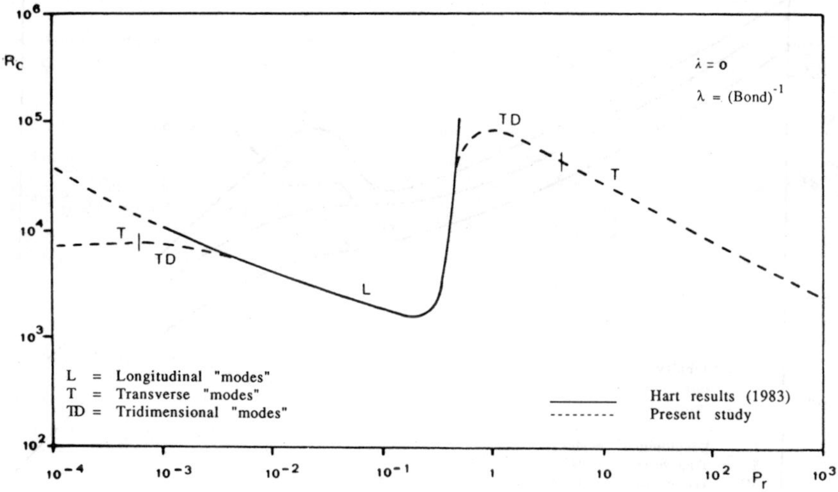

Fig. 6 Critical Reynolds number (R_c) versus the Prandtl number (Pr) for natural convection ($\lambda = 0$).

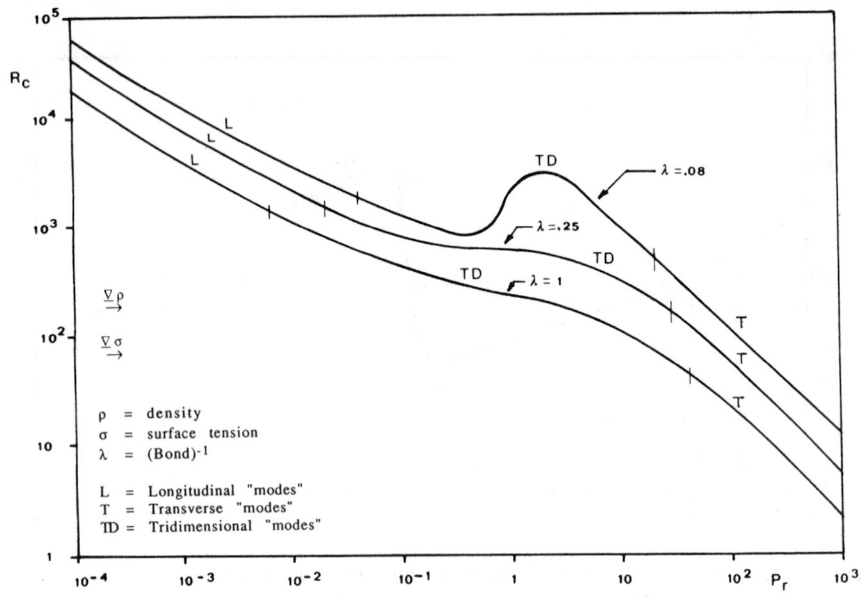

Fig. 7 Critical Reynolds number (R_c) versus the Prandtl number (Pr) for combined convection with $\lambda > 0$.

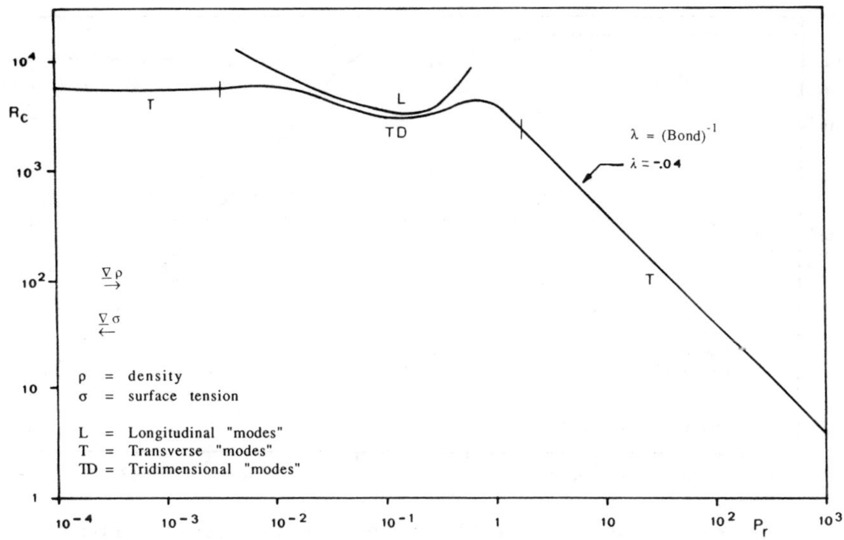

Fig. 8 Critical Reynolds number (R_c) versus the Prandtl number (Pr) for λ = - 0.4.

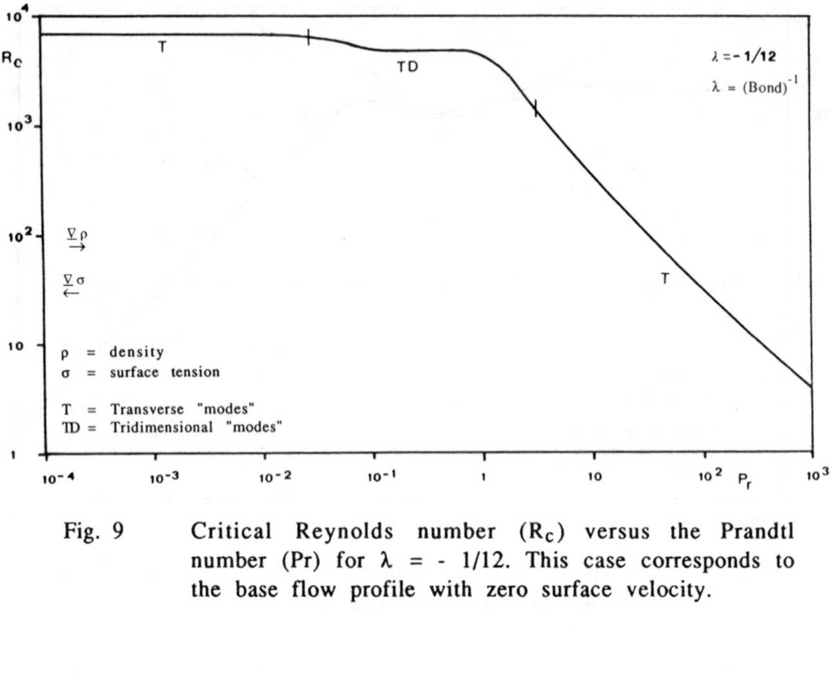

Fig. 9 Critical Reynolds number (R_c) versus the Prandtl number (Pr) for $\lambda = -1/12$. This case corresponds to the base flow profile with zero surface velocity.

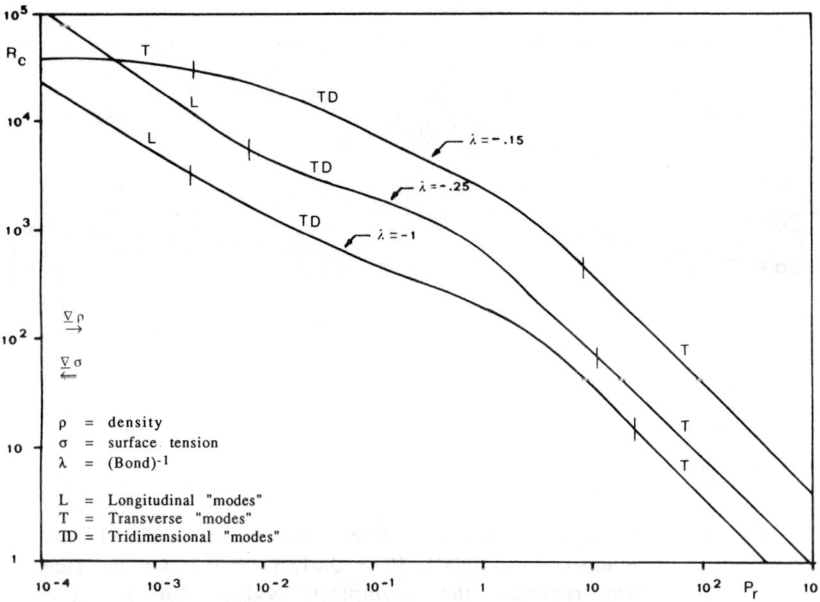

Fig. 10 Critical Reynolds number (R_c) versus the Prandtl number (Pr) for some values of $\lambda < 0$.

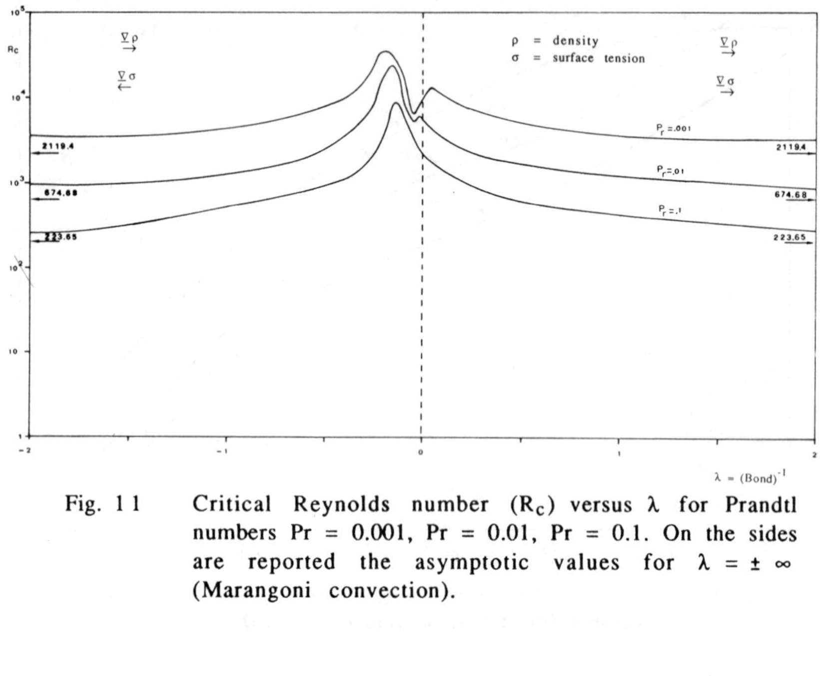

Fig. 1 1 Critical Reynolds number (R_c) versus λ for Prandtl numbers Pr = 0.001, Pr = 0.01, Pr = 0.1. On the sides are reported the asymptotic values for $\lambda = \pm \infty$ (Marangoni convection).

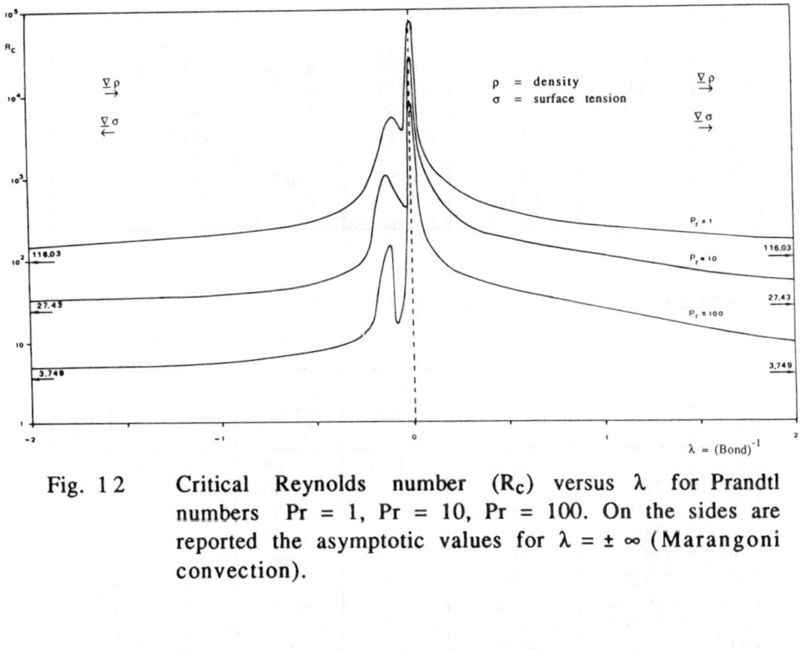

Fig. 12 Critical Reynolds number (R_C) versus λ for Prandtl numbers Pr = 1, Pr = 10, Pr = 100. On the sides are reported the asymptotic values for $\lambda = \pm\infty$ (Marangoni convection).

TABLE 1. Critical Reynolds number Rc(Pr,λ) and product Rc Pr for various values of λ for Pr=0, Pr=0.0001, Pr=0.001.

	Pr=0.001		Pr=0.0001		Pr=0	
λ	Rc	Rc Pr	Rc	Rc Pr	Rc	Rc Pr
+∞	2119	2.119	13041	1.3041	28208	0
1	3769	3.769	22714	2.2714	71769	0
0.5	5089	5.089	30285	3.0285	186112	0
0.25	7026	7.026	40960	4.0960	327768	0
0.08	10911	10.911	63628	6.3628	97271	0
0.04	13134	13.134	33552	3.3552	33564	0
0.01	10112	10.112	10015	1.0015	10009	0
0	7743	7.743	7694	0.7694	7690	0
-0.01	6603	6.603	6581	0.6581	6579	0
-0.04	5364	5.364	5351	0.5351	5348	0
-1/12	6212	6.212	6202	0.6202	6201	0
-0.125	22631	22.631	22595	2.2595	22593	0
-0.15	29176	29.176	29105	2.9105	29098	0
-0.25	21708	21.708	121927	12.1927	124910	0
-0.5	8588	8.588	57339	5.7339	88751	0
-1	4880	4.880	30850	3.0850	52691	0
-∞	2119	2.119	13041	1.3041	28208	0

TABLE 2. Critical Reynolds number Rc(Pr,λ) and product Rc Pr for various values of λ for Pr=100, Pr=1000, Pr=∞.

	Pr = 100		Pr = 1000		Pr = ∞	
λ	Rc	Rc Pr	Rc	Rc Pr	Rc	Rc Pr
+∞	3.749	3749	0.395	395	0	398
1	21.501	21501	2.165	2165	0	2267
0.5	32.282	32282	3.249	3249	0	3256
0.25	53.621	53621	5.390	5390	0	5391
0.08	130.132	130132	13.060	13060	0	13053
0.04	223.653	223653	22.450	22450	0	22458
0.01	644.412	644412	66.840	66840	0	64881
0	8729.301	8729301	2813.100	2813100	0	∞ (*)
-0.01	198.591	198591	19.870	19870	0	19872
-0.04	44.371	44371	4.436	4436	0	4428
-1/12	32.023	32023	3.164	3164	0	3159
-0.125	140.352	140352	14.160	14160	0	14184
-0.15	44.463	44463	4.490	4490	0	4490
-0.25	9.531	9531	0.958	958	0	960
-0.5	5.323	5323	0.536	536	0	537
-1	4.143	4143	0.419	419	0	420
-∞	3.749	3749	0.395	395	0	398

(*) For λ=0 (natural convection) is $Rc\,Pr^{1/2} \rightarrow$ const

NON LINEAR STABILITY RESULTS FOR A DISCRETE VELOCITY MODEL OF A GAS IN A SLAB.

Mariarosaria Padula

Dipartimento di Matematica, via Machiavelli 35, 44100 Ferrara, Italy.

The discrete velocity model in the kinetic theory of gases, has been used by many authors, see Gatignol (1975) for a clear description of the fundamental statements and Illner & Platowski (1988) for a recent survey. In this note we are concerned with the study of non-linear stability for a threedimensional model with 2n velocities in which at least two velocities have different modulus.Moreover, we shall assume the density function to depend only on one spatial variable, say x in a fixed direction **i**. Physically this last assumption reflects the behaviour of a gas between two parallel plates orthogonal to the direction **i**. As a consequence the most interesting mathematical well-posedness to be analized here appears to be the initial boundary value problem. Only recently Kawashima (preprint) proved an existence theorem for global solutions of a general discrete velocity model satisfying quite general boundary value conditions, as those of pure diffuse reflection or of general reflection. The main objective of this work is just connected with the initial boundary value problem in the sense that we prove that the above mentioned boundary conditions are actually *stabilizing*. This scope is achieved by analizing the stability of a particular Maxwellian state compatible with the boundary data. Also the method here used seems to be new in this field. Precisely, we employ the Ljapunov method in the manner suggested by Galdi & Padula (1990), in which an algorithm for the construction of the most appropriate Liapunov functional is furnished. This approach was then, in a series of successive papers, successfully applied to several stability problems of the classical Fluid-dynamics. However, the equations governing the evolution of the density probability for a discrete velocity model are quite different from the equations governing the flow of a fluid and several modifications of the method are needed. The stability result we achieve here reads as follows

Stability theorem Any regular perturbation $\{f_i\}$ to an equilibrium state F_i=s, either verifying diffuse boundary conditions or partially reflecting boundary conditions or unperturbed inflow data or subject to periodicity and symmetry conditions at the boundary, decays in the norm of $W^{1,2}(0,1)$ exponentially in time to zero once the initial data are "*sufficiently small*".

As it can be physically expected, the size of the initial data and the rate of decay for the perturbations depend on the basic state s, the number of velocities fixed by the model and the moduli v_i of such velocities. We also like to stress that the role played by the nonlinearities becomes crucial in limiting the size of the initial data and the distance between the planes as well.

Moreover, as obvious consequence of the theorem also the exponential decay of the supremum of the f_i is true.

By necessity, the proofs of calculations and estimates needed in this paper will be not given here. The results represent joint work with H. Babovski, and the detailed results and proofs may be found in Babovski & Padula (preprint).

n.1 Mathematical preliminaries.

In the sequel we will denote by $\|\cdot\|$ the usual L^2–norm and by $\phi_{,x}$ the partial derivative with respect to x.The following theorem constitues a generalization of the usual Poincare' inequality.

Lemma Let X and Y with $X \subseteq Y$ be two Banach spaces and let $|\cdot|_K$ denote a seminorm in X which becomes a norm in a close subspace $K \subseteq X$. Moreover, assume the set $\{u \in K: \|u\|_Y + \|u\|_K \leq 1\}$ is compact in Y. Then, there exists a constant $\gamma_o > 0$ such that for all $u \in K$

(1) $\|u\|_Y \leq \gamma_o |u|_K$; $\sup|\varphi(x)|^2 \leq (2\|\varphi_{,x}\| + \|\varphi\|)\|\varphi\|$.

The discrete velocity model which will be considered throughout the paper is described by the the 2n velocities $v_{\pm i}$, i=1,...n. The evolution of the density probability $F_i(x,t)$ to find a particle of the gas at the point x, at thc time t with velocity v_i is governed by the following system

(2) $F_{i,t} + v_i F_{i,x} = \sum_{j=0}^{n} F_j F_{-j} - nF_i F_{-i}$, $i=\pm1,\ldots,\pm n$,

where $v_i = v_1 \cdot i$.To system (2) will be appended the following boundary conditions:

(a) Unperturbed inflow at the planes x=0 and x=1 :

$F_{+i}(0,t)=s, \quad F_{-i}(1,t)=s;$

(b) Periodicity at the planes x=0,1,and symmetry inside the slab together with a compatibility condition

$F_{\pm i}(0,t)=F_{\pm i}(1,t); \quad F_{\pm i}(x,t)=F_{\mp i}(1-x,t), \qquad x \in (0,1);$

$\sum_{i=1}^{n} \int_0^1 (F_i(x,t)+F_{-i}(x,t))dx = 2ns;$

(c) Diffuse boundary conditions with the compatibility condition $(b)_3$ for the four velocity model;

$$F_i(0,t)=c\sum_{j=1}^{n} v_j F_{-j}(0,t); \quad F_{-i}(1,t)=c\sum_{j=1}^{n} v_j F_j(1,t);$$

(d) Partially reflecting boundary data :

$$F_i(0,t) = \sum_{j=1}^{n} a_{ij} F_{-j}(1,t)+b_i;$$

$$F_{-i}(1,t)= \sum_{j=1}^{n} a_{ij} F_j(1,t)+b_i;$$

where the matrix $A=A(v_i-v_j,x,t)=a_{i,j}$ represents the probability density that a molecule striking the wall at velocity v_i at the point x and time t reemerges with velocity v_{-j} and the vector b denotes the flux of particles absorbed by the walls. In the sequel we shall assume that A defines a positive definite quadratic form, and that all numbers b_i are positive (absorption only).

Denote by $F_i(x,t)=s+f_i(x,t)$ any flow satisfying one of the boundary conditions (a)-(d), we want to prove that the $W^{1,2}$norm of the perturbation $f_{\pm i}(x,t)$ to s or, equivalently, of the new variables $\sigma_i=f_i+f_{-i}$, $\delta_i=f_i-f_{-i}$ decays exponentially to zero in time. As it can be easily checked the new variables $\sigma_i(x,t)$, $\delta_i(x,t)$ verify the following system

$$\sigma_{i,t}+v_i\delta_{i,x}=2s\sum_{j=1}^{n}(\sigma_j-\sigma_i)+(1/2)\sum_{j=1}^{n}(\sigma_j^2-\sigma_i^2)$$

$$-(1/2)\sum_{j=1}^{n}(\delta_j^2-\delta_i^2)$$

(3)

$$\delta_{i,t}+v_i\sigma_{i,x} = 0, \quad i=1,\dots,n$$

with the following boundary conditions:

(aa) $\sigma_i(0,t)+\delta_i(0,t) = 0; \quad \sigma_i(1,t)-\delta_i(1,t) = 0;$

(bb) $\sigma_i(0,t) = \sigma_i(1,t); \quad \delta_i(0,t) = \delta_i(1,t);$

(cc) $\sigma_i(0,t)+\delta_i(0,t) = c\sum_{j=1}^{n} v_j(\sigma_j(0,t)-\delta_j(0,t));$

$\sigma_i(1,t)-\delta_i(1,t) = c\sum_{j=1}^{n} v_j(\sigma_j(1,t)+\delta_j(1,t));$

(dd) $\sigma_i(0,t)+\delta_i(0,t) = \sum_{j=1}^{n} a_{ij}(\sigma_j(0,t)-\delta_j(0,t));$

$\sigma_i(1,t)-\delta_i(1,t) = \sum_{j=1}^{n} a_{ij}(\sigma_j(1,t)+\delta_j(1,t)).$

Moreover, the compatibility condition $(b)_3$ reads

(4) $\sum_{i=1}^{n}\int_0^1 \sigma_i(x,t)dx = 0.$

In the sequel, in order to deal with the nonlinear terms we shall need the Lemma above stated concerning a generalization of the Poincare' inequality. To this end, let us use the symbol u to denote a 2n vector with components (σ_i,δ_i), and let us fix $X=W^{1,2}$, $Y=L^2$. Moreover,the subspace K will take into account the boundary conditions (a)-(d) in the following manner:

(Ka) $K = \{ u\in X : u \text{ verifies (aa)}, \quad |u|_K^2 \equiv \|u_{,x}\|^2 \}$

(Kb) $K = \{ u\in X : u \text{ verifies (bb), (4)}, \ \delta_i(1/2)=0,$

$|u|_K^2 \equiv \|u_{,x}\|^2+\sum_{i,j=1}^{n}\|\varphi_i-\varphi_j\|^2 \}$

(Kcd) $K = \{ u\in X : u \text{ verifies either (cc) or (dd), (4)},$

$|u|_K \equiv \|u_{,x}\|^2 \}$

It is not difficult to prove that K and $|\cdot|_K$ satisfy all conditions of the Lemma.

n.2 Stability results

It turn useful to write the stability problem in the following abstract form

(5) $$u_{,t} = (S+M)u + Nu$$

where $u \equiv (\sigma_1,\ldots,\sigma_n,\delta_1,\ldots,\delta_n)$ is a 2n vector function of L^2, moreover the linear symmetric bounded operator S, bounded in L^2, the linear operator M bounded in $W^{1,2}$, and the non linear operator N are given respectively by

$$Su=\Big(-2s\sum_{j=1}^{n}(\sigma_1-\sigma_j),\ldots,-2s\sum_{j=1}^{n}(\sigma_n-\sigma_j),0,\ldots,0\Big);$$

$$Mu=-(v_1\delta_{1,x},\ldots,v_n\delta_{n,x},v_1\sigma_{1,x},\ldots,v_n\sigma_{n,x});$$

$$Nu=\Big(\sum_{j=1}^{n}\{(\delta_1^2-\delta_j^2)-(\sigma_1^2-\sigma_j^2)\},..,\sum_{j=1}^{n}\{(\delta_n^2-\delta_j^2)-(\sigma_n^2-\sigma_j^2)\},0,$$

If we compute the evolution of the L^2 norm of u, E say, we obtain

(6) $$E_{,t} = (Su,u) + (Mu,u) + (Nu,u),$$

where (.,.) denotes the usual L^2 scalar product. A first lecture of (6) shows that S plays a non destabilizing role, moreover, in Galdi & Padula it has been shown that M may influence the stability only if the so called "coupling functional" $F \equiv (Su,Mu)$ is not zero. In this connection it was suggested to add F to the classical energy functional E in this way we are lead to the following Ljapunov functional

$$E \equiv \| u \|^2 + \mu \| u_{,x} \|^2 + F$$

In the case at hand we have $F = \sum_{i=1}^{n}\{(\delta_{i,x},\sigma_i)+(\sigma_{i,x},\delta_i)\}$ this motivates the following choice of $FF \equiv$

$$\sum_{i,j=1}^{n}\{\lambda(((\sigma_i/v_i)-(\sigma_j/v_j)),(\delta_{i,x}-\delta_{j,x}))$$
$$+\lambda\tau(((\delta_i/v_i)-(\delta_j/v_j)),(\sigma_i-\sigma_j))\}$$

with μ,λ and τ arbitrary parameters. The request for E to be positive definite imposes $(\lambda\tau/\sqrt{\mu})(1+\tau) \le (v_m/2n)$, $v_m \equiv \min\{v_i, i=1,\dots,n\}$. A carefull and long calculation ofthe time derivative of E along the solutions of (3) reveals that the boundary conditions (aa)-(dd) provide negative terms for $E_{,t}$ (in particular,the boundary data (aa),(cc),(dd) are *strictly dissipative*).

Bibliography

Babovski, H.& M. Padula, A new approach to nonlinear stability of a 2n velocity model of the kinetic theory ry, forthcoming

Galdi G.P.& M. Padula, A new approach to energy theory in the stability of fluid motion, Arch. for Ratl. Mech. Anal. and Mech., 1990 in the Press

Gatignol, R., Theorie Cinetique des Gaz a Repartition Discrete de Vitesses,Lecture Notes in Physics, Springer-Verlag,Berlin Heidelbewrg New York, 1975

Kawashima,S., Global solutions to the initial-boundary value problems for the discrete Boltzmann equation, preprint, 1989.

Platkovski, T. & R. Illner, Discrete velocity models of the Boltzmann equation: a survey on the mathematical aspects of the theory, SIAM Review, 30, 1988

STATIC AND DYNAMIC PROBLEMS FOR AXISYMMETRIC ELASTIC SHELLS

Franco PASTRONE, Maria Luisa TONON

Dipartimento di Matematica
Università degli Studi di Torino
Via Carlo Alberto, 10
10123 TORINO
ITALIA

ABSTRACT.The nonuniqueness of static solutions for an axisymmetric non-linearly elastic Kirchhoff shell is proved; in the dynamic case, appropriate field equations are derived, such that wave propagation and free oscillations can be studied.

1. INTRODUCTION

In this paper we discuss the problem of the behaviour of a generalized axisymmetric Kirchhoff shell, both in the static and dynamic case, within the framework of the non linear theory of elasticity. In the static problem we formulate the governing boundary problem when the shell is subjected to a wrench; we prove a local nonuniqueness theorem for the corresponding solution, i.e. the existence of buckled states. In the dynamic problem the field equations for an hyperelastic shell are obtained via a variational principle and natural boundary conditions are also given.Moreover, explicit values for the speeds of acceleration and shock waves can be found if we assigne an uncoupled strain energy function. Finally, we consider periodic oscillations of a closed circular cylindrical Kirchhoff shell.

2. GEOMETRY AND KINEMATICS

We consider an axisymmetric generalized Kirchhoff shell $\boldsymbol{S}$, i.e. a material surface of revolution about an axis a. An arbitrary configuration of $\boldsymbol{S}$ is given by the parametric equations:

$$0P = \mathbf{r} = r(S,t)\mathbf{e}_r(S, \Phi, t) + z(S, t)\mathbf{k}, \tag{1}$$

where 0 is a fixed point of a, $P \in \boldsymbol{S}$, $\mathbf{k}$ is the unit vector of the axis a, $\mathbf{e}_r$ the unit vector orthogonal to $\mathbf{k}$ in the plane of the meridian containing P;S and Φ are material coordinates on $\boldsymbol{S}$, specified in the following, while t is the time.

Since we allow bending, stretching, expansion, extension, torsion and shearing of a sector of the shell, we define a reference configuration $\boldsymbol{S}^*$ by:

$$0P_0 = R(S)\,\mathbf{E}_R(\Phi) + Z(S)\,\mathbf{k} \tag{2}$$

and we introduce the orthogonal group $\{Q_\beta\} \equiv SO(2)$, such that $Q_\beta \mathbf{k} = \mathbf{k}$, where $\beta = \beta(S, t)$ is an angle of rotation; then we can write $\mathbf{e}_r = Q_\beta \mathbf{E}_R(\Phi)$. As usual, capital letters refer to $\mathcal{S}^*$. The variables S and Φ are the arc length of the meridian containing the point $P_0 \in \mathcal{S}^*$ and the longitude of P_0, respectively; moreover, we assume $0 \leq S \leq 1$.

We introduce also the unit normal to $\mathcal{S}$ in P:

$$\mathbf{d} = \cos\theta(S,t)\,\mathbf{e}_r + \sin\theta(S,t)\,\mathbf{k}, \tag{3}$$

where θ is the angle between the tangent $\mathbf{t}$ to the meridian containing P and the axis a.

Let $\mathbf{e}_\Phi(S, \Phi, t) = Q_\beta \mathbf{E}_\Phi$ be the unit vector orthogonal to $\mathbf{k}$ and $\mathbf{e}_r$; then we have:

$$\begin{cases} \mathbf{e}'_r = -\beta' \mathbf{e}_\Phi, \quad \mathbf{r}' = r'\mathbf{e}_r + z'\mathbf{k} - r\beta'\mathbf{e}_\Phi, \quad \mathbf{r}_{,\Phi} = r\,\mathbf{e}_\Phi \\ \mathbf{d}' = -\sin\theta\theta'\mathbf{e}_r - \cos\theta\beta'\mathbf{e}_\Phi + \cos\theta\theta'\mathbf{k}, \quad \mathbf{d}_{,\Phi} = \cos\theta\,\mathbf{e}_\Phi \\ \dot{\mathbf{e}}_r = -\dot\beta\mathbf{e}_\Phi, \quad \dot{\mathbf{r}} = \dot r\mathbf{e}_r + \dot z\mathbf{k} - r\dot\beta\mathbf{e}_\Phi, \quad \dot{\mathbf{d}} = -\sin\theta\dot\theta\mathbf{e}_r - \cos\theta\dot\beta\mathbf{e}_\Phi + \cos\theta\dot\theta\mathbf{k}, \end{cases} \tag{4}$$

where the prime, the subscript Φ and the dot denote differentiation with respect to S, Φ and t, respectively.

If the shell is hyperelastic, there exists a strain energy density function:

$$W = W(r, \theta, r', \theta', z', \beta'; S, \Phi). \tag{5}$$

In the following the arguments S and Φ are omitted, in virtue of the hypothesis of homogeneity.

Finally we introduce the kinetic energy density $\hat{T} = \rho_0(\dot{\mathbf{r}} \cdot \dot{\mathbf{r}} + \alpha\,\dot{\mathbf{d}} \cdot \dot{\mathbf{d}})/2$, where ρ_0 is the 2-dimensional mass density and α an inertia term; in virtue of $(4)_{7,8}$ $\hat{T}$ can be expressed as:

$$\hat{T} = \rho_0[\dot r^2 + \dot z^2 + \alpha\dot\theta^2 + (r^2 + \alpha\cos^2\theta)\dot\beta^2]/2. \tag{6}$$

3. THE STATIC PROBLEM

We briefly formulate the problem of an axisymmetric generalized Kirchhoff shell subjected to a wrench. We only expose the problem and the final results, because this case has been extensively studied in [2].

We introduce the set of deformation variables $\epsilon_\kappa = \{\sigma, \rho, \chi, \theta', \beta'\}$ defined as:

$$\sigma = r/R - 1, \quad \chi = \cos(\theta + \Theta) - \cos\Theta, \quad \rho = (r'^2 + z'^2)^{1/2}, \quad \theta' = d\theta/dS, \quad \beta' = d\beta/dS. \tag{7}$$

The equations of equilibrium can be written as:

$$\bar h' = n_\Phi, \quad \bar p' = 0, \quad \bar v' = 0, \quad \bar m' + m_\Phi\sin\theta - \bar q\rho = 0, \tag{8}$$

where we have set $\bar h = hr$, $\bar p = pr$, $\bar v = vr$, $\bar m = mr$; h, p and v are the components along $\mathbf{e}_r$, $\mathbf{e}_\Phi$, $\mathbf{k}$, respectively, of the stress force resultant acting on the parallel; n_Φ is the component along $\mathbf{e}_\Phi$ of the stress resultant acting on the meridian, m and m_Φ are the components, one along $\mathbf{e}_\Phi$, the other along **t**, of the moment resultants acting on parallel and meridian, respectively.

If we assume the Kirchhoff shell be homogeneous and elastic, we can add the constitutive equations:

$$m_\Phi = \mathcal{L}(\epsilon_\kappa),\ \overline{m} = \mathcal{M}(\epsilon_\kappa),\ n_\Phi = \mathcal{N}(\epsilon_\kappa),\ \overline{p} = \mathcal{P}(\epsilon_\kappa),\ \overline{v} = \mathcal{V}(\epsilon_\kappa); \tag{9}$$

$\mathcal{L}$, $\mathcal{M}$, $\mathcal{N}$, $\mathcal{P}$, $\mathcal{V}$ are smooth functions of their arguments and must satisfy suitable physically meaningful restrictions (see [2]).

The following boundary conditions are prescribed:

(i) the edges of the shell are circles of constant radius R_0, i.e. $r(0) = r(1) = R_0$;

(ii) the edge corresponding to $S = 0$ is fixed, the edge corresponding to $S = 1$ remains orthogonal to the axis a;

(iii) the moving edge is acted upon by a vertical edge load **R** and a terminal wrench **M**:

$$\mathbf{R} = R\mathbf{k}\ ,\ \mathbf{M} = -M\mathbf{k}\ ; \tag{10}$$

if we introduce the forces per unit edge length $\mathbf{F} = F\mathbf{k}$ (traction or compression) and $\mathbf{K} = K\,\mathbf{e}_\Phi$ (shearing force), then the resultant force and moment (10) become:

$$\mathbf{R} = 2\pi R_0 \mathbf{F}\ ,\ \mathbf{M} = 2\pi R_0 \mathbf{K} \times \mathbf{e}_r\ ;$$

(iv) the meridians are free to rotate in their plane at both edges: $\overline{m}(0) = \overline{m}(1) = 0$, or equivalently $\theta'(0) = \theta'(1) = 0$.

If we take as reference configuration $\mathcal{S}^*$ a circular cylinder of radius R_0, corresponding to zero load ($F = K = 0$), we have from (7): $\sigma = r/R_0 - 1$, $\chi = \cos\theta - 1$.

It is easy to see that the b.v.p. described above has the solution (trivial solution or trivial state):

$$\sigma = 0,\ \theta = 0,\ \rho = E = \text{const.},\ \beta' = T = \text{const.}, \tag{11}$$

if and only if the algebraic system:

$$\mathcal{P}(0, E, 0, 0, T) = R_0 K,\ \mathcal{V}(0, E, 0, 0, T) = R_0 F, \tag{12}$$

which arises from $(8)_{2,3}$, is satisfied.

As shown in [4], one can prove the existence of non trivial solutions (buckled states) of the b.v.p. formulated above, i.e. functions:

$$r = r(S;\ F,\ K),\ \theta = \theta(S;\ F,\ K),\ \rho = \rho(S;\ F,\ K),\ \beta = \beta(S;\ F,\ K), \tag{13}$$

in addition to the horizontal reaction $\overline{h} = \overline{h}(S; F, K)$, which are expected to arise from critical values of the parameters F and K. We indicate such critical values F_n^C and K_n^C, where $n \in Z$. Let us remark that if $F>0$, $K=0$, $\beta=0$, the critical load F_n^C is in agreement with the value obtained in [4]; if $F<0$, $K=0$, $\beta=0$, $E=1$, we find the same critical value F_n^C given in [1].

Finally, the following local nonuniqueness theorem holds: under suitable hypotheses on the constitutive functions, the b.v.p. stated above has a trivial solution (namely a cylindrical configuration of equilibrium with constant extension and twist) and non trivial solutions which bifurcate when $F=F_n^c$ and $K=K_n^c$. The proof is given in [2].

If the shell is hyperelastic, a global theorem of nonuniqueness of solution can be proved (see [2]).

4. THE FIELD EQUATIONS

We suppose that the shell be hyperelastic; we define an energy functional:

$$\hat{E} = \int_{S^*} [\hat{T} - W - \mu(r'\cos\theta + z'\sin\theta)] \, d\Sigma, \qquad (14)$$

where $d\Sigma$ is the element of area in the referential configuration S^*, $\hat{T}$ and W are given by (6) and (5), respectively, μ is a Lagrange multiplier introduced because of the geometrical constraint of axial symmetry $r'\cos\theta + z'\sin\theta = 0$.

The total energy in absence of distributed loads and body forces is: $\mathcal{E} = (\hat{E} + B)/2\pi$, where B is a line integral including possible external forces acting on the boundaries.

The field equations, obtained via a variational principle, are:

$$\begin{cases} \rho_0\ddot{r} - \rho_0 r\dot{\beta}^2 = (\partial W/\partial r')' - \partial W/\partial r + (\mu\cos\theta)' \\ \rho_0\ddot{z} = (\partial W/\partial z')' + (\mu\sin\theta)' \\ \rho_0\alpha\ddot{\theta} + \rho_0\alpha\sin\theta\cos\theta\,\dot{\beta}^2 = (\partial W/\partial\theta')' - \partial W/\partial\theta + \mu(r'\sin\theta - z'\cos\theta) \\ \rho_0(r^2 + \alpha\cos^2\theta)\ddot{\beta} + 2\rho_0(r\dot{r} - \alpha\sin\theta\cos\theta\,\dot{\theta})\dot{\beta} = (\partial W/\partial\beta')', \end{cases} \qquad (15)$$

with the side equation:

$$r'\cos\theta + z'\sin\theta = 0. \qquad (16)$$

When B=0, the natural boundary conditions are:

$$\left(\partial W/\partial r' + \mu\cos\theta\right)\Big|_{S=0}^{S=1} = 0, \quad \left(\partial W/\partial z' + \mu\sin\theta\right)\Big|_{S=0}^{S=1} = 0, \quad \left(\partial W/\partial\theta'\right)\Big|_{S=0}^{S=1} = 0, \quad \left(\partial W/\partial\beta'\right)\Big|_{S=0}^{S=1} = 0. \qquad (17)$$

If we suppose, as in the previous section, that the boundaries remain circles of constant radius R_0, condition $(17)_1$ is substituted by the geometrical conditions $r(0) = r(1) = R_0$. If the edges are free to rotate in their plane, as supposed in section 2, condition $(17)_3$ becomes $\theta'(0) = \theta'(1) = 0$.

We suppose now $B \neq 0$. For instance, if a vertical load (traction or compression) $\mathbf{F} = \pm F\mathbf{K}$ is applied on the edge S=1, while the other one is supposed fixed, condition $(17)_2$ must be modified as follows:

$$\left((\partial W/\partial z') + \mu\sin\theta\right)\Big|_{S=1} = \pm F;$$

on the other hand, if a shearing force $\mathbf{K} = K\mathbf{e}_\Phi$ is applied on the edge S = 1, condition $(17)_4$ is substituted by: $(\partial W/\partial\beta')\big|_{S=1} = K$.

5. WAVES AND FREE OSCILLATIONS

In order to obtain explicit values of the wave speeds, we suppose the strain energy density (5) uncoupled in its arguments. Applying the usual kinematic conditions of compatibility for acceleration waves to the equations of motion (15), we obtain the propagation condition:

$$(Q - \nu^2 \mathcal{K})\, \mathcal{A} = 0, \tag{18}$$

where $Q = \mathrm{diag}\,(Q_{rr}, Q_{zz}, Q_{\theta\theta}, Q_{\beta\beta})$ is the acustic tensor determined by the constitutive equation (5), $\mathcal{K} = \mathrm{diag}\,(\rho_0, \rho_0, \rho_0\alpha, \rho_0(r^2 + \alpha\cos^2\theta))$, ν^2 the square of the wave speed and $\mathcal{A}$ the amplitude vector.

The wave speeds for acceleration waves are given by:

$$\begin{cases} \nu_r = (Q_{rr}/\rho_0)^{1/2}, \quad \nu_z = (Q_{zz}/\rho_0)^{1/2} \\ \nu_\theta = (Q_{\theta\theta}/\rho_0\alpha)^{1/2}, \quad \nu_\beta = [Q_{\beta\beta}/(\rho_0(r^2+\alpha\cos^2\theta))]^{1/2}. \end{cases} \tag{19}$$

According to the literature, we can call ν_r and ν_θ bending wave speeds, ν_z longitudinal - extensional wave speed and ν_β transverse - twisting wave speed.

The same values of wave speeds can be found applying the kinematic conditions of compatibility for shock waves to the balance equation of linear momentum in integral form. This result is well-known in the linear theory, but in our case we linearize only the strain energy function, while the left-sides of field equations (15) are non linear.

We consider periodic oscillations of a closed circular cylindrical Kirchhoff shell, with negligible thickness, hence $\alpha = 0$, $\mu = 0$, $\theta = \Theta = 0$, and we assume constant axial extension and twist, hence z' and ß' do not depend on S. Then $W = W(r)$ and (15) provides:

$$\rho_0\ddot{r} - \rho_0 r\,\dot{\beta}^2 = -\partial W/\partial r, \qquad r^2\dot{\beta} = R_0^2\,\dot{\beta}_0. \tag{20}$$

We choose as reference configuration a cylinder of radius R_0. By coupling equations $(20)_1$ and $(20)_2$ we can see that there exists the first integral of motion:

$$\dot{r}^2 = 2\,[R_0^2\,\dot{\beta}_0^2\rho_0/2 + W(R_0) - W(r) - R_0^4\rho_0\,\dot{\beta}_0^2/(2r^2)]/\rho_0 = G(r). \tag{21}$$

An oscillatory periodic motion exists if the equation $G(r) = 0$ has two strictly potive real roots $r_1 < r_2$ and the period of the oscillations is $T = \int_{r_1}^{r_2} G^{-1/2}\,dr$.

If $\dot{\beta}_0 = 0$, equations $(20)_2$ and $(20)_1$ yield:

$$\dot{\beta} = 0, \quad \rho_0\ddot{r} = -\partial W/\partial r;$$

in this case we have free radial periodical vibrations if we choose as initial configuration a cylindrical configuration with radius $R_i \neq R_0$. In particular, if $W = \rho_0 c^2 (r - R_0)^2/2$, the period is $T = 2\pi/c$.

If we want to seek the existence of extensional - axial free vibrations, we relax the assumption that z' does not depend on S and assume $r = R_0$.

It follows that $\dot{\beta} = 0$ and the only equation of motion reads:

$$\ddot{z} = (\partial W / \partial z'^2) z'' / \rho_0 = W_2 (z')\ z'', \qquad (22)$$

since the strain energy function, restricted to this class of deformations, is $W = W(z')$. We can have periodical solutions if and only if $W_2(z')$ is a constant positive: $W_2 = \omega^2$, $\omega \in R$, namely if $W = \rho_0 \omega^2 z'^2/2$. The period is $T = 4\pi / 2n+1$, i.e. we have an infinite number of vibration modes.

The model herein used, as well as the results obtained, can be compared with the model discussed in [3], where it is studied the problem of deformations for an elastic Cosserat surface, that is a more general scheme than a Kirchhoff shell. On the other hand, in [3] Crochet and Naghdi assume from the beginning the extension and twist be constant, while in our approach we allow also such deformations. Hence we obtain not only radial free vibrations, but also the other cases discussed above.

Acknowledgment: this work was supported by a grant of the Ministero della Pubblica Istruzione (Italy) and G.N.D.T. of C.N.R..

REFERENCES

1. Cohen, H., and Pastrone, F.,"Axisymmetric equilibrium states of nonlinear elastic cylindrical shells". Int. J. Non-Linear Mechanics **21**(1), 37-50 (1986).

2. Cohen, H., and Pastrone, F.,"Buckling of a cylindrical shell subjected to a wrench", to appear.

3. Crochet, M.J., and Naghdi, P.M., "Large deformation solutions for an elastic Cosserat surface", Int. J. Engng. Sci. **7**, 309-335 (1969).

4. Pastrone, F., and Tonon, M.L., "Nonuniqueness of equilibrium states for axisymmetric elastic shells in tension", J. of Elasticity **21**, 43-59 (1989).

A DERIVATION OF EQUATIONS OF BALANCE FOR MICROSTRUCTURED CONTINUA

Mario Pitteri

Dipartimento di Metodi e Modelli Matematici per le Scienze Applicate
Università di Padova 35131 Padova, Italy

ABSTRACT

The equations of balance for mass, linear momentum, microinertia and generalized rotational momentum are directly derived from a statistical mechanical description of a system of ideally constrained Newtonian mass-points. The equations are shown to be consistent with those obtained in a previous paper of the author from an equation of balance of energy and the proved transformation properties of gross fields under a change of observer.

1. Introduction

In this paper I present a direct derivation of equations of balance for mass, linear momentum, microinertia and generalized rotational momentum for a continuum with lagrangian internal structure. For the equation of balance of energy we refer the reader to PITTERI [1990]. These equations are derived from a description of the body as being made of a finite number of molecules, regarded as ideally constrained systems of mass points, under the assumption that the initial conditions are assigned by means of a probability density function which is convected by the system of points along its motion. The gross fields are constructed as expectations of microscopic functions with respect to the aforementioned probability density, and the equations of balance are consequence of an equation of evolution for the probability density which generalizes the so-called Liouville equation of classical statistical mechanics.

We assume both external and internal forces to be present, and allow internal forces to have the most general form compatible with Newtonian mechanics and the physical indistinguishability of inertial frames. In this case, according to BRESSAN [1962], the internal force $\mathbf{F}_{ij}$ that the point P_j exerts upon P_i has necessarily the constitutive equation

$$\mathbf{F}_{ij} = \mathbf{n}_{ij}\, \phi_{ij}(\mathrm{r}_{ij}, \|\mathbf{v}_j - \mathbf{v}_i\|, (\mathbf{v}_j - \mathbf{v}_i)\cdot \mathbf{n}_{ij}), \tag{1.1}$$

for some choice of functions ϕ_{ij} such that $\phi_{ij} \equiv \phi_{ji}$. Here $\mathbf{n}_{ij}$ [r_{ij}] is the unit vector [is the norm] of $\mathrm{P}_i\mathrm{P}_j$, and $\mathbf{v}_i$ [$\mathbf{v}_j$] is the velocity of P_i [of P_j] .

The general framework and the notations are the same used in PITTERI [1986] and PITTERI [1990]. We assume the reader to be already familiar with both papers, and use their notation without additional comments. The usual summation convention over repeated indices, one subscript and one superscript, is understood.

Equations of balance for mass, linear momentum, microinertia and rotational momentum for microstructured continua are derived in PITTERI [1990] from an equation of balance of energy which holds with respect to any observer ϕ, and from the computed transformation laws gross fields obey when ϕ is replaced by any other observer ϕ^* which is rigidly moving with respect to ϕ. We show that the forms of the equations of balance we obtain here for the aforementioned quantities are equivalent to the analogues in PITTERI [1990]. As is pointed out there, if one believes that the equation of balance for rotational momentum, which is a skew vector, should only be the skew part of a more general equation of balance for a tensorial quantity, then he must suitably generalize the idea that all the relevant equations of balance should be obtainable from an equation of balance of energy and from an assumption of invariance with respect to change of observer.

2. Equations of balance

Let us first recall the form of the Liouville equation considered in PITTERI [1990]

$$\begin{aligned} 0 = \frac{\partial W}{\partial t} &+ \sum_r \frac{\partial}{\partial \mathbf{x}_{r0}} \cdot \Big(W \frac{\partial \mathcal{H}}{\partial \mathbf{p}_{r0}}\Big) + \sum_r \frac{\partial}{\partial \mathbf{p}_{r0}} \cdot \Big(W(\hat{\mathbf{R}}_r - \frac{\partial \mathcal{H}}{\partial \mathbf{x}_{r0}})\Big) \\ &+ \sum_r \frac{\partial}{\partial \mathrm{q}^i_r} \Big(W \frac{\partial \mathcal{H}}{\partial \mathrm{p}_{ri}}\Big) + \sum_r \frac{\partial}{\partial \mathrm{p}_{ri}} \Big(W(\hat{\mathrm{Q}}_{ri} - \frac{\partial \mathcal{H}}{\partial \mathrm{q}^i_r}\Big), \end{aligned} \tag{2.1}$$

and the definitions of the gross fields

$$\rho = \sum_{r,\alpha} \mathrm{m}_{r\alpha} \langle\, W |\, \mathbf{x}_{r0} = \mathbf{x} \rangle = \sum_r \mathrm{m}_r \langle\, W |\, \mathbf{x}_{r0} = \mathbf{x} \rangle \quad \text{and} \tag{2.2}$$

$$\rho \mathbf{u} = \sum_{r,\alpha} \mathrm{m}_{r\alpha} \langle\, \mathbf{v}_{r\alpha}\, W |\, \mathbf{x}_{r0} = \mathbf{x} \rangle = \sum_r \mathrm{m}_r \langle\, \mathbf{v}_{r0}\, W |\, \mathbf{x}_{r0} = \mathbf{x} \rangle\,. \tag{2.3}$$

By differentiating under the integral sign and using (2.1) and the divergence theorem to integrate by parts, we obtain *the equation of balance for mass* :

$$\begin{aligned} \frac{\partial \rho}{\partial t} &= \sum_r \mathrm{m}_r \langle \frac{\partial W}{\partial t} \mid \mathbf{x}_{r0} = \mathbf{x} \rangle = -\sum_r \mathrm{m}_r \langle \frac{\partial W}{\partial \mathbf{x}_{r0}} \cdot \mathbf{v}_{r0} \mid \mathbf{x}_{r0} = \mathbf{x} \rangle \\ &= -\operatorname{div}(\rho\, \mathbf{u})\,. \end{aligned} \tag{2.4}$$

By introducing the additional gross fields

$$\mathbf{T} = \mathbf{T}_{\mathrm{K}} + \mathbf{T}_{\mathrm{V}}\,,\ \boldsymbol{\eta}_{r0} = \mathbf{v}_{r0} - \mathbf{u}\,,\ \mathbf{T}_{\mathrm{K}} = -\sum_r \mathrm{m}_r \langle\, \boldsymbol{\eta}_{r0} \otimes \boldsymbol{\eta}_{r0}\, W |\ \mathbf{x}_{r0} = \mathbf{x} \rangle\,, \tag{2.5}$$

$$\mathbf{T}_{\mathrm{V}} = \sum_{\substack{r,\alpha \\ s,\beta}}{}' \int_{\mathbf{R}^3} d^3\mathbf{z} \int_0^{+\infty} d\theta\, \langle \frac{\mathrm{V}'_{r\alpha s\beta}}{\|\mathbf{x}_{r\alpha} - \mathbf{x}_{s\beta}\|} (\mathbf{x}_{r\alpha} - \mathbf{x}_{s\beta})\, W \mid \mathbf{x}_{r0} = \mathbf{x} + \theta\mathbf{z}, \tag{2.6}$$
$$\mathbf{x}_{s0} = \mathbf{x} - (1-\theta)\theta\mathbf{z} \rangle \otimes \mathbf{z},$$

$$\mathbf{b} = \sum_r \langle\, \hat{\mathbf{R}}_r\, W |\ \mathbf{x}_{r0} = \mathbf{x} \rangle\,, \tag{2.7}$$

by differentiating under the integral sign and using (2.1) and the divergence theorem to integrate by parts, we obtain *the equation of balance for linear momentum*:

$$\begin{aligned} \frac{\partial}{\partial t}(\rho\mathbf{u}) &= \sum_r \mathrm{m}_r\, \langle \mathbf{v}_{r0} \frac{\partial W}{\partial t} \mid \mathbf{x}_{r0} = \mathbf{x} \rangle \\ &= -\sum_{r,s} \mathrm{m}_r \langle \mathbf{v}_{r0} \Big[\frac{\partial W}{\partial \mathbf{x}_{s0}} \cdot \mathbf{v}_{s0} + \frac{\partial}{\partial \mathbf{p}_{s0}} \cdot \Big(W (\hat{\mathbf{R}}_s - \frac{\partial \mathcal{H}}{\partial \mathbf{x}_{s0}}) \Big) \\ &\quad + \frac{\partial}{\partial \mathrm{q}_s^i} \Big(W \frac{\partial \mathcal{H}}{\partial \mathrm{p}_{si}} \Big) + \frac{\partial}{\partial \mathrm{p}_{si}} \Big(W (\hat{\mathrm{Q}}_{si} - \frac{\partial \mathcal{H}}{\partial \mathrm{q}_s^i}) \Big) \Big] \mid \mathbf{x}_{r0} = \mathbf{x} \rangle \\ &= -\operatorname{div}\,(\rho\, \mathbf{u} \otimes \mathbf{u} - \mathbf{T}_{\mathrm{K}}) + \mathbf{b} + \operatorname{div} \mathbf{T}_{\mathrm{V}}\ . \end{aligned} \tag{2.8}$$

By introducing the gross fields

$$\mathcal{I}_r = \sum_\alpha \mathrm{m}_{r\alpha}\, \boldsymbol{\xi}_{r\alpha} \otimes \boldsymbol{\xi}_{r\alpha}\,, \quad \mathcal{K}_{ri} = \sum_\alpha \mathrm{m}_{r\alpha}\, \boldsymbol{\xi}_{r\alpha} \otimes \frac{\partial \boldsymbol{\xi}_{r\alpha}}{\partial \mathrm{q}_r^i}\,, \tag{2.9}$$

$$\mathbf{I} = \sum_r \langle \mathcal{I}_r\, W \mid \mathbf{x}_{r0} = \mathbf{x} \rangle\,, \quad \mathbf{K} = \sum_r \langle (a_r^{-1})^{ij}\, \mathrm{p}_{ri}\, \mathcal{K}_{rj}\, W \mid \mathbf{x}_{r0} = \mathbf{x} \rangle\,, \tag{2.10}$$

$$\varXi_{\mathbf{I}} = \frac{1}{2} \sum_r \langle\, \mathcal{I}_r \otimes \boldsymbol{\eta}_{r0}\, W \mid \mathbf{x}_{r0} = \mathbf{x} \rangle\,, \tag{2.11}$$

by differentiating under the integral sign and using (2.1), $(2.5)_2$ and the divergence theorem to integrate by parts, we deduce *the equation of balance for microinertia*:

$$\begin{aligned} \frac{\partial \mathbf{I}}{\partial t} &= -\sum_{r,\alpha,s} \mathrm{m}_{r\alpha} \langle \boldsymbol{\xi}_{r\alpha} \otimes \boldsymbol{\xi}_{r\alpha} \Big[\frac{\partial W}{\partial \mathbf{x}_{s0}} \cdot \mathbf{v}_{s0} + \frac{\partial}{\partial \mathbf{p}_{s0}} \cdot \Big(W (\hat{\mathbf{R}}_s - \frac{\partial \mathcal{H}}{\partial \mathbf{p}_{s0}}) \Big) \\ &\quad + \frac{\partial}{\partial \mathrm{q}_s^i} \Big(W \frac{\partial \mathcal{H}}{\partial \mathrm{p}_{si}} \Big) + \frac{\partial}{\partial \mathrm{p}_{si}} \Big(W (\hat{\mathrm{Q}}_{si} - \frac{\partial \mathcal{H}}{\partial \mathrm{q}_s^i}) \Big) \Big] \mid \mathbf{x}_{r0} = \mathbf{x} \rangle \\ &= -\operatorname{div}\,(\mathbf{I} \otimes \mathbf{u}\,) - \operatorname{div} \varXi_{\mathbf{I}} + \mathbf{K} + \mathbf{K}^T\ . \end{aligned} \tag{2.12}$$

Notice that by integrating by parts the terms $\partial/\partial \mathbf{p}_{s0}(\ldots)$ and $\partial/\partial \mathrm{p}_{si}(\ldots)$ we obtain zero. Also, the equations of balance we have obtained so far coincide with the ones bearing the same name in PITTERI [1990].

We introduce the gross fields

$$\mathcal{K}_r = (a_r^{-1})^{kj}\, \mathrm{p}_{rk}\, \mathcal{K}_{rj}\,, \quad \boldsymbol{\Xi}_{\mathbf{K}} = \sum_r \langle\, \mathcal{K}_r \otimes \boldsymbol{\eta}_{r0}\, W \mid \mathbf{x}_{r0} = \mathbf{x} \rangle\,, \quad \text{and} \tag{2.13}$$

$$\boldsymbol{\Theta} = \sum_r \langle\, \Big(\frac{\partial \mathcal{K}_r}{\partial \mathrm{q}_r^i} \frac{\partial \mathcal{H}}{\partial \mathrm{p}_{ri}} + \frac{\partial \mathcal{K}_r}{\partial \mathrm{p}_{ri}} (\hat{\mathrm{Q}}_{ri} - \frac{\partial \mathcal{H}}{\partial \mathrm{q}_r^i}) \Big)\, W \mid \mathbf{x}_{r0} = \mathbf{x} \rangle\,. \tag{2.14}$$

We notice that this definition of $\boldsymbol{\Xi}_{\mathbf{K}}$ does not coincide with the one given in PITTERI [1990]. Indeed the tensor $\boldsymbol{\Xi}_{\mathbf{K}}$ there is the skew part of the one in (2.13), and shall be denoted here by $\boldsymbol{\Xi}'_{\mathbf{K}}$. We obtain *the equation of balance of generalized moment of momentum*:

$$\begin{aligned}
\frac{\partial \mathbf{K}}{\partial t} = -\sum_{r,s} \langle \mathcal{K}_r \Big[\frac{\partial W}{\partial \mathbf{x}_{s0}} \cdot \mathbf{v}_{s0} + \frac{\partial}{\partial \mathbf{p}_{s0}} \cdot \Big(W\, (\hat{\mathbf{R}}_s - \frac{\partial \mathcal{H}}{\partial \mathbf{x}_{s0}}) \Big) + \frac{\partial}{\partial \mathrm{q}_s^i} \Big(W \frac{\partial \mathcal{H}}{\partial \mathrm{p}_{si}} \Big) \\
+ \frac{\partial}{\partial \mathrm{p}_{si}} \Big(W\, (\hat{\mathrm{Q}}_{si} - \frac{\partial \mathcal{H}}{\partial \mathrm{q}_s^i}) \Big) \Big] \mid \mathbf{x}_{r0} = \mathbf{x} \rangle \qquad (2.15) \\
= -\operatorname{div}(\mathbf{K} \otimes \mathbf{u}) - \operatorname{div} \boldsymbol{\Xi}_{\mathbf{K}} + \boldsymbol{\Theta}.
\end{aligned}$$

For consistency, we have to show that the skew part of this equation, which is the balance of rotational momentum, is equivalent to the one obtained in PITTERI [1990], that is,

$$\begin{aligned}
\frac{\partial \mathbf{K}'}{\partial t} + \operatorname{div}(\mathbf{K}' \otimes \mathbf{u}) + \operatorname{div} \boldsymbol{\Xi}'_{\mathbf{K}} - \operatorname{div}(\boldsymbol{\Psi} + \boldsymbol{\Phi}) + \frac{1}{2}(\mathbf{T} - \mathbf{T}^T) = \qquad (2.16) \\
\frac{1}{2} \sum_r \langle\, \boldsymbol{\xi}_{r\alpha} \wedge (\hat{\mathbf{f}}_{r\alpha} + \mathbf{g}_{r\alpha})\, W \mid \mathbf{x}_{r0} = \mathbf{x} \rangle
\end{aligned}$$

where, as we mentioned above, $\mathbf{K}'$ and $\boldsymbol{\Xi}'_{\mathbf{K}}$ are obtained by skewsymmetrizing the tensors $\mathbf{K}$ and $\boldsymbol{\Xi}_{\mathbf{K}}$ with respect to the first pair of indices, so, for instance,

$$\mathbf{K}' = \frac{1}{2} \sum_r \langle\, (a_r{}^{-1})^{ij}\, \mathrm{p}_{ri} \sum_\alpha \mathrm{m}_{r\alpha}\, \boldsymbol{\xi}_{r\alpha} \wedge \frac{\partial \boldsymbol{\xi}_{r\alpha}}{\partial \mathrm{q}_r^j}\, W \mid \mathbf{x}_{r0} = \mathbf{x} \rangle\,. \tag{2.17}$$

In addition

$$\begin{aligned}
\boldsymbol{\Psi} = \frac{1}{4} \sum_{\substack{r,\alpha \\ s,\beta}}{}' \int_{\mathbf{R}^3} d^3\mathbf{z} \int_0^{+\infty} d\theta\, \langle (\mathbf{x}_{r0} + \mathbf{x}_{s0} - 2\mathbf{x}) \wedge (\mathbf{x}_{r\alpha} - \mathbf{x}_{s\beta})\, \frac{V'_{r\alpha s\beta}}{\|\mathbf{x}_{r\alpha} - \mathbf{x}_{s\beta}\|} \qquad (2.18) \\
W \mid \mathbf{x}_{r0} = \mathbf{x} + \theta \mathbf{z}, \mathbf{x}_{s0} = \mathbf{x} - (1-\theta)\mathbf{z} \rangle \otimes \mathbf{z}\,,
\end{aligned}$$

and

$$\boldsymbol{\Phi} = \frac{1}{4}\sum_{\substack{r,\alpha\\ s,\beta}}{}' \int_{\mathbf{R}^3} d^3\mathbf{z} \int_0^{+\infty} d\theta \, \langle (\boldsymbol{\xi}_{r\alpha} + \boldsymbol{\xi}_{s\beta}) \wedge (\mathbf{x}_{r\alpha} - \mathbf{x}_{s\beta}) \frac{V'_{r\alpha s\beta}}{\|\mathbf{x}_{r\alpha} - \mathbf{x}_{s\beta}\|} W \mid \mathbf{x}_{r0} =$$

$$\mathbf{x} + \theta\mathbf{z}, \mathbf{x}_{s0} = \mathbf{x} - (1-\theta)\mathbf{z}\rangle \otimes \mathbf{z} \,. \quad (2.19)$$

By comparing (2.15) and (2.16), we have to prove that

$$\boldsymbol{\Theta}' = \operatorname{div}(\boldsymbol{\Psi} + \boldsymbol{\Phi}) - \tfrac{1}{2}(\mathbf{T} - \mathbf{T}^T) + \tfrac{1}{2}\textstyle\sum_r \langle \boldsymbol{\xi}_{r\alpha} \wedge (\hat{\mathbf{f}}_{r\alpha} + \mathbf{g}_{r\alpha}) \, W | \, \mathbf{x}_{r0} = \mathbf{x}\rangle \,, \quad (2.20)$$

where the skew part $\boldsymbol{\Theta}'$ of $\boldsymbol{\Theta}$ is given by

$$\boldsymbol{\Theta}' = \sum_r \langle \Big(\frac{\partial \mathcal{K}'_r}{\partial \mathrm{q}^i_r} \frac{\partial \mathcal{H}}{\partial \mathrm{p}_{ri}} + \frac{\partial \mathcal{K}'_r}{\partial \mathrm{p}_{ri}} (\hat{\mathrm{Q}}_{ri} - \frac{\partial \mathcal{H}}{\partial \mathrm{q}^i_r}) \Big) W \mid \mathbf{x}_{r0} = \mathbf{x} \rangle \,, \quad (2.21)$$

$$\mathcal{K}'_r = (a_r^{-1})^{kj} \, \mathrm{p}_{rk} \, \mathcal{K}'_{rj} \,, \quad \mathcal{K}'_{rj} = \frac{1}{2} \sum_\alpha \mathrm{m}_{r\alpha} \, \boldsymbol{\xi}_{r\alpha} \wedge \frac{\partial \boldsymbol{\xi}_{r\alpha}}{\partial \mathrm{q}^j_r} \,. \quad (2.22)$$

By the semi-Hamiltonian equations of motion

$$\dot{\mathbf{x}}_{r0} = \frac{\partial \mathcal{H}}{\partial \mathbf{p}_{r0}} \,, \qquad \dot{\mathbf{p}}_{r0} = \hat{\mathbf{R}}_r - \frac{\partial \mathcal{H}}{\partial \mathbf{x}_{r0}} \,, \quad (2.23)$$

$$\dot{\mathrm{q}}^i_r = \frac{\partial \mathcal{H}}{\partial \mathrm{p}_{ri}} \,, \qquad \dot{\mathrm{p}}_{ri} = \hat{\mathrm{Q}}_{ri} - \frac{\partial \mathcal{H}}{\partial \mathrm{q}^i_r} \,, \quad (2.24)$$

presented in **Pitteri** [1990], we recognize that, since $\mathcal{K}'_r$ depends neither on t nor on $\mathbf{x}_{r0}$ nor on $\mathbf{p}_{r0}$,

$$\boldsymbol{\Theta}' = \sum_r \langle (\mathcal{K}'_r)^{\cdot} \, W \mid \mathbf{x}_{r0} = \mathbf{x} \rangle \,, \quad (2.25)$$

where the superimposed dot means total differentiation with respect to time along the microscopic motion, that is, along a solution of the semi-Hamiltonian equations (2.23), (2.24). By (2.22) and the Newtonian equations of motion

$$\mathrm{m}_{r\alpha} \, \mathbf{a}_{r\alpha} = \mathbf{f}_{r\alpha} + \boldsymbol{\phi}_{r\alpha} \quad (2.26)$$

for the constrained system of mass points, presented in **Pitteri** [1990] together with the assumption that the constraints are ideal and “internal” in a suitable sense, we deduce

$$\begin{aligned}
\boldsymbol{\Theta}' &= \frac{1}{2}\sum_r \langle \Big(\sum_\alpha \mathrm{m}_{r\alpha} \, \boldsymbol{\xi}_{r\alpha} \wedge \dot{\boldsymbol{\xi}}_{r\alpha} \Big)^{\cdot} \, W \mid \mathbf{x}_{r0} = \mathbf{x} \rangle \\
&= \frac{1}{2}\sum_{r,\alpha} \langle \boldsymbol{\xi}_{r\alpha} \wedge \mathrm{m}_{r\alpha} \, \ddot{\boldsymbol{\xi}}_{r\alpha} \, W \mid \mathbf{x}_{r0} = \mathbf{x} \rangle \qquad (2.27) \\
&= \frac{1}{2}\sum_{r,\alpha} \langle \boldsymbol{\xi}_{r\alpha} \wedge \mathrm{m}_{r\alpha} \, \ddot{\mathbf{x}}_{r\alpha} W \mid \mathbf{x}_{r0} = \mathbf{x} \rangle = \frac{1}{2}\sum_{r,\alpha} \langle \boldsymbol{\xi}_{r\alpha} \wedge \mathbf{f}_{r\alpha} W \mid \mathbf{x}_{r0} = \mathbf{x} \rangle
\end{aligned}$$

since $\sum_\alpha \boldsymbol{\xi}_{r\alpha} \wedge \boldsymbol{\phi}_{r\alpha} = \mathbf{0}$. Indeed, since the constraints are ideal, $\sum_{r\alpha} \boldsymbol{\phi}_{r\alpha} \cdot \delta \mathbf{x}_{r\alpha} = \mathbf{0}$ for any choice of the virtual displacements $\delta \mathbf{x}_{r\alpha}$, and the condition in PITTERI [1990] that the constraints be internal implies that such are the displacements $\mathbf{A}\boldsymbol{\xi}_{r\alpha}$ for any skew double tensor $\mathbf{A}$. Then the assertion above on the reactions easily follows.

As in PITTERI [1990],

$$\mathbf{f}_{r\alpha} = \mathbf{g}_{r\alpha} + \hat{\mathbf{f}}_{r\alpha} - \sum_{s,\beta} \frac{V'_{r\alpha s\beta}}{\|\mathbf{x}_{r\alpha} - \mathbf{x}_{s\beta}\|} (\mathbf{x}_{r\alpha} - \mathbf{x}_{s\beta}) . \tag{2.28}$$

Therefore (2.20) is equivalent to

$$-\frac{1}{2} \sum_{\substack{r,\alpha \\ s,\beta}} \langle \frac{V'_{r\alpha s\beta}}{\|\mathbf{x}_{r\alpha} - \mathbf{x}_{s\beta}\|} \boldsymbol{\xi}_{r\alpha} \wedge (\mathbf{x}_{r\alpha} - \mathbf{x}_{s\beta}) \, W \mid \mathbf{x}_{r0} = \mathbf{x} \rangle \tag{2.29}$$
$$= \operatorname{div}(\boldsymbol{\Psi} + \boldsymbol{\Phi}) - \frac{1}{2}(\mathbf{T} - \mathbf{T}^T) .$$

On the other hand, by differentiating under the integral sign and applying PITTERI's [1987] version of NOLL's [1955] lemma, we can prove that

$$\operatorname{div}(\boldsymbol{\Psi} + \boldsymbol{\Phi}) = -\frac{1}{2} \sum_{\substack{r,\alpha \\ s,\beta}}{}' \int_{\mathbf{R}^3} d^3\mathbf{z} \int_0^{+\infty} d\theta \, \langle \mathbf{z} \wedge (\mathbf{x}_{r\alpha} - \mathbf{x}_{s\beta}) \frac{V'_{r\alpha s\beta}}{\|\mathbf{x}_{r\alpha} - \mathbf{x}_{s\beta}\|} W \mid \mathbf{x}_{r0} = \mathbf{x} + \theta \mathbf{z}, \mathbf{x}_{s0} = \mathbf{x} - (1-\theta)\mathbf{z} \rangle$$
$$-\frac{1}{4} \sum_{\substack{r,\alpha \\ s,\beta}}{}' \langle (\mathbf{x}_{r0} + \mathbf{x}_{s0} - 2\mathbf{x} + \boldsymbol{\xi}_{r\alpha} + \boldsymbol{\xi}_{s\beta}) \wedge (\mathbf{x}_{r\alpha} - \mathbf{x}_{s\beta}) \frac{V'_{r\alpha s\beta}}{\|\mathbf{x}_{r\alpha} - \mathbf{x}_{s\beta}\|} W \mid \mathbf{x}_{r0} = \mathbf{x} \rangle$$
$$= -\tfrac{1}{2}(\mathbf{T}^T - \mathbf{T}) \tag{2.30}$$
$$-\frac{1}{4} \sum_{\substack{r,\alpha \\ s,\beta}}{}' \langle (\mathbf{x}_{s0} - \mathbf{x}_{r0} + \boldsymbol{\xi}_{r\alpha} + \boldsymbol{\xi}_{s\beta}) \wedge (\mathbf{x}_{r\alpha} - \mathbf{x}_{s\beta}) \frac{V'_{r\alpha s\beta}}{\|\mathbf{x}_{r\alpha} - \mathbf{x}_{s\beta}\|} W \mid \mathbf{x}_{r0} = \mathbf{x} \rangle .$$

But the identity $(\mathbf{x}_{r\alpha} - \mathbf{x}_{s\beta}) \wedge (\mathbf{x}_{r\alpha} - \mathbf{x}_{s\beta}) = \mathbf{0}$ implies that

$$\boldsymbol{\xi}_{r\alpha} \wedge (\mathbf{x}_{r\alpha} - \mathbf{x}_{s\beta}) = (\mathbf{x}_{s0} - \mathbf{x}_{r0} + \boldsymbol{\xi}_{s\beta}) \wedge (\mathbf{x}_{r\alpha} - \mathbf{x}_{s\beta}) \tag{2.31}$$

so indeed (2.29) holds, hence the equation of balance of generalized rotational momentum (2.15) has the skew part which is equivalent to the equation of balance of rotational momentum obtained in PITTERI [1990].

ACKNOWLEDGEMENTS: This work was performed as part of the program of the 40% research project of the italian M.P.I. on Continuum Thermomechanics and of the C.N.R. research project on Mathematical Models for Modern Continua.

REFERENCES

A. BRESSAN [1962], "Metodo di assiomatizzazione in senso stretto della meccanica classica. Applicazione di esso ad alcuni problemi di assiomatizzazione non ancora completamente risolti", *Rendiconti del Seminario Matematico Università di Padova* **32**, 55-212.

W. NOLL [1955], "Die Herleitung der Grundgleichungen der Thermomechanik der Kontinua aus der statistischen Mechanik", *Journal of Rational Mechanics and Analysis* **4**, 627-646.

M. PITTERI [1986], "Continuum equations of balance in classical statistical mechanics", *Archive for Rational Mechanics and Analysis* **94**, 291-305 and Corrigendum, *Archive for Rational Mechanics and Analysis* **100** (1988), 315-316.

M. PITTERI [1987], "On the objectivity of certain gross fields constructed within classical statistical mechanics", *Proceedings of the ISIMM Symposium on Kinetic Theory and Extended Thermodynamics* (I. Müller and T. Ruggeri Ed.), Pitagora, Bologna.

M. PITTERI [1990], "On a statistical-kinetic model for generalized continua", appearing in *Archive for Rational Mechanics and Analysis.*

HYPERBOLIC EVOLUTION OF PHASE INTERFACES

Paolo Podio-Guidugli
Dipartimento di Ingegneria Civile Edile
Università di Roma 2

ABSTRACT - This expository lecture consists of three parts. In Part I, the ideas involved in the formulation of evolution laws for phase interfaces are briefly reviewed, and the differences between various formulations are pointed out; here I mostly draw from [1], [2] and [3]. In Part II, a short account is given, based essentially on [4], [5] and [6], of the phenomenology of melting-freezing waves in He^4. The oscillatory character of this interfacial motion motivates the search for hyperbolic theories of the evolution of phase interfaces; such a theory has been constructed in [3]. In Part III, the distinctive features of the hyperbolic theory of [2] for the evolution of plane curves are presented, and application to melting-freezing waves is sketched.

PART I - EVOLUTION OF PHASE INTERFACES

Consider a continuous medium, modelled as $\mathbb{R}^2$, consisting of two *phases*: a solid phase, the *crystal*, a compact subset $\mathcal{C}$ of $\mathbb{R}^2$; and a liquid phase, the *melt*, the interior $\mathcal{M}$ of the complement of $\mathcal{C}$ in $\mathbb{R}^2$. The *phase interface* is the crystal's boundary $\partial\mathcal{C}$, a smooth, simple closed curve. Our purpose here is to review various formulations of the evolution of phase interfaces.

a) Geometrical Theory

This is based on the *curve-shortening equation* [6]:

$$v = \kappa \quad , \tag{I.1}$$

where v = *normal velocity* and κ = *curvature*. This is a *parabolic* PDE, as one can see when the phase interface has a local representation as the graph of a function $y = h(x,t)$,

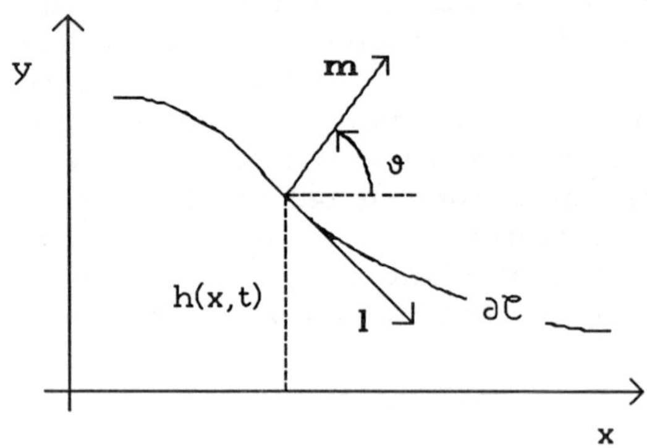

and (I.1) takes the simplified form

$$h_t = \sin^2\vartheta \; h_{xx} \tag{I.2}$$

(*cf.* the accompanying figure, where $\mathbf{l}$ is the unit tangent and $\mathbf{m}$ is the unit normal to $\partial\mathcal{C}$, pointing to the melt).

b) Mechanical Theories with Immobile Crystal and Melt

There are two theories of this type, presented in [1] and [2], respectively. In both theories neither crystal nor melt move, and melt is regarded as a source of atoms for cristallization processes. Both theories are based on a *balance law of interfacial momentum*. In [1], this law leads to the following major modification of the curve-shortening equation:

$$\beta(\vartheta)\,v = \kappa\,[\psi(\vartheta) + \psi''(\vartheta)] - F \quad , \tag{I.3}$$

where $\beta(\vartheta)v$ = *drag force*; ψ = *interfacial energy*, with the symbol $'$ denoting differentiation; and F = *driving force* (the difference in chemical potentials of crystal and melt). Remarkably, (I.3) is parabolic whenever $[\psi(\vartheta) + \psi''(\vartheta)]$ is positive. In [2], balance of interfacial momentum and an additional *balance law of interfacial mass* imply the *hyperbolic* equation

$$\rho(\vartheta)\,v^{\circ} + \beta(\vartheta)\,v = \kappa\,[\psi(\vartheta) + \psi''(\vartheta)] - F \quad , \tag{I.4}$$

where ρ = *effective interfacial inertia* and the symbol $^{\circ}$ denotes the operation of *normal* time-differentiation. We postpone a more detailed introduction of these balance laws until Part III.

c) Mechanical Theory with Mobile Melt

A common feature of all theories reviewed so far is that only the phase interface is endowed with mathematical structure. This is not the case with the theory of [3], where the interface is assumed to be massless, and thus has less structure than in [2], but the melt has more, being modelled as an inviscid incompressible fluid capable only of irrotational motions. Both theories do account for inertial effects in the evolution of phase interfaces, although in a different way. In [3] the governing equations for the combined motion of melt and interface are

$$\Delta u = 0 \quad \text{in } \mathcal{M} \quad , \tag{I.5}$$

and, on $\partial\mathcal{C}$,

$$\partial_m u = \alpha v, \quad \beta(\vartheta)v = \kappa[\psi(\vartheta)+\psi''(\vartheta)] + u_t + (2\alpha)^{-1}(\mathrm{grad}\,u)^2, \qquad (I.6)$$

where u = *velocity potential*, the symbol ∂_m denotes directional differentiation, and $\alpha = (\rho_{\mathcal{C}} - \rho_{\mathcal{M}})^2/\rho_{\mathcal{M}}$ is a density ratio. Equations (I.5), (I.6) integrate a new type of free-boundary problem; when linearized about a flat equilibrium interface, they support oscillatory solutions and so have, in the words of M.E. Gurtin, "a flavor of hyperbolicity"; an undesirable feature is perhaps that disturbances propagate with infinite speed.

PART II - MELTING-FREEZING WAVES IN HELIUM-4

By melting-freezing waves one refers to a nonstandard type of wave motion where the atoms located at the interface of a crystal of He^4 with its melt apparently switch back and forth from the solid to the liquid phase so that oscillatory motions of the phase interface are observed even if the individual atoms nearby (especially so those in the solid phase) seem to undergo no substantial motion. The phenomenology of melting-freezing waves in He^4 is better grasped if one has in mind the *phase diagram* sketched below (*cf.* [4]).

Crystal growth (by *supercooling*) is obtained in a cryostat in the .4 - .9 °K range at sufficiently high pressure: a H(exagonal) C(lose-)P(acked) solid phase is found in the presence of a *superfluid* liquid phase. Superfluids have viscosity close to zero and thermal conductivity close to infinity. The latter property, together with the evidence that the freezing pressure is almost constant and therefore the latent heat of fusion is almost null in the indicated temperature range, suggests that freezing of He^4 may be regarded as a *mechanical* more than *thermal* process [4], and modelled consequently, just as is done in [2] and [3].

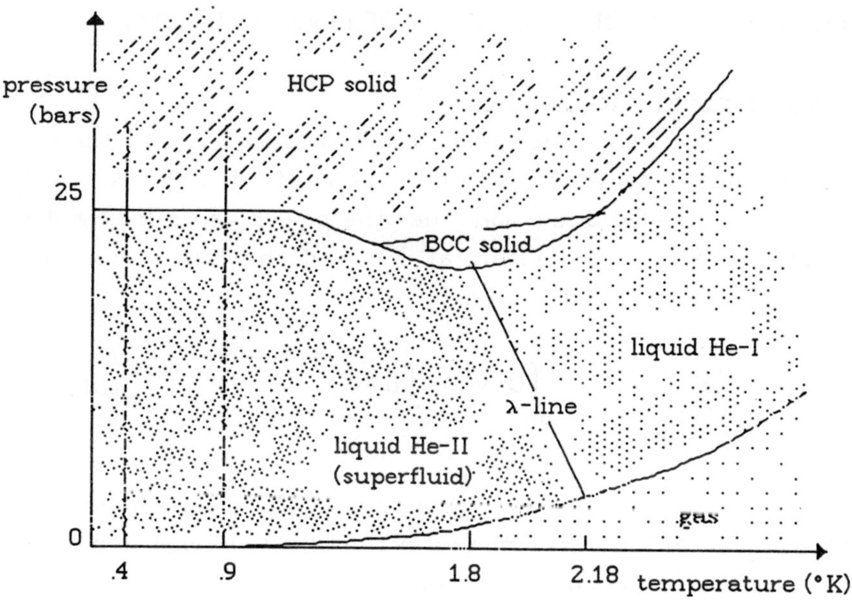

Tapping the cryostat's wall, one observes almost by the naked eye oscillatory motions of the phase interface, in the following ranges : temperature .4 - .6 °K; amplitude 1 - 2 mm; frequency 3 - 10^7 kHz; typical values of surface tension are around .023 gr/cm (*cf.* [6]). Oscillations dampen away with temperature, so much so that at .8 °K none persists, the quantum mechanical explanation being that at higher temperatures the thermal phonons and rotons interact with the interface and eventually inhibit its motion [4]. For oscillatory motions of the form

$$h(x,t) = h_0 \, e^{i(\lambda x - \omega t)} \qquad \text{(II.1)}$$

the observed *dispersion relation* is

$$\omega^2 \approx \lambda^3 \ . \tag{II.2}$$

PART III - HYPERBOLIC EVOLUTION OF PLANE CURVES

a) Deduction of Model Equation

Let $\mathcal{S} \subset \partial\mathcal{C}$ be an interfacial chunk, and let $\partial\mathcal{S}$ be the set of its boundary points. We assume (the reader is referred to [3] for a more detailed discussion) that the *power expended* by the *interfacial stress* $\mathbf{C}$ (acting at $\partial\mathcal{S}$) and the *bulk-interface interaction force* $\mathbf{b}$ (acting on $\mathcal{S}$) is

$$\int_{\partial\mathcal{S}} \mathbf{C}\cdot\mathbf{v} + \int_{\mathcal{S}} \mathbf{b}\cdot\mathbf{v} \tag{III.1}$$

(here $\mathbf{v} = v\,\mathbf{m}$). We also assume that two scalar fields are defined over $\partial\mathcal{C}$, namely, the *effective inertial density* ρ and the *interfacial mass supply* r , and write

$$\int_{\mathcal{S}} \frac{1}{2}\rho\, v^2 \quad \text{and} \quad \int_{\mathcal{S}} \frac{1}{2} r\, v^2$$

for the *kinetic energy* and the *kinetic energy supply*, respectively, of the interfacial chunk $\mathcal{S}$. With this, we let the *mechanical energy-production* in the chunk $\mathcal{S}$, be defined as the difference between the *rate of change of kinetic energy*

$$\frac{d}{dt}\Big(\int_{\mathcal{S}} \frac{1}{2}\rho v^2\Big) - \int_{\mathcal{S}} \frac{1}{2} r\, v^2$$

and the power expended on $\mathcal{S}$ by $\mathbf{C}$ and $\mathbf{b}$. We denote such difference by $\mathcal{E}(\mathcal{S})$:

$$\mathcal{E}(\mathcal{S}) = [\ \frac{d}{dt}\ (\ \int_{\mathcal{S}} \frac{1}{2}\rho\, v^2\ ) - \int_{\mathcal{S}} \frac{1}{2} r\, v^2\ ] - [\ \int_{\partial\mathcal{S}} \mathbf{C}\cdot\mathbf{v} + \int_{\mathcal{S}} \mathbf{b}\cdot\mathbf{v}\ ]\ . \qquad \text{(III.2)}$$

We then call the transformation

$$x \mapsto x^{+} = x + ct \quad \text{, with } c = \text{a constant vector,} \qquad \text{(III.3)}$$

a *Galilean change in lattice observer*, and note that the normal velocity transforms in the following way under such changes in observer:

$$v \mapsto v^{+} = v + c\ . \qquad \text{(III.4)}$$

Our basic axiom is that, for each interfacial chunk $\mathcal{S}$, *the mechanical energy-production* $\mathcal{E}(\mathcal{S})$ *be invariant under Galilean changes in lattice observer*. This axiom yields local balance laws for interfacial mass

$$\rho^{\circ} - \rho\kappa v = r \qquad \text{(III.5)}$$

and momentum

$$\rho\,\mathbf{v}^{\circ} = \mathbf{C}_s + \mathbf{b} \qquad \text{(III.6)}$$

(here a subscript s indicates differentiation with respect to arc lenght on $\partial\mathcal{C}$). Only the normal component of (III.6), namely,

$$\rho\, v^{\circ} = \xi_s + \sigma\kappa + b\ ,\quad \xi = \mathbf{C}\cdot\mathbf{m}\ ,\ \sigma = \mathbf{C}\cdot\mathbf{l}\ ,\ b = \mathbf{b}\cdot\mathbf{m}\ , \qquad \text{(III.7)}$$

restrict the motion of $\partial\mathcal{C}$; indeed, (III.5) determines the interfacial mass supply r, and the tangential component of (III.6) determines the tangential component of the bulk-interface interaction force $\mathbf{b}$. We then supplement (III.7) with constitutive prescriptions for the *surface tension* σ and the *surface shear* ξ :

there is a interfacial energy $\psi(\vartheta) > 0$ such that

$$\sigma(\vartheta) = \psi(\vartheta) \quad \text{and} \quad \xi(\vartheta) = \psi'(\vartheta) \; ; \qquad \text{(III.8)}_1$$

for the normal component b of the interaction force :

$$b = -F + \beta(\vartheta)v \;, \quad \beta(\vartheta) > 0 \; ; \qquad \text{(III.8)}_2$$

and for the effective inertial density

$$\rho(\vartheta) > 0 \; . \qquad \text{(III.8)}_3$$

Finally, combining $(\text{III.7})_1$ and (III.8) we arrive at (I.4).

b) When Interface is Locally a Graph

In this case (I.4) yields the following broad generalization of (I.2):

$$h_{tt} + B(\vartheta)\,h_t + \sin 2\vartheta\, h_t h_{tx} = \sin^2\vartheta\,[G(\vartheta) - (h_t \cos\vartheta)^2]\,h_{xx} - D(\vartheta)\,, \qquad \text{(III.9)}$$

$$h_x \tan\vartheta = -1\,, \quad B(\vartheta) = \frac{\beta(\vartheta)}{\rho(\vartheta)}\,, \quad G(\vartheta) = \frac{\psi(\vartheta) + \psi''(\vartheta)}{\rho(\vartheta)}\,, \quad D(\vartheta) = \frac{F}{\rho(\vartheta)\sin\vartheta}\,.$$

As $(\text{III.8})_3$ prevails, it can be shown that equation (III.9) is hyperbolic if and only if $[\psi(\vartheta) + \psi''(\vartheta)]$ is positive , *i.e.*, as we have already pointed out, if and only if equation (I.3) is parabolic.

c) Small Oscillations about a Flat Interface

With a view towards modelling melting-freezing waves, assume now that $F = 0$, so that (I.4) admits flat interfaces ($\kappa \equiv 0$) as equilibrium solutions, and linearize (III.9) about such an equilibrium solution. The result is

$$h_{tt} + B_0 h_t = G_0 h_{xx} \ , \quad B_0 = B(\pi/2) \ , \ G_0 = G(\pi/2) \ ,$$

and this equation supports damped oscillations of the form

$$h(x,t) = h_0 \, e^{i(\lambda x - \omega t) - \varsigma t} \ , \quad \varsigma = B_0/2 \ ,$$

whose dispersion relation is

$$\omega^2 = G_0 \lambda^2 - \varsigma^2 \ . \tag{III.10}$$

Comparison of (III.10) with (I.2) shows that, as far as modelling melting-freezing waves is concerned, our present theory gives only qualitative agreement with experiments [5]. This is not the case with the theory of [3], where a more sophisticated modelling of inertial effects leads to a dispersion relation in complete agreement with (I.2).

REFERENCES

[1] Angenent, S. & M.E. Gurtin, Multiphase thermomechanics with interfacial structure. 2. Evolution of an isothermal interface. Arch. Rational Mech. Anal., forthcoming.

[2] Gurtin, M.E. & P. Podio-Guidugli, A hyperbolic theory for the evolution of plane curves. SIAM J. Math. Anal., forthcoming.

[3] Gurtin, M.E., A mechanical theory for crystallization of a rigid solid in a liquid melt: melting-freezing waves. Arch. Rational Mech. Anal., forthcoming.

[4] Maris, H.J. & A.F. Andreev, The surface of crystalline helium-4. Phys. Today, 25-30 (Feb. 1987).

[5] Andreev, A.F. & A. Ya. Parshin, Equilibrium shape and oscillations of the surface of quantum crystals. Sov. Phys. JETP 48, 763-766 (1978).

[6] Keshishev, K.O., A. Ya. Parshin & A.V. Babkin, Experimental detection of crystallization waves in He^4. Sov. Phys. JETP Lett. 30, 56-59 (1979).

[7] Brakke, K. A., *The Motion of a Surface by its Mean Curvature.* Princeton University Press (1978).

LINEAR TRANSPORT AND KINETIC THEORY IN STOCHASTIC MEDIA

G.C.Pomraning*

Department of Mathematics
University of Bologna
Piazza di Porta S.Donato, 5
40127 Bologna, Italy

*Permanent address: School of Engineering and Applied Science, University of California, Los Angeles, Los Angeles, CA 90024-1597, U.S.A.

ABSTRACT

We discuss the formulation of linear transport and kinetic theory describing particle flow in a random mixture of two or more immiscible fluids. Included in this discussion are: (1) the method of smoothing and the Liouville master equation approaches in the case of Markovian statistics as the description of the fluid mixing; (2) the use of asymptotics to derive various limiting descriptions of the Markov model; (3) the use of the theory of alternating renewal processes to obtain an integral equation formulation for non-Markovian statistics; and (4) the reduction of this non-Markovian integral formulation to differential equations with statistical effects represented by pseudo-scattering terms.

INTRODUCTION

In the last few years several papers have been written concerned with the problem of describing linear particle transport in a stochastic medium consisting of two randomly mixed immiscible fluids.[1–21] The goal in these works has been to develop a relatively simple and accurate description for the ensemble-averaged solution of the stochastic transport problem. In this paper we will attempt to summarize the salient features of these papers and present an overall view of the status of this area of research.

The generic linear transport equation we will be concerned with is written

$$\frac{1}{v}\frac{\partial\psi}{\partial t} + \boldsymbol{\Omega}\cdot\nabla\psi + \sigma\psi = \frac{\sigma_s}{4\pi}\varphi + S, \tag{1}$$

where

$$\varphi = \int_{4\pi} d\mathbf{\Omega}\psi(\mathbf{\Omega}). \tag{2}$$

In writing Eqs.(1) and (2) we have used the notation of neutron transport theory, but our considerations are equally applicable in any linear transport setting. The dependent variable in Eq.(1) is the angular flux $\psi(\mathbf{r}, \mathbf{\Omega}, t)$, with $\mathbf{r}, \mathbf{\Omega}$, and t denoting the spatial, angular (neutron flight direction), and time variables, respectively. The quantity $\varphi(\mathbf{r}, t)$ is the scalar flux, v is the neutron speed, $\sigma(\mathbf{r}, t)$ is the macroscopic total cross section, $\sigma_s(\mathbf{r}, t)$ is the macroscopic scattering cross section, and $S(\mathbf{r}, \mathbf{\Omega}, t)$ is the external source of neutrons. We have assumed isotropic and coherent (no energy exchange) scattering in Eq.(1), but this simplification is not necessary for the essentials of the considerations to follow. Thus Eq.(1) is a monoenergetic (one group) transport equation, and there is no need to display the independent energy variable which is simply a parameter.

To treat the case of a binary statistical mixture, the quantities σ, σ_s, and S in Eq.(1) are considered as discrete random variables, each of which assumes, at any $\mathbf{r}, \mathbf{\Omega}$, t, one of two sets of values characteristic of the two fluids constituting the mixture. We denote the two fluids by an index i, with $i = 0, 1$, and in the ith fluid these three quantities are denoted by $\sigma_i(\mathbf{r}, t)$, $\sigma_{si}(\mathbf{r}, t)$, and $S_i(\mathbf{r}, \mathbf{\Omega}, t)$. That is, as a neutron traverses the mixture along any path, it encounters alternating segments of the two fluids, each of which has known deterministic values of σ, σ_s, and S. The statistical nature of the problem enters through the statistics of the fluid mixing, i.e., through the knowledge as to what fluid is present in the mixture at space point $\mathbf{r}$ and time t. Since σ, σ_s and S in Eq. (1) are (two state, discrete) random variables, the solution of Eqs.(1) and (2) for ψ and φ is stochastic, and we let $\bar{\psi}$ and $\bar{\varphi}$ denote the ensemble-averaged angular and scalar fluxes, respectively. The goal is to develop accurate and, hopefully, simple models for these ensemble averages.

To discuss this problem in a qualitative way, it is convenient to consider the simple case of time-independent transport in a purely absorbing ($\sigma_s = 0$) medium. Then Eq.(1) reduces to

$$\frac{d\psi}{ds} + \sigma\psi = S, \tag{3}$$

where s denotes the spatial variable in the direction of propagation. If we write any random variable Q as the sum of its ensemble average $\bar{Q}$ and the deviation from this average $\tilde{Q}$, an ensemble averaging of Eq.(3) gives

$$\frac{d\bar{\psi}}{ds} + \bar{\sigma}\bar{\psi} + (\overline{\tilde{\sigma}\tilde{\psi}}) = \bar{S}. \tag{4}$$

It is clear from Eq. (4) that one needs to calculate (or approximate) the cross correlation term involving the ensemble average of the product of $\tilde{\sigma}$ and $\tilde{\psi}$ to obtain a formulation for the quantity of interest, namely $\bar{\psi}$.

On physical grounds, and it can also be shown mathematically by using asymptotics,[14] it is clear that this term is very small and can be neglected when $\sigma_i l_i << 1$, where l_i is a characteristic chord length for the fluid packets of fluid i. However, when this inequality is not satisfied, the neglect of the cross correlation term in Eq. (4) can lead to large errors in $\bar{\psi}$. In particular, the attenuation of a beam of particles can be grossly overestimated by using Eq. (4) without the cross correlation term.[11] Simply stated then, the challenge is to compute the cross correlation term in Eq. (4). To do this, one must specify the statistics of the fluid mixing, and we discuss both Markovian and non-Markovian mixtures.

MARKOVIAN MIXTURES

We initially consider the simple transport equation given by Eq. (3), under the assumption of Markovian mixing. These mixing statistics are defined by a no-memory statement. Given that point s is in fluid i, the probability of point $s + ds$ being in fluid $j \neq i$ is given by

$$Prob(i \rightarrow j) = \frac{ds}{\lambda_i(s)}, \; j \neq i, \tag{5}$$

where the $\lambda_i(s)$ are the prescribed Markovian transition probabilities. The probability $p_i(s)$ of finding fluid i at position s, and the correlation length $\lambda_c(s)$ for the Markov mixture are uniquely and simply related to the $\lambda_i(s)$.[13] With the mixing statistics defined by Eq. (5), one can obtain an exact description for $\bar{\psi}$ corresponding to the transport problem given by Eq. (3). This is given by the two coupled equations [1–3,11]

$$\frac{d}{ds}\begin{bmatrix} \bar{\psi} \\ \chi \end{bmatrix} + \begin{bmatrix} \bar{\sigma} & \nu \\ \nu & \hat{\sigma} \end{bmatrix}\begin{bmatrix} \bar{\psi} \\ \chi \end{bmatrix} = \begin{bmatrix} \bar{S} \\ T \end{bmatrix}, \tag{6}$$

where

$$\nu\chi = (\overline{\tilde{\sigma}\tilde{\psi}}), \quad \nu = (p_0 p_1)^{1/2}(\sigma_0 - \sigma_1), \tag{7}$$

$$\bar{\sigma} = p_0\sigma_0 + p_1\sigma_1, \quad \hat{\sigma} = p_1\sigma_0 + p_0\sigma_1 + \frac{1}{\lambda_c}, \tag{8}$$

$$\bar{S} = p_0 S_0 + p_1 S_1, \quad T = (p_0 p_1)^{1/2}(S_0 - S_1). \tag{9}$$

Two derivations of Eq. (6) can be given. The first is based upon the so-called method of smoothing.[1,2,11] Subtraction of Eq. (4) from Eq. (3) gives

$$\frac{d\tilde{\psi}}{ds} + \bar{\sigma}\tilde{\psi} + [\tilde{\sigma}\tilde{\psi} - (\overline{\tilde{\sigma}\tilde{\psi}})] = \tilde{S} - \tilde{\sigma}\bar{\psi}. \tag{10}$$

Inserting a smallness parameter in front of the bracketted term in Eq. (10) and seeking a solution as a power series in this smallness parameter gives the Neumann series

$$\tilde{\psi} = \sum_{n=0}^{\infty} (-1)^n [G(I-P)\tilde{\sigma}]^n G(\tilde{S} - \tilde{\sigma}\bar{\psi}), \tag{11}$$

where I is the identity and G and P are Green's and projection operators defined by

$$Gf = \int_0^s ds' \left[exp - \int_{s'}^s d\xi \bar{\sigma}(\xi)\right] f(s'), \quad Pf = \bar{f}. \tag{12}$$

For Markovian statistics, all of the terms in the sum of Eq. (11) can be evaluated and the infinite summation performed. The result of these algebraic manipulations is just Eq. (6).

The second derivation of Eq. (6) is based upon the observation that the transport problem given by Eq. (3) describes a Markov process; it is an initial value problem with the spatial variable s playing the role of time. Thus, together with the assumed Markovian mixing, we have a joint Markov process and we can write a Liouville master equation for the joint probability $P_i(s,\psi)$, defined such that $P_i d\psi$ is the probability of finding fluid i and intensity ψ lying between ψ and $\psi + d\psi$ at position s.[3,11,20] This master equation is given by

$$\frac{\partial P_i}{\partial s} - \frac{\partial}{\partial \psi}(\psi\sigma_i P_i - S_i P_i) = \frac{P_j}{\lambda_j} - \frac{P_i}{\lambda_i}, \; j \neq i. \tag{13}$$

The probability $p_i(s)$ is just the integral of $P_i(s,\psi)$ over ψ, and the conditional ensemble-averaged intensity given that position s is in fluid $i, \psi_i(s)$, is the integral of $\psi P_i(s,\psi)$ over ψ, divided by p_i. Multiplying Eq. (13) by ψ and integrating over ψ gives two coupled equations for the ψ_i. Making the change of variables

$$\bar{\psi} = p_0\psi_0 + p_1\psi_1, \quad \chi = (p_0 p_1)^{1/2}(\psi_0 - \psi_1) \tag{14}$$

once again leads to Eq. (6).

No such exact results are available when time-dependence and scattering are present in the transport problem, i.e., when Eq. (1) is the underlying stochastic transport equation. In this case, the algebra of the method of smoothing cannot be carried through. Further, with scattering the transport problem is not a Markov process (it is a boundary value problem) and hence the master equation approach cannot rigorously be used.[9] Nevertheless, it has been suggested[11] that the use of the master equation might lead to a useful, albeit approximate, model. This model is

$$\left(\frac{1}{v}\frac{\partial}{\partial t} + \boldsymbol{\Omega}\cdot\nabla + \boldsymbol{\Sigma}\right)\begin{bmatrix}\bar{\psi}\\ \chi\end{bmatrix} = \frac{1}{4\pi}\boldsymbol{\Sigma}_s\begin{bmatrix}\bar{\varphi}\\ \eta\end{bmatrix} + \begin{bmatrix}\bar{S}\\ T\end{bmatrix}, \tag{15}$$

where

$$\boldsymbol{\Sigma} = \begin{bmatrix}\bar{\sigma} & \nu\\ \nu & \hat{\sigma}\end{bmatrix}, \quad \boldsymbol{\Sigma}_s = \begin{bmatrix}\bar{\sigma}_s & \nu_s\\ \nu_s & \hat{\sigma}_s\end{bmatrix}, \tag{16}$$

where, in addition to Eqs. (7) through (9), we have

$$\bar{\varphi} = \int_{4\pi} d\Omega\bar{\psi}, \quad \eta = \int_{4\pi} d\Omega\chi, \quad \nu_s = (p_0 p_1)^{1/2}(\sigma_{s0} - \sigma_{s1}), \tag{17}$$

$$\bar{\sigma}_s = p_0\sigma_{s0} + p_1\sigma_{s1}, \quad \hat{\sigma}_s = p_1\sigma_{s0} + p_0\sigma_{s1}. \tag{18}$$

Several other derivations of this model are available. Sahni[16] has shown that the assumption of independent particle flight paths leads to Eq. (15), and he has also shown[17] that nuclear reactor noise techniques can be used to obtain this model. It has also been demonstrated[12] that one can derive exact particle balance equations in this stochastic transport setting, but involving two ensemble averages conditioned upon being in fluid i, namely a volumetric average ψ_i and a surface (interface between fluids) average ψ_i^s. Making the approximation that these two ensemble averages are equal again yields the model given by Eq. (15).[12] Numerical comparisons with exact Monte Carlo results[12] shows the accuracy of this model to be of the order of ten percent or so, with the error being the largest, as expected, for problems with multiple scattering interactions.

Although no exact transport-like equation(s) formulation is available in general for $\bar{\psi}$ in the presence of time dependence and scattering, a few exact results have been reported for certain time-independent problems including the scattering interaction in rod (particles are restricted to move along a line) and planar geometry.[6–10,20] Vanderhaegen and Deutsch[9] considered the sourcefree purely-scattering rod problem and used a Riccati transformation to convert the boundary value transport problem for the intensity to an initial value problem for the rod transmission. They then applied the master equation approach to this initial value problem. Their results were confirmed in a recent paper [20] devoted to numerical and analytical solution methods for the Liouville master equation applicable to a large class of initial value problems. An exact formalism in rod and planar geometry has been developed[6–8,10] for a restricted class of problems, in which σ_s/σ and S/σ are nonstochastic; the only stochasticity is the optical depth variable. The results of Vanderhaegen and Deutsch[9] were again confirmed[6], and results for several classical halfspace and infinite medium transport problems were developed.[7,8,10]

If one accepts Eq. (15) as a reasonable model of particle transport in a binary Markovian mixture, two simplifications involving certain asymptotic limits have been reported.[14] The first of these corresponds to a small amount, of $0(\epsilon^2)$, of one fluid admixed with an $0(1)$ amount of the second fluid. Further, the first fluid has large, of $0(1/\epsilon^2)$, source and cross sections compared to $0(1)$ quantities for the second fluid. In this asymptotic limit, one finds a renormalized transport equation of the standard form given by[14]

$$\left(\frac{1}{v}\frac{\partial}{\partial t} + \mathbf{\Omega}\cdot\nabla + \sigma_{eff}\right)\bar{\psi} \;=\; \frac{\sigma_{s,eff}}{4\pi}\bar{\varphi} + S_{eff} + 0(\epsilon^2), \tag{19}$$

with

$$\sigma_{eff} = \bar{\sigma} - \frac{\nu^2}{\hat{\sigma}}, \quad \sigma_{s,eff} = \sigma_{eff} - \sigma_{a,eff}, \tag{20}$$

and

$$\sigma_{a,eff} = (\bar{\sigma} - \bar{\sigma}_s) - \frac{(\nu - \nu_s)^2}{(\hat{\sigma} - \hat{\sigma}_s)}, \quad S_{eff} = \bar{S} - \frac{(\nu - \nu_s)}{(\hat{\sigma} - \hat{\sigma}_s)}T. \tag{21}$$

Equation (21) for S_{eff} assumes that the sources are isotropic, i.e., $S_i \neq S_i(\mathbf{\Omega})$. The corresponding expression for anisotropic sources is more complex.[14] The result given by Eqs. (19) through (21) is robust in that all of these effective cross sections and source are always nonnegative, even when one is far from the asymptotic limit under consideration. This derivation of a renormalized transport equation as an asymptotic limit has been generalized to a mixture containing an arbitrary number of fluid components, and leads to an entire class of reduced (in complexity) transport descriptions.[19]

The second asymptotic limit of Eq. (15) is one in which the scattering interaction is dominant, of $0(1)$. Absorption, sources, and all derivatives are assumed small, of $0(\epsilon^2)$, except for the spatial derivative which is scaled to be $0(\epsilon)$. This scaling eliminates the angular variable in the transport problem, and is analogous to the Hilbert expansion which gives the Euler fluid equations from the Boltzmann equation. If we restrict ourselves to isotropic sources as well as an isotropic correlation length, one finds in this asymptotic limit two coupled diffusion equations given by[14]

$$\left(\frac{1}{v}\frac{\partial}{\partial t} - \nabla \cdot \mathbf{D}\nabla + \mathbf{\Sigma}_a\right)\begin{bmatrix} \bar{\varphi} \\ \eta \end{bmatrix} = 4\pi \begin{bmatrix} \bar{S} \\ T \end{bmatrix} + 0(\epsilon^2), \tag{22}$$

where $\mathbf{\Sigma}_a = \mathbf{\Sigma} - \mathbf{\Sigma}_s$ and

$$\mathbf{D} = \frac{1}{3}\mathbf{\Sigma}^{-1} = \frac{1}{3(\bar{\sigma}\hat{\sigma} - \nu^2)}\begin{bmatrix} \hat{\sigma} & -\nu \\ -\nu & \bar{\sigma} \end{bmatrix}. \tag{23}$$

It can easily be shown that the diffusion coefficient matrix $\mathbf{D}$ is positive definite under all choices of the physical parameters, thus assuring robustness in this diffusion approximation. Since the scaling which led to Eq. (22) corresponds to a singular perturbation problem, both boundary and temporal layers exist. Performing the appropriate asymptotic matching analyses, one obtains boundary and initial conditions for Eq. (22). The boundary condition is of the mixed (Robbin) type, involving the number 0.710446... , the classic linear extrapolation distance for the Milne problem.[14]

NON-MARKOVIAN MIXTURES

We consider a general class of statistics describing alternating segments on a line, and which includes Markovian statistics as a special case,[13] defined by

$$Q_i(y,x) = Prob\{[y,x] in\ i | x\ in\ i\ and\ x+dx\ not\ in\ i\}, \tag{24}$$

$$R_i(y,x) = Prob\{[y,x] in\ i | x\ in\ i\}. \tag{25}$$

These two probabilities are not independent, but are related to the fundamental unconditional probability that the interval $[y,x]$ is in fluid i.[13] Restricting ourselves to time independent transport with no scattering as described by Eq.(3), the theory of alternating renewal processes has been used to obtain $\bar{\psi} = p_0\psi_0 + p_1\psi_1$, where the ψ_i are obtained as the solution of the coupled integral equations[4,5,13,18,21]

$$\psi_0(x) = G_0[0,x;\hat{\psi}]R_0(0,x) + \int_0^x d_y R_0(y,x,)G_0[y,x;\phi_1(y)], \tag{26}$$

$$\phi_0(x) = G_0[0,x;\hat{\psi}]Q_0(0,x) + \int_0^x d_y Q_0(y,x)G_0[y,x;\phi_1(y)], \tag{27}$$

where

$$G_0[y,x;b] = b\ exp[-\tau_0(y,x)] + \int_y^x dx' S_0(x')exp[-\tau_0(x',x)], \tag{28}$$

and

$$\tau_0(y,x) = \int_y^x d\xi\sigma_0(\xi). \tag{29}$$

The interval of interest has been taken as $x \geq 0$, with $\hat{\psi}$ denoting the (nonstochastic) boundary condition at $x = 0$. A similar set of equations is obtained by interchanging the indices 0 and 1. Physically, $\psi_i(x)$ denotes the ensemble-averaged solution given that position x is in fluid i, and $\phi_i(x)$ is the ensemble-averaged solution given that position x is an interface between fluids, with fluid i to the left.

In general, these four coupled integral equations are only a model, but for a certain subclass of the mixing statistics defined by Eqs. (24) and (25), they constitute an exact description.[18,21] Included in this class is the case of homogeneous statistics, i.e., the statistics are independent of position, in which case Q_i and R_i depend only upon the single variable $z \equiv x - y$, and for which Q_i and R_i are given by

$$Q_i(z) = \int_z^\infty dz' f_i(z'), \quad R_i(z) = \frac{1}{\lambda_i}\int_z^\infty dz' Q_i(z'). \tag{30}$$

Here $f_i(z)$ is the density function for chord length z in fluid i, and λ_i is the mean chord length in fluid i. Further, in this case we have $p_i = \lambda_i/(\lambda_0 + \lambda_1)$. Homogeneous Markov statistics is a special case of these statistics,[13] with

$$\lambda_i f_i = Q_i = R_i = exp(-z/\lambda_i). \tag{31}$$

An interesting subclass of these homogeneous statistics is that for which $Q_i(z)$ can be represented as a Laplace transform according to

$$Q_i(z) = \int_0^\infty du q_i(u) e^{-z/u} \tag{32}$$

for some class of functions $q_i(u)$. Then one deduces

$$R_i(z) = \frac{1}{\lambda_i}\int_0^\infty du u q_i(u) e^{-z/u}, \quad \lambda_i = \int_0^\infty du u q_i(u), \tag{33}$$

and it can be shown[13] that the four coupled renewal integral equations reduce to the two equation integro-differential description

$$\frac{\partial\psi_0(x,u)}{\partial x} + [\sigma_0(x) + 1/u]\psi_0(x,u) = S_0(x) + \int_0^\infty du' \frac{q_1(u')}{u}\,\psi_1(x,u'), \tag{34}$$

with a second equation obtained by interchanging the indices 0 and 1. The ψ_i needed to compute the ensemble-averaged intensity according to $\bar{\psi}(x) = p_0\psi_0(x) +$

$p_1\psi_1(x)$ are given by quadrature as

$$\psi_i(x) = \frac{1}{\lambda_i} \int_0^\infty du u q_i(u) \psi_i(x,u). \tag{35}$$

Recalling that the underlying transport process is one with no scattering as described by Eq. (3), we see that the statistical nature of the problem has introduced a scattering-like term into the formalism, as described by the integral term in Eq. (34). This term can be interpreted as pseudo-scattering from one fluid to the other due to statistical fluctuation "scattering centers". In the special case of homogeneous fluids, i.e., σ_i independent of position, a singular eigenfunction analysis of Eq. (34), and its complement found by interchanging indices, has recently been given.[15]

Sanchez[18] has generalized the integral equation formulation given by Eqs. (26) and (27) to include time dependence and scattering for general statistics. The treatment is very involved and quite abstract, involving an infinite set of very complex coupled integral equations. Some method, unknown at this time, is needed to truncate and simplify this set of equations for this treatment to be useful in applications. A completely different approach to treating non-Markovian mixing has also been suggested.[5] In this approach, Eq. (15), the Markovian master equation model, is modified by replacing the correlation length λ_c contained in $\hat{\sigma}$ [see Eq. (8)] with an effective correlation length to account in an approximate way for the non-Markovian statistics of the mixing. It was shown that Eq. (15) with this simple modification preserves certain properties of the more fundamental integral equations. In particular, the single change from λ_c to an effective correlation length allows one to preserve, for time-independent problems with homogeneous statistics and σ_i, σ_{si}, and S_i independent of position, both the infinite medium solution and the mean distance to collision. Finally, we note that numerical calculations have suggested[5] that in the non-Markovian case, the information required from the mixing statistics is the mean and variance of the chord length distribution for each fluid component. The mean is not sufficient, and moments higher than the variance have little effect on the transport solutions, at least for the unimodal distributions tested.

ACKNOWLEDGEMENTS

The author wishes to express his gratitude to the Italian National Research Council (C.N.R.) for the opportunity to visit the University of Bologna under C.N.R.'s Visiting Professor Program. The preparation of this paper was also partially supported by the U.S. Department of Energy under grant DE-FG03-89ER14016.

REFERENCES

1. G. C. Pomraning, *Transport Th. Statis. Phys.* **15**, 773 (1986).
2. C. D. Levermore, G. C. Pomraning, D. L. Sanzo, and J. Wong, *J. Math. Phys.* **27**, 2526 (1986).
3. D. Vanderhaegen, *J. Quant. Spectros. Radiat. Transfer* **36**, 557 (1986).
4. D. Vanderhaegen, *J. Quant. Spectros. Radiat. Transfer* **39**, 333 (1988).
5. C. D. Levermore, G. C. Pomraning, and J. Wong, *J. Math. Phys.* **29**, 995 (1988).
6. G. C. Pomraning, *J. Quant. Spectros. Radiat. Transfer* **40**, 479 (1988).
7. G. C. Pomraning, *Transport Th. Statis. Phys.* **17**, 595 (1988).
8. F. Malvagi and G. C. Pomraning, *J. de Physique, Coll. C7* **49**, 321 (1988).
9. D. Vanderhaegen and C. Deutsch, *J. Statis. Phys.* **54**, 331 (1989).
10. G. C. Pomraning, *J. Quant. Spectros. Radiat. Transfer* **41**, 103 (1989).
11. G. C. Pomraning, C. D. Levermore, and J. Wong, *Lecture Notes in Pure and Applied Mathematics* **Vol. 115**, P. Nelson et al., ed. Marcel Dekker, New York, pp. 1-35 (1989).
12. M. L. Adams, E. W. Larsen, and G. C. Pomraning, *J. Quant. Spectros. Radiat. Transfer* **42**, 253 (1989).
13. G. C. Pomraning, *J. Quant. Spectros. Radiat. Transfer* **42**, 279 (1989).
14. F. Malvagi, C. D. Levermore, and G. C. Pomraning, "Asymptotic Limits of a Statistical Transport Description," *Transport Th. Statis. Phys.*, in press.
15. H. Frisch, C. G. Pomraning, and P. F. Zweifel, "An Exact Analytical Solution of a Radiative Transfer Problem in a Binary Mixture," *J. Quant. Spectros. Radiat. Transfer*, in press.
16. D. C. Sahni, "Equivalence of Generic Equation Method and the Phenomenological Model for Linear Transport Problems in a Two State Random Scattering Medium," *J. Math. Phys.* (submitted 1988).

17. D. C. Sahni, "An Application of Reactor Noise Techniques to Neutron Transport Problems in a Random Medium," *Annals Nucl. Energy* (submitted 1988).

18. R. Sanchez, "Linear Kinetic Theory in Stochastic Media," *J. Math. Phys.* (submitted 1989).

19. F. Malvagi and G. C. Pomraning, "Renormalized Equations for Linear Transport in Stochastic Media," *J. Math. Phys.* (submitted 1989).

20. V. C. Boffi, F. Malvagi, and G. C. Pomraning, "Solution Methods for Discrete State Markovian Initial Value Problems," *J. Statis. Phys.* (submitted 1989).

21. G. C. Pomraning and R. Sanchez, "The Use of Renewal Theory for Stochastic Transport," *J. Quant. Spectros. Radiat. Transfer* (submitted 1989).

INHOMOGENEOUS DEFORMATIONS IN NON-LINEARLY ELASTIC MATERIALS

K.R. Rajagopal
Department of Mechanical Engineering
University of Pittsburgh
Pittsburgh, PA 15261

ABSTRACT

We discuss a systematic method for studying inhomogeneous deformations of isotropic non-linearly elastic materials.

1. INTRODUCTION

In most technical applications wherein solids undergo large deformations, the deformation is primarily non-homogeneous. Even so, few non-homogeneous deformations have been subject to the intense scrutiny which they deserve. One reason for this might be the fact that such deformations are not "universal"[1] within the context of general compressible elastic materials and thus have to be studied within the context of special constitutive theories rather than in full generality. Ericksen [1], [2] established two important results which have wide ramifications. Ericksen [1] proved that all "universal deformations"[1] in isotropic compressible materials, in the absence of body forces, are homogeneous. He had also shown that in the case of incompressible isotropic elastic materials five classes of non-homogeneous universal deformations are possible [2]. This result was subsequently extended to more general simple materials by Carroll [3], Wineman [4] and Fosdick [5]. Universal deformations and motions play a special role in both fluid and solid mechanics in that they suggest experiments in which the deformation of the body under consideration is known from the outset. Also, many of the simple and elegant exact solutions in both solid and fluid theories are universal. However, universal solutions have limited impact when it comes to considering boundary value problems which have relevance to every day applications wherein non-homogeneous deformations are the order of the day. More importantly, there is no reason to expect all materials, say for instance rubber and steel, to respond in an identical manner. Hence, it would seem more reasonable for one to ask the following question: given a non-homogeneous deformation what class of constitutive equations can support such a deformation? We shall see that this question in general leads to determining the existence of solutions to a set of partial differential or ordinary differential equations subject to the appropriate initial and/or

[1]A deformation or motion is said to be "universal" for a particular class of materials subject to a body force, if it is possible in all materials belonging to the class.

boundary conditions. In the light of the above remarks, it is not surprising that a considerable amount of interest has been evinced recently in the study of non-homogeneous deformations of isotropic non-linear elastic materials (cf. Sensening [6], Holden [7], Varley and Cumberbatch [8], Ogden [9], Isherwood and Ogden [10], Currie and Hayes [11], Rajagopal and Wineman [12], Rajagopal, Troy and Wineman [13], Chao, Rajagopal and Wineman [14], McLeod, Rajagopal and Wineman [15], and Fu, Rajagopal and Szeri [16]).[2]

It would be useful for the purpose of modeling if we could determine the largest class of constitutive equations which would support a given inhomogeneous deformation. Thus, starting with a specific non-homogeneous deformation which is observed, we could delineate the class of constitutive equations which would be appropriate for modeling such deformations. We shall now discuss a few non-homogeneous deformations of non-linear elastic materials.

The recent work of Rajagopal and Wineman [20] on the non-homogeneous twisting of a non-linearly elastic slab of a Mooney-Rivlin material presents a novel feature which seems to have been overlooked in previous studies, namely the existence of solutions that are not symmetric. Rajagopal and Wineman [20] show that the classical axisymmetric solution is imbedded in a one-parameter family, and is the only one of the class that possesses axisymmetry. Similarly, in their study of the non-homogeneous extension of an elastic layer, Rajagopal and Wineman [20] show that the classical homogeneous uniaxial extension is imbedded in a much larger class of solutions, all but the classical solution being non-homogeneous. Rajagopal, Troy, and Wineman [13] have established similar results in the case of a class of non-homogeneous shearing, which they call proportional shearing. A question that comes to mind immediately is whether these non-homogeneous solutions are stable. The stability or instability of these solutions has a great deal of significance with regard to the status and use of the classical solution.

Recently, Chen and Rajagopal [21] studied the stability of the inhomogeneous deformations obtained by Rajagopal and Wineman [20] in the case of a Neo-Hookean material. Since the problems under consideration were in unbounded domains, Chen and Rajagopal [21] established a stability criterion based on disturbances which have compact support with fixed value for the displacements on the boundary. Using this stability criterion, Chen and Rajagopal [27] obtained sufficient conditions for stability, i.e., they showed that the inhomogeneous deformations are stable provided that the inhomogeneity, measured in some manner, is sufficiently small. In the case of some other models the ideas of Chen and Rajagopal[21] can be extended.

---

[2]Recently, Carroll [17], [18], [19] has exhibited several exact solutions within the context of elastodynamics for isotropic compressible elastic materials.

2. PRELIMINARIES

Let Ω denote the reference configuration of the body and let Ω_t denote the current configuration of the body at time t. By the motion of the body we mean a one-to-one invertible map

$$\mathbf{x} = \chi\,(\mathbf{x},t)) \tag{1}$$

which assigns to each $\mathbf{x} \in \Omega$ at time t, a point $\mathbf{x} \in \Omega_t$.

The stress $\mathbf{T}$ in an isotropic compressible elastic material has the form

$$\mathbf{T} = \alpha_1 \mathbf{1} + \alpha_2 \mathbf{B} + \alpha_3 \mathbf{B}^2\,, \tag{2}$$

where

$$\alpha_i = \alpha_i\,(\mathrm{I,II,III})\,, \quad i = 1,2,3 \tag{3}$$

$$\mathrm{I} = \mathrm{tr}\,\mathbf{B}\,, \tag{4}$$

$$\mathrm{II} = \frac{1}{2}\left[(\mathrm{tr}\,\mathbf{B})^2 - \mathrm{tr}\,\mathbf{B}^2\right], \tag{5}$$

$$\mathrm{III} = \det \mathbf{B}\,. \tag{6}$$

In the case of an isotropic incompressible material the stress takes the form

$$\mathbf{T} = -\,p\mathbf{1} + \widehat{\alpha}_1 \mathbf{B} + \widehat{\alpha}_2 \mathbf{B}^2\,, \tag{7}$$

where $-\,p\mathbf{1}$ is the indeterminate part of the stress due to the constraint of incompressibility, and

$$\widehat{\alpha}_i = \widehat{\alpha}_i\,(\mathrm{I,II})\,, \quad i = 1,2\,, \tag{8}$$

and

$$\mathrm{III} = 1.$$

If the material is hyperelastic, then the representation for the stress $\mathbf{T}$ in an isotropic compressible elastic material, and an isotropic incompressible elastic material take the form

$$\mathbf{T} = \frac{2}{\sqrt{\mathrm{III}}}\left[\left(\mathrm{II}\frac{\partial W}{\partial \mathrm{II}} + \mathrm{III}\frac{\partial W}{\partial \mathrm{III}}\right)\mathbf{1} + \frac{\partial W}{\partial \mathrm{I}}\mathbf{B} - \mathrm{III}\frac{\partial W}{\partial \mathrm{II}}\mathbf{B}^{-1}\right], \tag{9}$$

$$\mathbf{T} = -\,p\mathbf{1} + \frac{\partial \widehat{W}}{\partial \mathrm{I}}\mathbf{B} - \frac{\partial W}{\partial \mathrm{II}}\mathbf{B}^2\,, \tag{10}$$

respectively, where

$$W = W\,(\mathrm{I,II,III})\,, \tag{11}$$

and

$$\widehat{W} = \widehat{W}\,(\mathrm{I,II})\,, \tag{12}$$

are the stored energy function for the respective materials.

3. INHOMOGENEOUS DEFORMATIONS

In the introduction we mentioned several recent investigations into the non-homogeneous deformations of non-linearly elastic materials (cf. [6] - [14]). All these investigations involved specific non-linear constitutive theories. However, recently McLeod, Rajagopal and Wineman [15] investigated a non-homogeneous deformation within the spirit of the comments in the introduction. They tried to determine a class of constitutive equations which could support such a deformation. Since their problem best exemplifies the kind of research proposed here, we shall briefly discuss their work.

Consider a deformation of the form

$$x = \frac{1}{\sqrt{\lambda}} X + f(Z) , \tag{13}$$

$$y = \frac{1}{\sqrt{\lambda}} Y + g(Z) , \tag{14}$$

$$z = \lambda Z , \tag{15}$$

where (x,y,z) represent reference coordinates of a material point and (x,y,z) its current coordinates and λ is a constant.

The above deformation is a generalization of the classical uniaxial extension or compression wherein material filaments parallel to the z-axis are allowed to buckle (cf. [15]).

In the case of an incompressible isotropic hyperelastic material the problem reduces to solving (cf. [15])

$$\frac{d^2}{dZ^2}\left\{\frac{df}{dZ}\left(\lambda\frac{\partial W}{\partial I} + \frac{\partial W}{\partial II}\right)\right\} = 0 , \tag{16}$$

$$\frac{d^2}{dZ^2}\left\{\frac{df}{dZ}\left(\lambda\frac{\partial W}{\partial I} + \frac{\partial W}{\partial II}\right)\right\} = 0 , \tag{17}$$

where

$$I = \frac{2}{\lambda} + \lambda^2 + f'^2 + g'^2 , \tag{18}$$

$$II = 2\lambda + \frac{1}{\lambda^2} + \frac{f'^2 + g'^2}{\lambda} . \tag{19}$$

Defining

$$W\left(f'^2 + g'^2\right) = W(I,II) , \tag{20}$$

where the prime denotes differentiation with respect to Z, the problem reduces to solving

$$f'\overline{W}\left(f'^2 + g'^2; \lambda\right) = c_1 Z + d_1 , \tag{21}$$

$$g'\overline{W}\left(f'^2 + g'^2; \lambda\right) = c_2 Z + d_2 . \tag{22}$$

The appropriate boundary conditions are

$$f(0) = m_x, \quad f(H) = l_x , \tag{23}$$

$$g(0) = m_y, \quad g(H) = l_y . \tag{24}$$

Let

$$t \equiv f'^2 + g'^2 . \tag{25}$$

McLeod, Rajagopal and Wineman [15] use degree theory to establish that the equations (21), (22) subject to (23), (24) have a solution provided

$$t\,\overline{W}(t;\lambda) \to 0 \quad \text{as } t \to 0 \tag{26}$$

$$t\,\overline{W}(t;\lambda) \to \infty \quad \text{as } t \to \infty \tag{27}$$

and $t\,\overline{W}(t;\lambda) > 0$ whenever $t > 0$.

Note that the above conditions are satisfied for both neo-Hookean and Mooney-Rivlin materials. Of course, a much larger class of constitutive equations are possible, in fact precisely those which satisfy the conditions (26) - (27). Thus, for instance, even if $t\,\overline{W}(t;\lambda) \equiv F(t)$ is of the form shown in Figure 1, solutions would exist for the problem under consideration.

Thus, starting with an assumed form for the deformation, i.e., equations (13) - (15), we obtain the equations of equilibrium (16) - (17) in terms of the stored energy function which depends in an arbitrary manner on the specific funtions used in the assumed deformation (13) - (15). Then, we determine the form of the stored energy function for which there would be existence of solutions to the equations of equilibrium under appropriate boundary conditions. Hence, we delineate the class of constitutive equations (or materials) for which a deformation of the assumed form is possible. In the example shown above, the problem reduces to a system of non-linear ordinary differential equations. In general, the problem could be a system of partial differential equations, this depends on the assumed form for the deformation.

In the case of the Neo-Hookean and Mooney-Rivlin materials, we can exhibit explicit exact solutions for the functions f(Z) and g(Z). Since the solution is not "universal" in general (unless special forms are assumed for f and g) the form of f and g are different for different materials.

ACKNOWLEDGEMENTS

Dr. K.R. Rajagopal would like to thank the National Science Foundation for its support.

REFERENCES

[1] J.L. Ericksen, ZAMP, 5, 466 (1954).

[2] J.E. Ericksen, J. Math. Phys., 34, 126 (1955).

[3] M.M. Carroll, Intl. J. Eng. Sci., 5, 515 (1967).

[4] A.S. Wineman, University of Michigan Techical Report (1967).

[5] R.L. Fosdick, Arch. Ratl. Mech. Anal., 29, 272 (1968).

[6] C.B. Sensing, Comm Pure and Appl. Math., 18, 147 (1965).

[7] J.T. Holden, Appl. Sci. Research, 19, 171 (1968).

[8] E. Varley and E. Cumberbatch, Finite Elasticity (R.S. Rivlin, Editor), ASME, New York (1977).

[9] R.W. Ogden, Mechanics Research Communications, 4, 347 (1977).

[10] D.A. Isherwood and R.W. Ogden, Intl. J. Solids and Structures, 13, 105 (1977).1

[11] P.K. Currie and M. Hayes, Proceedings of the IUTAM Symposium on Finite Elasticity (D.E. Carlson and R.T. Shield, Editors), Martinus Nijhoff Publishers, The Hagne-Boston-London (1981).

[12] K.R. Rajagopal and A.S. Wineman, J. of Structural Mechanics, 4, 471 (1984).

[13] K.R. Rajagopal, A.S. Wineman and W. Troy, A Note on Non-Universal Deformations of Non-Linear Elastic Layers, submitted for publication.

[14] R.M. Chao, K.R. Rajagopal and A.S. Wineman, Archives of Mechanics, 39, 551 (1987).

[15] J.B. McLeod, K.R. Rajagopal and A.S. Wineman, Proc. Roy. Irish Academy, 88A, 91 (1988).

[16] D. Fu, K.R. Rajagopal and A.S. Wineman, In Press, Intl. J. Non-Linear Mechanics.

[17] M.M. Carroll, J. of Elasticity, 8, 323 (1978).

[18] M.M. Carroll, ZAMP, 25, 83 (1974).

[19] M.M. Carroll, J. of Appl. Mechanics, 1 (1983).

[20] K.R. Rajagopal and A.S. Wineman, Intl. J. Eng. Sci., 23, 217 (1985).

[21] Y.C. Chen and K.R. Rajagopal, Archives Rational Mech. Analysis, 108, 1 (1989).

NONLINEAR CONVECTION PROBLEMS INVOLVING INTERNAL HEAT GENERATION AND VARIABLE GRAVITY EFFECTS

S.Rionero[1] & B.Straughan[2]

[1]Dipartimento di Matematica e Applicazioni,
Università di Napoli.

[2]Department of Mathematics,
University of Glasgow.

Abstract.

Nonlinear convection thresholds are determined for the problem of a layer of fluid heated internally, with zero heat flux on the lower boundary and constant temperature upper surface, and for the problem of convection in a fluid saturated porous layer which is heated internally and where the gravitational field varies with distance through the layer.

1. Introduction. The goal of this paper is to present results for two problems. The first is that of finding the critical Rayleigh number for nonlinear energy stability in a fluid saturated porous layer in the presence of an internal heat source while subject to the effect of a variable gravity field; further details are given in Rionero & Straughan[4)]. The second determines the critical Rayleigh number for nonlinear energy stability for convection in a fluid layer with internal heat generation when the lower boundary is thermally insulated; further details may be found in Straughan[6)].

The effect of a heat source on convection in a plane layer has been studied by Joseph & Shir[2)], Roberts[5)] and by Tritton & Zarraga[7)]. Roberts[5)] found the linear instability boundary and studied the solution behaviour after the onset of convection, for a constant heat source, Q. We here present a nonlinear stability boundary by using energy theory: Joseph & Shir[2] investigated the same problem for constant Q, by energy theory, although they used different boundary conditions to those of Roberts[5]. In fact, Roberts[5], observes that in the experiments of Tritton & Zarraga[7] it may not be appropriate to have Q constant: we allow for the possibility of a non-uniform heat source.

2. Convection in a porous medium with internal heat source and variable gravity effects.

The non-dimensional perturbation equations governing the stability of convection in a porous layer with internal heat source and variable gravity effects are:

$$p_{,i} = RH(z)\theta k_i - u_i, \tag{1}$$

$$u_{i,i} = 0, \tag{2}$$

$$Pr(\theta_{,t} + u_i\theta_{,i}) = RN(z)w + \Delta\theta, \tag{3}$$

where $\mathbf{u} = (u, v, w)$, $N(z) = 1 + \delta q(z)$ and $H(z) = 1 + \epsilon h(z)$ with $\epsilon h(z)$ and $\delta q(z)$ representing the gravity variation and internal heat source effects, respectively. These equations hold on the region $\{(x, y) \in \mathbf{R}^2\} \times \{z \in (0, 1)\}$ and the boundary conditions to be satisfied are:

$$\theta = 0, \quad z = 0, 1; \qquad w = 0, \quad z = 0, 1; \tag{4}$$

we also assume the functions $\mathbf{u}, \theta, p$ are spatially periodic in the sense that they satisfy

$$\Delta^*\theta = -a^2\theta,$$

for a wave number a, where

$$\Delta^* \equiv (\partial^2/\partial x^2 + \partial^2/\partial y^2).$$

We develop an energy analysis for (1)–(4). The energy identities are derived as:

$$\frac{1}{2}Pr\,\frac{d}{dt}\,\|\theta\|^2 = R < Nw\theta > -D(\theta),$$

$$R < Hw\theta >= \|\mathbf{u}\|^2,$$

where $< . >$ and $\|.\|$ denote integration over, and the L^2-norm on V, respectively; V here is the cell of solution periodicity.

For $\lambda(> 0)$ to be chosen we form the identity:

$$\frac{dE}{dt} = R < M(z)w\theta > -D(\theta) - \lambda\|\mathbf{u}\|^2,$$

where

$$E = \frac{1}{2} Pr\|\theta\|^2,$$

$$M(z) = 1 + \delta q + \lambda(1 + \epsilon h).$$

If we define R_E by

$$R_E^{-1} = \max_{\mathcal{H}} \frac{I}{\mathcal{D}}, \tag{5}$$

with

$$I =< Mw\theta >, \qquad \mathcal{D} = D(\theta) + \lambda\|\mathbf{u}\|^2,$$

where, in (5), $\mathcal{H}$ is the space of admissible functions, then it is straightforward to show $E(t) \to 0$ at least exponentially provided $R < R_E$, cf. Rionero[3)]. The existence of a maximizing solution to problems like (5) is discussed in Rionero[3)]: an explicit proof may be derived by a slight modification of the analysis in Galdi *et al.*[1)].

The maximization problem (5) is solved numerically; in fact we determine

$$\max_{\lambda} \min_{a^2} R_E(a^2; \lambda). \tag{6}$$

Some values of the critical Rayleigh and wave numbers for nonlinear energy stability, R_E^2, a_E, and for linear instability, R_L^2, a_L, for various ϵ, δ, are now given when $h = -z$, $q = z$. Further values, including those for other choices of h and q are given in Rionero & Straughan[4)].

R_L^2	R_E^2	a_L^2	a_E^2	ϵ	δ	$\bar{\lambda}$
39.478	39.478	9.870	9.870	0.0	0.0	1.000
77.020	74.971	10.209	10.178	1.0	0.0	1.500
132.020	122.374	12.314	11.928	1.5	0.0	2.565
63.114	61.563	10.026	10.110	1.0	0.5	2.318
111.930	101.513	11.887	11.562	1.5	0.5	3.201
35.384	35.092	9.835	9.870	0.5	1.0	2.000
53.375	51.715	9.918	10.032	1.0	1.0	2.797

Table 1. Critical Rayleigh and wave numbers for $h = -z$, $q = z$. ($\bar{\lambda}$ denotes the best value of λ in (6).)

3. Nonlinear stability for convection in a fluid with internal heat generation.

We now obtain *quantitative nonlinear* stability estimates which guarantee nonlinear stability for the problem of convection in a plane layer with a non-uniform heat source, a constant temperature upper surface, and thermally insulated from below. This is the problem studied by Roberts[5] for constant heat source; he used the methods of linear instability theory, truncated expansions and mean field theory. Roberts[5] is primarily concerned with the structure of the finite amplitude solution and attempting to explain the experimental findings of Tritton & Zarraga[7]. It is known that for the internally heated convection problem sub-critical instabilities may occur and so our nonlinear results which establish a critical Rayleigh number below which convection cannot occur complement the linear value of Roberts[5] above which convection occurs. The parameter region of possible sub-critical instability is small. Roberts[5], pp. 34–35, remarks that the temperature dependence of the electrical resistivity of the working fluid in the experiments of Tritton & Zarraga[7] prevents the experiment attaining a *constant* heat source. Hence, we also investigate the non-uniform heat source problem and obtain the nonlinear energy stability – linear instability critical Rayleigh numbers.

Consider then a layer of heat-conducting linear viscous fluid occupying the horizontal layer $z \in (0, d)$ with the lower boundary $z = 0$ thermally insulated (i.e. zero heat flux through $z = 0$) and with the temperature scale selected so that the temperature at $z = d$ remains zero. This is the situation described in Roberts[5]. The equations are

$$v_{i,t} + v_j v_{i,j} = -\frac{1}{\rho} p_{,i} + \nu \Delta v_i + g k_i \alpha T,$$

$$v_{i,i} = 0,$$

$$T_{,t} + v_i T_{,i} = \kappa \Delta T + Q,$$

where $\nu, \rho, g, \alpha, \kappa$ are viscosity, density, gravity, thermal expansion coefficient and thermal diffusivity. These equations are defined on the spatial region $\mathbf{R}^2 \times \{z \in (0, d)\}$. The boundary conditions are

$$\mathbf{v} = \mathbf{0}, \quad z = 0, d; \qquad T = 0, \quad z = d; \qquad \frac{\partial T}{\partial z} = 0, \quad z = 0. \tag{7}$$

We here consider three heat supply functions:

$$\text{I.} \quad Q = Q\,(\text{constant}),$$

$$\text{II.} \quad Q = Q\,(e^{z/d} - 1),$$

$$\text{III.} \quad Q = Q \sin \frac{2\pi z}{d}.$$

Note that the last choice heats the lower half of the layer while cooling the rest.

The steady solutions $(\bar{\mathbf{v}}, \bar{T})$ corresponding to boundary conditions (7) are

$$\bar{\mathbf{v}} \equiv \mathbf{0}; \qquad \bar{T} = \frac{Q}{2\kappa}(d^2 - z^2) \qquad \text{(case I)}$$

$$\bar{\mathbf{v}} \equiv \mathbf{0}; \qquad \bar{T} = \frac{Qd^2}{\kappa}\Big(e - \frac{3}{2} - e^{z/d} + \frac{z}{d} + \frac{z^2}{2d^2}\Big) \qquad \text{(case II)}$$

$$\bar{\mathbf{v}} \equiv \mathbf{0}; \qquad \bar{T} = \frac{Qd^2}{2\pi\kappa}\Big(1 + \frac{1}{2\pi}\sin\frac{2\pi z}{d} - \frac{z}{d}\Big) \qquad \text{(case III)}$$

the hydrostatic pressure being determined from the momentum equation.

To investigate the stability of these solutions we introduce perturbations

$$(\mathbf{u}, \theta, \pi)$$

by

$$v_i = \bar{v}_i + u_i, \qquad T = \bar{T} + \theta, \qquad p = \bar{p} + \pi.$$

Then, the perturbation equations are non-dimensionalized according to the scales (stars denote dimensionless quantities)

$$t = t^* \frac{d^2}{\nu} \qquad U = \frac{\nu}{d} \qquad \mathbf{x} = \mathbf{x}^* d \qquad T = T^* T^\#$$

$$Pr = \frac{\nu}{\kappa} \qquad T^\# = \frac{U}{\kappa}\sqrt{\frac{Qd\nu}{\alpha g}} \qquad R^2 = \frac{Qd^5 g\alpha}{\kappa^2 \nu}.$$

Here Pr is the Prandtl number and $R^2 = Ra$ is the Rayleigh number (as defined by Roberts[5]).

The *non-dimensional* equations are (*after omitting all stars*):

$$\begin{aligned} u_{i,t} + u_j u_{i,j} &= -\pi_{,i} + \Delta u_i + R\theta k_i; \\ u_{i,i} &= 0; \\ Pr(\theta_{,t} + u_i \theta_{,i}) &= Rf(z)w + \Delta\theta; \end{aligned} \tag{8}$$

where $\mathbf{k} = (0,0,1), w = u_3$, and

$$f(z) = \begin{cases} z, & \text{case I;} \\ e^z - 1 - z, & \text{case II;} \\ \frac{1}{2\pi}(1 - \cos 2\pi z), & \text{case III.} \end{cases}$$

Equations (8) hold on $\mathbf{R}^2 \times (0,1)$ and the boundary conditions adopted are:

$$\left. \begin{array}{lll} u_i = 0, & z = 0, 1; & \\ \theta = 0, & z = 1; & \dfrac{\partial\theta}{\partial z} = 0, \quad z = 0; \\ \mathbf{u}, \theta, p & \text{have a periodic structure in } x, y. & \end{array} \right\} \tag{9}$$

Conditions (9) are appropriate to two fixed surfaces. For two stress free surfaces it can be shown that exchange of stabilities holds. Our nonlinear energy results are close to those of stationary convection and so we limit consideration to this case for the linear theory.

To obtain the nonlinear stability limit we multiply $(8)_1$ by u_i, $(8)_3$ by $\lambda\theta$, add, and integrate over a period cell V, for $\lambda(>0)$ a coupling parameter to be chosen. This yields, after use of (9),

$$\frac{dE}{dt} = RI - \mathcal{D}, \tag{10}$$

where

$$E = \frac{1}{2}\|\mathbf{u}\|^2 + \frac{1}{2}Pr\lambda\|\theta\|^2,$$

$$I =< (1+\lambda f)\theta w >,$$

$$\mathcal{D} = \|\nabla\mathbf{u}\|^2 + \lambda\|\nabla\theta\|^2.$$

From (10) we now deduce

$$\frac{dE}{dt} \leq -\mathcal{D}\left(\frac{R_E - R}{R_E}\right), \tag{11}$$

where

$$R_E^{-1} = \max_{\mathcal{H}} \frac{I}{\mathcal{D}}, \tag{12}$$

$\mathcal{H}$ being the space of admissible functions. If now

$$R < R_E$$

use of Poincaré's inequality in (11) quickly leads to exponential decay of $E(t)$ and so to global nonlinear stability. The value of R_E is obtained from (12) where it is convenient to set $\phi = \lambda^{\frac{1}{2}}\theta$ and then the Euler-Lagrange equations for R_E are:

$$\left.\begin{aligned} \Delta u_i + MR_E k_i \phi &= \varpi_{,i}, \\ \Delta\phi + MR_E w &= 0, \end{aligned}\right\} \tag{13}$$

where ϖ is a Lagrange multiplier and M is given by

$$M = \frac{1+\lambda f}{2\sqrt{\lambda}}.$$

To find R_E from (13) we take curlcurl $(13)_1$ and then take the third component of the resulting equation. Normal modes are introduced, i.e.

$$w = W(z)G(x,y), \quad \phi = \Phi(z)G(x,y),$$

where $\Delta^* G = -a^2 G$, $\Delta^* = \partial^2/\partial x^2 + \partial^2/\partial y^2$ and a is a horizontal wavenumber. The object is then to solve the system

$$\left.\begin{aligned}(D^2 - a^2)^2 W &= R_E M a^2 \Phi,\\ (D^2 - a^2)\Phi &= -M R_E W,\end{aligned}\right\} \qquad (14)$$

$D = d/dz$, on $(0,1)$ with

$$W = DW = 0, \quad z = 0,1; \qquad \Phi = 0, \quad z = 1; \qquad D\Phi = 0, \quad z = 0; \qquad (15)$$

for the lowest eigenvalue $R_E(a^2;\lambda)$ and then find

$$Ra_E = \max_{\lambda} \min_{a^2} R_E^2(a^2;\lambda).$$

Here Ra_E is the critical Rayleigh number of energy theory.

By comparison with (14), the eigenvalue problem of linear theory for stationary convection is

$$\left.\begin{aligned}(D^2 - a^2)^2 W &= R a^2 \Theta,\\ (D^2 - a^2)\Theta &= -R f(z) W,\end{aligned}\right\} \qquad (16)$$

together with (15) with Θ replacing Φ. For (16) we find

$$Ra_L = \min_{a^2} R^2(a^2),$$

where Ra_L is the critical Rayleigh number of linear theory (stationary convection).

Numerical results are given below.

f	Ra_L	a_L^2	Ra_E	a_E^2	$\bar{\lambda}$
z	2772.275	6.913	2737.162	6.946	2.068
$e^z - 1 - z$	8382.810	7.377	8008.576	7.485	5.725
$\frac{1-\cos 2\pi z}{2\pi}$	5335.721	6.661	5312.097	6.678	4.077

Table 2. Critical Rayleigh numbers of linear theory, Ra_L, and energy theory, Ra_E, together with the respective critical wavenumbers a_L, a_E, and the best value of λ, $\bar{\lambda}$.

Two facts are immediately evident from the above table. The first is that the linear and energy results are very close: certainly close enough to be useful from a practical viewpoint. The second is that a variation in Q can influence the critical Rayleigh number enormously: this indicates that the experiment needs to be performed very carefully to determine the correct form for Q.

References.

1. Galdi, G.P., Payne, L.E., Proctor, M.R.E. & Straughan, B. Convection in thawing subsea permafrost. *Proc. Roy. Soc. London A*, **414**, 83–102 (1987).

2. Joseph, D.D. & Shir, C.C. Subcritical convective instability. Part 1. Fluid layers. *J. Fluid Mech.*, **26**, 753–768 (1966).

3. Rionero, S. Metodi variazionali per la stabilità asintotica in media in magnetoidrodinamica. *Ann. Matem. Pura Appl.*, **78**, 339–364 (1968).

4. Rionero, S. & Straughan, B. Convection in a porous medium with internal heat source and variable gravity effects. *Int. J. Engng. Sci.*, to appear.

5. Roberts, P.H. Convection in horizontal layers with internal heat generation. Theory. *J. Fluid Mech.*, **30**, 33–49 (1967).

6. Straughan, B. Continuous dependence on the heat source and nonlinear stability for convection with internal heat generation, to appear.

7. Tritton, D.T. & Zarraga, M.N. Convection in horizontal layers with internal heat generation. Experiments. *J. Fluid Mech.*, **30**, 21–31 (1967).

LONG TIME BEHAVIOR OF INITIAL BOUNDARY VALUE PROBLEMS FOR A DISCRETE VELOCITY BOLTZMANN EQUATION

M. Slemrod*

Center for the Mathematical Sciences
University of Wisconsin
Madison, WI 53706

0. INTRODUCTION

In a series of recent papers S. Kawashima [1]-[6] has given very nice results on initial boundary value problems for discrete velocity Boltzmann equations. In particular, his results apply to the Broadwell model for rarefied gas dynamics. While Kawashima's results apply to various interesting cases, they do not include the long term trend to equilibria of the Broadwell model with specularly reflective boundary conditions. In this note I will give a short survey of Kawashima's results as they apply to the Broadwell model and then give a sketch of my approach to the specularly reflective boundary condition problem [7]. My purpose is to provide the reader with a thumb-nail sketch of the "state of the art" for initial-boundary value problems for the Broadwell model with the hope that the report will stimulate further research into new methods and techniques of analysis.

1. BROADWELL MODEL

The general Broadwell model describes a gas of particles with identical masses moving along three perpendicular coordinate axes with the same speed

* Sponsored in part by the Air Force Office of Scientific Research, Air Force Systems Command, USAF, under Contract/Grant No. AFOSR-87-0315. The United States government is authorized to reproduce and distribute reprints for Government purposes notwithstanding any copyright herein. Also supported in part by a Meyerhoff Visiting Professorship, Weizmann Institute of Science.

c. Results of a particular collision have the same probability and only binary collisions are considered. Let $N_i = N_i(x, y, z, t)$, $i = 1, \ldots, 6$, denote the densities of particles moving in the six allowed directions. Then $N_1, \ldots, N_6$ satisfy the Broadwell system

$$\begin{aligned}
\frac{\partial N_1}{\partial t} + c\frac{\partial N_1}{\partial x} &= \sigma(N_3N_4 + N_5N_6 - 2N_1N_2),\\
\frac{\partial N_2}{\partial t} - c\frac{\partial N_2}{\partial x} &= \sigma(N_3N_4 + N_5N_6 - 2N_1N_2),\\
\frac{\partial N_3}{\partial t} + c\frac{\partial N_3}{\partial y} &= \sigma(N_1N_2 + N_5N_6 - 2N_3N_4),\\
\frac{\partial N_4}{\partial t} - c\frac{\partial N_4}{\partial y} &= \sigma(N_1N_2 + N_5N_6 - 2N_3N_4),\\
\frac{\partial N_5}{\partial t} + c\frac{\partial N_5}{\partial z} &= \sigma(N_1N_2 + N_3N_4 - 2N_5N_6),\\
\frac{\partial N_6}{\partial t} - c\frac{\partial N_6}{\partial z} &= \sigma(N_1N_2 + N_3N_4 - 2N_5N_6),
\end{aligned}$$

where $\sigma/2c$ is the cross section for binary collisions.

For flows which are independent of y, z and for which $N_3 = N_4 = N_5 = N_6$ the above six velocity Broadwell model reduces to the simpler form

$$\begin{aligned}
\frac{\partial N_1}{\partial t} + c\frac{\partial N_1}{\partial x} &= N_3^2 - N_1N_2,\\
\frac{\partial N_2}{\partial t} - c\frac{\partial N_2}{\partial x} &= N_3^2 - N_1N_2,\\
\frac{\partial N_3}{\partial t} &= \frac{1}{2}(N_1N_2 - N_3^2).
\end{aligned} \tag{1}$$

(As the results given here are independent of σ the convenient choice $\sigma = 1/2$ is made.)

Initial conditions are

$$N_1(x,0) = N_{10}(x), \quad N_2(x,0) = N_{20}(x), \quad N_3(x,0) = N_{30}(x). \tag{2}$$

2. KAWASHIMA'S RESULTS

In Kawashima's work, the following boundary conditions are considered. At $x = 0$ either

$$N_1(0,t) = B_1^0 \quad \text{or} \tag{3}$$

$$N_1(0,t) = B^0 N_2(0,t) \tag{3'}$$

and at $x = 1$ either

$$N_2(1,t) = B_2^1 \quad \text{or} \tag{4}$$

$$N_2(1,t) = B^1 N_1(1,t) \; . \tag{4'}$$

The boundary data B_1^0, B_2^1 are positive constants, the coefficients B^0, B^1 are nonnegative constants. For (4′) Kawashima requires

$$1 \geq B^1 \; , \qquad M_2 = B^1 M_1 \; ,$$

where $M = (M_1, M_2, M_3)$ is the same constant Maxwellian, i.e., (M_1, M_2, M_3) are positive constants satisfying

$$M_1 M_2 - (M_3)^2 = 0 \; .$$

We now summarize Kawashima's results:

Existence and uniqueness of solutions.

Define	Problem (I)	by	(1), (2), (3), (4)
	Problem (II)	by	(1), (2), (3), (4′)
	Problem (III)	by	(1), (2), (3′), (4′) .

Then for $N_{10}(x), N_{20}(x), N_{30}(x)$ positive and continuously differentiable on $[0,1]$, Problems (I), (II) and (III) have unique continuously differentiable solutions if the initial data satisfy appropriate compatibility conditions up to order one.

Asymptotic behavior as $t \to \infty$.

As to asymptotic behavior Kawashima has proven: Consider Problem (I) and let M be any fixed Maxwellian and suppose $N_0 - M \in H^1(0,1)$ and N_0 satisfies the compatibility conditions up to order zero. If $\|N_0 - M\|_1$ is small enough, then there exists a unique global solution N satisfying $N - M \in C^0([0,\infty); H^1(0,1)) \cap C^1([0,\infty);\; L^2(0,1))$. Moreover this solution converges uniformly in x to the solution $N^\infty(x)$ of the corresponding stationary problem at an exponential rate as $t \to \infty$.

Asymptotic behavior for Problems (II), (III) is unknown.

3. SPECULARLY REFLECTIVE BOUNDARY CONDITIONS

Kawashima's results do not cover specularly reflective boundary conditions. In [7] I considered (1), (2) with specularly reflective boundary conditions

$$N_1(x,t) = N_2(x,t) \ , \quad t > 0 \ , \quad x = 0,1 \ . \tag{5}$$

The initial data N_{10}, N_{20}, N_{30} are non-negative, continuously differentiable on $[0,1]$ and satisfy the consistency condition

$$\frac{d}{dx}N_{10} = -\frac{d}{dx}N_{20} \quad \text{at} \ x = 0,1 \ . \tag{6}$$

With regard to existence, I proved:

For data as given above, the initial boundary value problem possesses a unique non-negative continuously differentiable solution. (Such data will be called *admissible*.)

More interesting are the results on asymptotic behavior. In this case standard methods seem to fail and I introduced an approach based on the weak invariance principle (see [8]).

The main result is formulated in terms of some elementary concepts. First define for a triple $(f_1, f_2, f_3) : \ [0,1] \to \mathbf{R}$ the extension (f_{1e}, f_{2e}, f_{3e}) given by

$$\begin{aligned} &f_{ie}(x) = f_i(x) \ , \quad 0 \le x \le 1 \ , \quad i = 1,2,3 \ ; \\ &f_{1e}(x) = f_{2e}(-x) \ , \quad f_{2e}(x) = f_{1e}(-x) \ , \\ &f_{3e}(x) = f_{3e}(-x) \ , \quad -1 \le x < 0 \ ; \\ &f_{ie}(x) = f_{ie}(x+2) \ , \qquad x \in \mathbf{R} \ . \end{aligned}$$

Next we introduce the appropriate function spaces for our problem.

We follow the presentation for Orlicz spaces given in Adams [9]. A similar discussion may be found in Zygmund [10].

Let a be a real valued function on $[0,\infty)$ with the following properties:

(i) $a(0) = 0$, $a(t) > 0$ if $t > 0$, $\lim_{t\to\infty} a(t) = \infty$;

(ii) a is nondecreasing $\big(a(s) \ge a(t),\ s > t \ge 0\big)$;

(iii) a is right continuous ($\lim_{s\to t+} a(s) = a(t)$). The function $A(t)$ is defined on $[0,\infty)$ by

$$A(t) = \int_0^t a(\tau)d\tau.$$

Let Ω be a domain in $\mathbb{R}^n$. Denote the *Orlicz class* $K_A(\Omega)$ by the set of all (equivalence classes modulo equality a.e. in Ω) of measurable functions $u(x)$ defined on Ω satisfying

$$\int_\Omega A(|u(x)|)dx < \infty .$$

The *Orlicz space* $L_A(\Omega)$ is defined to be the linear hull of the Orlicz class $K_A(\Omega)$ i.e. the smallest vector space (under pointwise addition and scalar multiplication) containing $K_A(\Omega)$. We endow $L_A(\Omega)$ with the norm

$$\|u\|_A =: \inf\left\{ k > 0;\ \int_\Omega A\left(\frac{|u(x)|}{k}\right)dx \le 1\right\} .$$

Let $E_A(\Omega)$ denote the closure in $L_A(\Omega)$ of the space of functions u which are bounded on Ω and have bounded support in $\overline{\Omega}$.

Let $A(s) = (1+s)\ln(1+s) - s$ $(a(s) = \log(1+s))$ and $\widetilde{A}(s) = e^s - s - 1$ $(\widetilde{a}(s) = e^s - 1)$. Then

(i) $L_A(\Omega) = K_A(\Omega) = E_A(\Omega)$;

(ii) $L_A(\Omega)$ is a separable Banach space;

(iii) $E_{\widetilde{A}}(\Omega)$ is a separable Banach space;

(iv) $[E_{\widetilde{A}}(\Omega)]' = L_A(\Omega)$, i.e. $L_A(\Omega)$ is the dual of the separable Banach space $E_{\widetilde{A}}(\Omega)$.

Definition. Set $X = L_A(0,1)^3$, $E = E_{\widetilde{A}}(0,1)^3$ for A, $\widetilde{A}$ as above, where

$$\|(u_1,u_2,u_3)\|_X^2 = \sum_{i=1}^3 \|u_i\|_A^2, \qquad \|(u_1,u_2,u_3)\|_E^2 =: \sum_{i=1}^3 \|u_i\|_{\widetilde{A}}^2 .$$

Then X, E are separable Banach spaces, $E' = X$.

Definition. For $\boldsymbol{f} \in X$ define the *mass functional* M and the *discrete velocity Boltzmann H-functional* H by

$$M(\boldsymbol{f}) =: \int_0^1 f_1 + f_2 + 4f_3 dx$$

$$H(\boldsymbol{f}) =: \int_0^1 f_1 \log f_1 + f_2 \log f_2 + 4f_3 \log f_3 dx$$

It is easy to prove the following lemmas.

Lemma. $M(\boldsymbol{N}(\cdot,t)) = M(\boldsymbol{N}_0)$;

$$\frac{d}{dt} H(\boldsymbol{N}(\cdot,t)) = -\int_0^1 (N_3^2 - N_1N_2)(\log N_3^2 - \log N_1N_2) dx$$

$$\leq -4 \int_0^1 \left((N_3 - (N_1N_2)^{1/2}\right)^2 dx \leq 0.$$

Now fix an admissible $\boldsymbol{N}_0$ and let $\boldsymbol{N}(t;\boldsymbol{N}_0)$ denote the associated unique solution to the initial-boundary value problem (1), (2), (5). From the above lemmas we see that positive orbit

$$\gamma^+(\boldsymbol{N}_0) =: \bigcup_{t\geq 0} \boldsymbol{N}(t;\boldsymbol{N}_0)$$

lies in $B_R =: \{f \in X;\ \|f\|_X \leq R\}, 0 < R < \infty$. (Use for example the inequality $(1+s)\log(1+s) - s \leq c_0 + 4(s\log s + s)$, $s \geq 0$, where $c_0 = (1+s_0)\log(1+s_0) - s_0$, $s_0 = \frac{1}{2}(1+\sqrt{5})$.)

Definition. The *weak* ω-limit set* is defined as $\omega^*(\boldsymbol{N}_0) =: \{\widehat{\boldsymbol{N}} \in X$; there exists $\{t_n\}$, $t_n \to \infty$ as $n \to \infty$, so that $\boldsymbol{N}(t_n;\boldsymbol{N}_0) \overset{*}{\rightharpoonup} \widehat{\boldsymbol{N}}$ in X as $n \to \infty\}$.

Since $X = E'$ with E separable the weak* topology on $B_R \subset X$ is metrizable. Let d denote this metric.

Definition. The *weak* distance* between two sets B_1, B_2 in B_R is defined by

$$\text{weak* dist}(B_1, B_2) = \inf_{\substack{b_1 \in B_1 \\ b_2 \in B_2}} d(b_1, b_2)\ .$$

Theorem. (i) $\omega^*(\boldsymbol{N}_0)$ is non-empty subset of B_R and (ii) weak* dist$\big(\boldsymbol{N}(t;\boldsymbol{N}_0), \omega^*(\boldsymbol{N}_0)\big) \to 0$ as $t \to \infty$.

Definition. For $\boldsymbol{f} \in X$ define the one parameter family of isometries $\mathsf{T}_t\boldsymbol{f} =: \big(f_{1e}(\cdot - ct), f_{2e}(\cdot + ct), f_{3e}(\cdot)\big)$, $\mathsf{T}_t\colon X \to X$, $t \in \mathbb{R}$.

We can state the main result of [7].

Theorem. For any admissible initial data $\boldsymbol{N}_0$ there is an element $\widehat{\boldsymbol{N}} \in X$ satisfying $M(\mathsf{T}_t\widehat{\boldsymbol{N}}_e) = M(\boldsymbol{N}_0)$, $t \in \mathbb{R}_+$ (mass conservation) and $\big(\widehat{N}_{1e}(x - ct)\widehat{N}_{2e}(x + ct)\big)^{1/2} \geq \widehat{N}_{3e}(x)$ a.e. $(0,\infty) \times (0,1)$ (collision inequality) so that weak* dist$\big(\boldsymbol{N}(t;\boldsymbol{N}_0), \bigcup_{0\leq t\leq 2/c} \mathsf{T}_t\widehat{\boldsymbol{N}}_e\big) \to 0$ as $t \to \infty$.

This theorem does not prove decay to the associated equilibrium $\tilde{\boldsymbol{N}}_e = a(1,1,1)$ for some non-negative constant. It only asserts decay to some oscillating free motion $\mathsf{T}_t\widehat{\boldsymbol{N}}_e$ which is an orbital stability result. It remains to be seen whether this points out a failure of the analysis or a failure of model. While this is an open problem, my guess is the latter.

REFERENCES

[1] Adams, R.A., "Sobolev Spaces", Academic Press (1975).

[2] Kawashima, S., "Global existence and stability of solutions for discrete velocity models of the Boltzmann equation", Recent Topics in Nonlinear PDE, Lecture Notes in Num. Appl. Anal. 6, Kinokuniya, 59-85 (1983).

[3] Kawashima, S., "Initial-boundary value problem for the discrete Boltzmann equation", Journées *Equations aux Dérivées Partielles*, Ecole Polytechnique, 1988.

[4] Kawashima, S., "Global solutions to the initial-boundary value problems for the discrete Boltzmann equation", in preparation.

[5] Kawashima, S., "Existence and stability of stationary solutions to the discrete Boltzmann equation", in preparation.

[6] Kawashima, S., "On the initial-boundary value problems for the discrete velocity Boltzmann equation", preprint.

[7] Shizuta, Y. and Kawashima, S., "Systems of equations of hyperbolic-parabolic type with applications to the discrete Boltzmann equation", Hokkaido Math. J. 14, 249-275 (1985).

[8] Slemrod, M., "Large time behavior of the Broadwell model of a discrete velocity gas with specularly reflective boundary conditions", to appear Archive for Rational Mechanics and Analysis.

[9] Slemrod, M., "Weak asymptotic decay via a "relaxed invariance principle" for a wave equation with nonlinear, non-monotone damping", Proc. Royal Soc., Edinburgh (1989).

[10] Zygmund, M., "Trigonometric Series", Cambridge University Press (1968).

NONLOCAL VARIATIONAL MECHANICS AND EQUILIBRIUM OF SMALL CRYSTALS

by

Giulio Starita

Universitá degli Studi di Napoli
Dipartimento di Matematica e Applicazioni "R. Caccioppoli"
Via Mezzocannone, 8 - 80134 Napoli (Italy)

Abstract. The equilibrium configurations of a crystal with its melt are discussed in the framework of the nonlocal theories of Continuum Mechanics starting from a variational principle. The Gibbs relations and the Wulff relations are derived from the field equations.

1. INTRODUCTION AND PRELIMINARY CONSIDERATIONS.

The equilibrium of a small crystal with its melt is described, from a phenomenological point of view, by a very simple model (see [1–3]). Starting from the observation that the surface enclosing the crystal is composed by planar faces and limiting the analysis to very small crystals, we can assume the following basic hypotheses:

i) the stress in the crystal is given by a constant hydrostatic pressure p_1;

ii) the pressure in the melt has a constant value p_2 different from p_1, so that there is a jump in the pressure field across the interface separating the two phases;

iii) each face of the crystal is subjected to a constant surface tension γ_i depending on the face.

The conditions connecting, at equilibrium, the physical parameters p_1, p_2, γ_i and the geometrical shapes of the crystal was first derived, in 1878, by Gibbs ([4]) starting from the

Gibbs Principle. *A convex polyhedron having a fixed number of faces with fixed orientations represents an equilibrium configuration for the crystal if it is an extremum for the total free energy of the system, under the constraint that the total volume of the system is fixed.*

By applying this principle, Gibbs deduced the following set of equations

$$p_1 - p_2 = \sum_{j,i}{}' \frac{(\gamma_j \csc \alpha_{ij} - \gamma_i \cot \alpha_{ij}) l_{ij}}{s_i} \qquad (i = 1, \dots, n), \tag{1}$$

called Gibbs relations (see also [5]). In (1) s_i is the area of the i-th face, α_{ij} the angle between the i-th and the j-th faces, l_{ij} the lenght of the edge separating the same two faces and $\sum'_{j,i}$ denotes, for a fixed i, the sum extended to all the indexes j of the faces adjacent to the i-th.

In 1901 Wulff ([6]) proposed a construction of the equilibrium shapes of a crystal based on the assumption that the distance h_i of any face from an internal point O of the crystal is proportional to the surface tension acting on the same face (Wulff law). Simple thermodynamical arguments allow us to relate the proportionality factor with the pressure jump $p_1 - p_2$ (see, for instance, [2]); the Wulff law is then expressed by the relations

$$p_1 - p_2 = \frac{2\gamma_i}{h_i} \qquad (i = 1, \ldots, n) \tag{2}$$

When the relation between the area s_i of the i-th face and the h_i's and the α_{ij}'s is introduced in (1), the Gibbs relations assume the form

$$\sum_{j,i}{}' \left[\left(\frac{p_1 - p_2}{2} h_i - \gamma_i \right) \cot \alpha_{ij} - \left(\frac{p_1 - p_2}{2} h_j - \gamma_j \right) \csc \alpha_{ij} \right] l_{ij} = 0, \tag{3}$$

(see [7]).

In the sequel a configuration of a crystal will be called Gibbs configuration or Wulff configuration if it verifies conditions (1) or (2), respectively. Equation (3) shows that the Wulff configurations represent a subset of the Gibbs configurations.

In this paper we propose a more general formulation of the Gibbs principle in order to derive the field equations describing the equilibrium of a crystal with its melt when the pressure and the surface tension can depend on the position and the faces are not necessary planar. In the last section is showed that the Gibbs relations can be obtained from the equilibrium system under suitable physical assumptions.

2. GENERALIZED GIBBS PRINCIPLE.

Let $\mathcal{S}$ be a system composed by two bulk phases $\mathcal{B}_1$ (the crystal) and $\mathcal{B}_2$ (the melt) separated by an interface $\mathcal{I}$ which, in turn, is composed by n faces $\mathcal{I}_i$ $(i = 1, \ldots, n)$ separated by m edges $\mathcal{E}_k$ $(k = 1, \ldots, m)$ (the edge separating the i-th face by the j-th one will be also explicitely denoted by $\mathcal{E}_{ij}$). We describe the bulk phases, the interface and the edges as three-dimensional, two-dimensional and one-dimensional continuous systems, respectively.

The total free energy of $\mathcal{S}$ is given by

$$\Psi = \int_{v_1} \rho_{V_1} \psi_{V_1} \, dv + \int_{v_2} \rho_{V_2} \psi_{V_2} \, dv + \sum_{i=1}^{n} \int_{s_i} \rho_{S_i} \psi_{S_i} \, ds. \tag{4}$$

In (4) ρ_{V_1} and ρ_{V_2} are the volume mass densities of the bulk phases, ρ_{S_i} is the surface mass density of the i-th face, ψ_{V_1}, ψ_{V_2}, ψ_{S_i} are the specific free energies per unit mass. Moreover v_1, v_2 and s_i denote the volumes occupied by the crystal and the melt and the surface occupied by yhe i-th face of the crystal. We assume that $s = \bigcup_{i=1}^{n} \bar{s}_i$ is a closed surface enclosing v_1 ($s = \partial v_1$) and that the total volume $v = \bar{v}_1 \cup \bar{v}_2$ is assigned.

The free energy densities appearing in (4) must be related to the configurations of the system by constitutive equations. Near the melting point the solid and the interface can be considered liquidlike (see [3], chapter VII); therefore the more natural idea is to assume that ψ_{V_1}, ψ_{V_2}, ψ_{S_i} depend on ρ_{V_1}, ρ_{V_2}, ρ_{S_i}, respectively. Nevertheless, under such hypothesis, the jump of the pressure field at the interface disappears when the interface is planar.

A suitable generalization of the classical constitutive assumptions can be found in the framework of the nonlocal theories of continuum mechanics. We refer to [8] for a description and an extensive bibliography of the nonlocal field theories. Here we limit ourselves to recall that they are based on two assumptions:

a) the global balance laws are valid only for the whole system;
b) the physical response of a material particle depends on the state of the whole system.

The equilibrium of a small crystal in its melt has been already described in [7] assuming the hypothesis a). In the present paper we are chiefly interested to hypothesis b), which is directly referred to the constitutive equations. In accordance with it we assume the following expressions for the free energy densities:

$$\psi_{V_1} = \psi_{V_1}(\rho_{V_1}), \qquad \psi_{V_2} = \psi_{V_2}(\rho_{V_2}), \qquad \psi_{S_i} = \psi_{S_i}(k_i), \tag{5}$$

where k_i is defined by

$$\begin{aligned} k_i(\mathbf{r}) = & \int_{v_1} \rho^*_{V_1} \widehat{\phi}_i(|\mathbf{r}-\mathbf{x}^*|)dv^* + \int_{s_i} \rho^*_{S_i} \widetilde{\phi}_i(\rho_{S_i}, |\mathbf{r}-\mathbf{r}^*|)ds^* + \\ & + {\sum_{j,i}}' \int_{l_{ij}} \rho^*_{L_{ij}} \bar{\phi}_i(|\mathbf{r}-\mathbf{s}^*|)dl^*. \end{aligned} \tag{6}$$

In (6), $\mathbf{x}^*$ is the generic particle of the crystal, $\mathbf{r}$ and $\mathbf{r}^*$ are two generic particles of the i-th face, $\mathbf{s}^*$ is the generic particle of an edge; moreover $\rho_{L_{ij}}$ is the linear mass density of the edge between the i-th and the j-th faces whereas a star denotes the value of a function at a starred variable.

It appear that $k_i(\mathbf{r})$ depends, in general, on the values of the deformation at every point of the system. Only when $\widehat{\phi}_i = \bar{\phi}_i = 0$ and $\widetilde{\phi}_i$ is independent of $|\mathbf{r}-\mathbf{r}^*|$ the equation $(5)_3$ reduces to the local one.

The function $\widehat{\phi}_i$ represents the energy density due to the interactions between a surface particle and a crystal particle; in the same way $\bar{\phi}_i$ describes the interactions between the particles of the surface and of the edges. Finally, there is a contribution to the energy at $\mathbf{r}$ due to the deformation at this point and to the interactions between surface particles. We assume that the functions $\widehat{\phi}_i$, $\widetilde{\phi}_i$ and $\bar{\phi}_i$ vanish when the distance between the interacting particles is greather than a certain value which represents the range of the nonlocal interactions.

We can now state the

Generalized Gibbs Principle (GGP). *The equilibrium configurations for a crystal in its melt are the extrema for the total free energy under the constraint that the total mass of the system is fixed.*

If

$$\Phi = \int_{v_1} \rho_{V_1}\, dv + \int_{v_2} \rho_{V_2}\, dv \tag{7}$$

is the total mass of the bulk phases of $\mathcal{S}$ and λ is a constant Lagrange multiplier, the GGP is equivalent to the assumption that the equilibrium configurations of $\mathcal{S}$ are the solutions of the functional equation

$$\delta\Psi - \lambda\delta\Phi = 0, \tag{8}$$

where δ denotes the Fréchet differentiation.

The determination of the local conditions equivalent to (8) constitutes a nonlocal variational problem*. Particularizing the results obtained in [11], where we started from a more general form of the functional Ψ, we can prove the following

Theorem. *A configuration of $\mathcal{S}$ is an equilibrium configuration if and only if the Gibbs potential of the bulk phases is continuous on the interface and, moreover, the following system of local equations is verified:*

$$\begin{aligned}
-\operatorname{grad} p_1 + \rho_{V_1}\widehat{\mathbf{f}}_V &= 0 && \text{in } v_1,\\
\operatorname{grad} p_2 &= 0 && \text{in } v_2,\\
(p_1 - p_2)\mathbf{n} + \operatorname{grad}_S \gamma_i + 2H\gamma_i \mathbf{n}_i + \rho_{S_i}\widehat{\mathbf{f}}_{S_i} &= 0 && \text{on } s_i\ (i = 1, \dots, n),\\
-(\gamma_i \boldsymbol{\nu}_{ij} + \gamma_j \boldsymbol{\nu}_{ji}) + \rho_{L_k}\widehat{\mathbf{f}}_{L_k} &= 0 && \text{on } l_k\ (k = 1, \dots, m).
\end{aligned} \tag{9}$$

In (9) $\boldsymbol{\nu}_{ij}$ is a versor normal to l_{ij} and belonging to the tangent plane to s_i, grad_S is the surface gradient operator, $\mathbf{n}_i$ is the normal versor to the i-th face;

* See Edelen, [9], for the foundation of the nonlocal variational calculus and Cimmelli, Starita, [10], for his extension to systems having surface and line action densities.

moreover, we have put

$$p_1 = \rho_{V_1}^2 \frac{\partial \psi_{V_1}}{\partial \rho_{V_1}}, \quad p_2 = \rho_{V_2}^2 \frac{\partial \psi_{V_2}}{\partial \rho_{V_2}}, \quad \gamma_i = -\rho_{S_i}^2 \int_{s_i} \rho_{S_i}^* \frac{\partial \psi_{S_i}}{\partial k_i} \frac{\partial \widetilde{\phi}_i}{\partial \rho_{S_i}} \, ds^*,$$

$$\widehat{\mathbf{f}}_V = \sum_{i=1}^n \int_{s_i} \rho_{S_i}^* \frac{\partial \psi_{S_i}^*}{\partial k_{S_i}^*} \frac{\partial \widehat{\phi}_{S_i}^*}{\partial \mathbf{k}^*} \, ds^* = \sum_{i=1}^n \widehat{\mathbf{f}}_V^{(S_i)},$$

$$\widehat{\mathbf{f}}_{S_i} = -\int_{v_1} \rho_{V_1}^* \frac{\partial \psi_{S_i}}{\partial k_{S_i}} \frac{\partial \widehat{\phi}_{S_i}}{\partial \mathbf{k}} \, dv^* - \int_{s_i} \rho_{S_i}^* \left(\frac{\partial \psi_{S_i}}{\partial k_{S_i}} \frac{\partial \widetilde{\phi}_{S_i}}{\partial \mathbf{k}} - \frac{\partial \psi_{S_i}^*}{\partial k_{S_i}^*} \frac{\partial \widetilde{\phi}_{S_i}^*}{\partial \mathbf{k}^*} \right) ds^* - \tag{10}$$

$$-\sum_{j,i}{}' \int_{l_{ij(k)}} \rho_{L_k}^* \frac{\partial \psi_{S_i}}{\partial k_{S_i}} \frac{\partial \bar{\phi}_{S_{i(k)}}}{\partial \mathbf{k}} dl^* = \widehat{\mathbf{f}}_{S_i}^{(V)} + \widehat{\mathbf{f}}_{S_i}^{(S_i)} + \sum_{j,i}{}' \widehat{\mathbf{f}}_{S_i}^{(L_{ij})},$$

$$\widehat{\mathbf{f}}_{L_k} = \int_{s_i} \rho_{S_i}^* \frac{\partial \psi_{S_i}^*}{\partial k_{S_i}^*} \frac{\partial \widehat{\phi}_{S_{i(k)}}^*}{\partial \mathbf{k}^*} \, ds^* + \int_{s_j} \rho_{S_j}^* \frac{\partial \psi_{S_j}^*}{\partial k_{S_j}^*} \frac{\partial \widehat{\phi}_{S_{j(k)}}^*}{\partial \mathbf{k}^*} \, ds^* = \widehat{\mathbf{f}}_{L_k}^{(S_i)} + \widehat{\mathbf{f}}_{L_k}^{(S_j)}.$$

In the last expressions a star denotes an inversion between starred and unstarred variables.

It is worth to observe that the terms $\widehat{\mathbf{f}}_V$, $\widehat{\mathbf{f}}_{S_i}$, $\widehat{\mathbf{f}}_{L_k}$, which represent forces due to long range interactions, verify the identity

$$\int_{v_1} \rho_{V_1} \widehat{\mathbf{f}}_V \, dv + \sum_{i=1}^n \int_{s_i} \rho_{S_i} \widehat{\mathbf{f}}_{S_i} \, ds + \sum_{k=1}^m \int_{l_{ij(k)}} \rho_{L_k} \widehat{\mathbf{f}}_{L_k} \, dl = 0 \tag{11}$$

so that they can be interpreted as localization residuals.

3. GIBBS AND WULFF CONFIGURATIONS OF CRYSTAL.

In this section we consider only crystals with planar faces.

Equation $(9)_2$ implies that the pressure in the melt is constant. If we assume that nonlocal interactions have a short range with respect to the linear dimensions of the crystal, also the pressure of the crystal and the surface tensions are constant, except that in a thin layer near the interface, the first, and in a strip near the boundary of the faces, the other ones.

The normal component of $(9)_3$ and $(9)_4$ can be written in the form

$$\begin{aligned} (p_1 - p_2)\mathbf{n}_i &= \boldsymbol{\mu}_i && \text{on } s_i, \\ -(\gamma_i \boldsymbol{\nu}_{ij} + \gamma_j \boldsymbol{\nu}_{ji}) &= \mathbf{r}_{ij} && \text{on } l_{ij(k)}, \end{aligned} \tag{12}$$

where

$$\boldsymbol{\mu}_i = -\rho_{S_i} \widehat{\mathbf{f}}_{S_i}^{(V)}, \qquad \mathbf{r}_{ij} = -\rho_{L_k} (\widehat{\mathbf{f}}_{L_k}^{(S_i)} + \widehat{\mathbf{f}}_{L_k}^{(S_j)}). \tag{13}$$

It is easily seen that the terms $\boldsymbol{\mu}_i$ and $\mathbf{r}_{ij}$ defined in (13) verify the global condition

$$\sum_{i=1}^{n} \int_{s_i} \boldsymbol{\mu}_i \, ds + \sum_{k=1}^{m} \int_{l_{ij(k)}} \mathbf{r}_{ij} \, dl = 0. \tag{14}$$

Rearranging the line integrals in a suitable way and recalling (12), equation (14) assumes the form

$$\sum_{i=1}^{n} \left[(p_1 - p_2) s_i - {\sum_{j,i}}' (\gamma_j \csc \alpha_{ij} - \gamma_i \cot \alpha_{ij}) l_{ij} \right] \mathbf{n}_i = 0. \tag{15}$$

The last identity supplies a constraint between physical and geometrical parameters characterizing the equilibrium of a crystal in its melt. If we assume that the equilibrium of a face is not affected by the other faces, from (15) we deduce the Gibbs relations (1) and the Wulff relations (2) as a particular case.

REFERENCES.

[1] STRICKLAND CONSTABLE, R. F.: *Kinetics and Mechanism of Crystallization*, Academic Press (1968).

[2] DEFAY, R. and PRIGOGINE, I.: *Tension superficielle et adsorption*, Desoer (1951).

[3] ADAMSON, A.: *Physical Chemistry of Surfaces, IV ed.*, J. Wiley (1982).

[4] GIBBS, J.: *Collected works*, vol. 2, Longamans Green & Co. (1928).

[5] IANNECE D., ROMANO, A. and STARITA, G.: *The Gibbs Principle for the Equilibrium of Systems with two Simple or Composed Phases*, Meccanica (to appear).

[6] WULFF, G.: *Zur frage der Geschwindigkeit des Wachsthums und der auflosung der Kristallflachen*, Z. Krystallogr., **34**, 449 (1901).

[7] ROMANO, A. and SUHUBI, E.S.: *On Wulff's law about Equilibrium Configurations of Crystals*, Int. J. Engng Sci., **27**, 1135 (1989).

[8] EDELEN, D.G.B.: *Nonlocal Field Theories*, on *Continuum Physics*, ERINGEN, A.C. (ed.), vol. IV (1976), Academic Press.

[9] EDELEN, D.G.B.: *Nonlocal Variational Mechanics I–VII*, Int. J. Engng Sci., **7**, 269, 287, 373, 391, 401, 677, 843, (1969).

[10] CIMMELLI, V. A. and STARITA, G.: *Nonlocal Variational Theories for systems with an Interface*, Int. J. Engng Sci. (to appear).

[11] CIMMELLI, V. A., ROMANO, A. and STARITA, G.: *The Gibbs principle for the equilibrium of crystals*, Int. J. Engng Sci. (to appear).

SIMILARITY SOLUTIONS FOR ONE - DIMENSIONAL NONLINEAR ELASTODYNAMICS

Erdoğan S.Şuhubi
Faculty of Science, Istanbul Technical University

ABSTRACT

Isovector fields corresponding to one-dimensional motions of hyperelastic solids are investigated. The conditions to be satisfied by the strain energy function of the material in order that field equations admit certain class of symmetries are obtained. Some corresponding similarity solutions are given.

ONE - DIMENSIONAL MOTIONS OF HYPERELASTIC SOLIDS

We represent a one-dimensional motion by the following equations of motion

$$x_k(X_K,t) = X_K\delta_{kK} + u_k(X,t), \quad X_1 \equiv X\ ; \quad K,k=1,2,3 \tag{1}$$

where x_k and X_K are spatial and material cartesian coordinates and u_k are components of the displacement vector. Thus, equations of motion in reference configuration become[1)]

$$\frac{\partial T_{1k}}{\partial X} = \rho_o \frac{\partial^2 u_k}{\partial t^2} \tag{2}$$

where

$$T_{1k} = \frac{\partial \Sigma}{\partial p_k}$$

are the relevant components of Piola-Kirchhoff stress tensor T_{Kk}, $\Sigma = \Sigma(p_k)$ is the strain energy function of a *homogeneous* hyperelastic material and displacement gradients are defined by

$$p_k = \frac{\partial u_k}{\partial X} \tag{3}$$

ρ_o is the constant density in the reference configuration. We shall assume henceforth that Eqs. (2) are properly non-dimensionalized so that we take $\rho_o = 1$. Hence equations of motion should be written as

$$C_{k\ell}(p_m)\,\frac{\partial^2 u_\ell}{\partial X^2} = \frac{\partial^2 u_k}{\partial t^2} \tag{4}$$

where we defined

$$C_{k\ell} = C_{\ell k} = \frac{\partial^2 \Sigma}{\partial p_k \partial p_\ell}$$

An equivalent first order quasilinear system is

$$C_{k\ell}(p_m)\,\frac{\partial p_\ell}{\partial X} - \frac{\partial v_k}{\partial t} = 0, \qquad v_k = \frac{\partial u_k}{\partial t}, \quad p_k = \frac{\partial u_k}{\partial X}, \quad \frac{\partial p_k}{\partial t} - \frac{\partial v_k}{\partial X} = 0 \tag{5}$$

We now consider the manifold $M = R^{11}$ with a coordinat cover (X, t, u_k, p_k, v_k) and define the following forms of the graded exterior algebra $\Lambda(M)$:

$$\sigma_k = du_k - p_k dX - v_k dt \;\varepsilon \Lambda^1(M), \quad d\sigma_k = -dp_k \wedge dX - dv_k \wedge dt \;\varepsilon\; \Lambda^2(M)$$

$$\omega_k = d\left(\frac{\partial \Sigma}{\partial p_k}\right) \wedge dt - dv_k \wedge dX \;\varepsilon \Lambda^2(M) \tag{6}$$

Since $d\omega_k = 0$ the ideal $I(\omega_k, \sigma_k, d\sigma_k)$ is closed. Hence the system is integrable[2,3]. A map $\phi : S \to M$ where S is a two-dimensional submanifold of M defined by

$$\Phi : u_k = u_k(X,t), \quad p_k = p_k(X,t), \quad v_k = v_k(X,t) \tag{7}$$

is a solving map of the ideal I if and only if

$$\phi^*\sigma_k = \left(\frac{\partial u_k}{\partial X} - p_k\right)dX + \left(\frac{\partial u_k}{\partial t} - v_k\right)dt = 0$$

$$\phi^* d\sigma_k = d\phi^*\sigma_k = \left(\frac{\partial p_k}{\partial t} - \frac{\partial v_k}{\partial X}\right)dX \wedge dt = 0$$

$$\phi^*\omega_k = \left(C_{k\ell}\frac{\partial p_\ell}{\partial X} - \frac{\partial v_k}{\partial t}\right)dX \wedge dt = 0$$

where $\phi : \Lambda(M) \to \Lambda(S)$ is induced by ϕ. Obviously such a map ϕ is a solution of the original field equations (5).

ISOVECTOR FIELDS

Let us consider a vector field V in the tangent space T(M) of M represented by

$$V = \Phi\frac{\partial}{\partial X} - \Psi\frac{\partial}{\partial t} - U_k\frac{\partial}{\partial u_k} - P_k\frac{\partial}{\partial p_k} - V_k\frac{\partial}{\partial v_k}$$

All components are assumed to be smooth functions of 11 variables X, t, u_k, p_k and v_k. V is *an isovector field* of the ideal I if and only if[3,4]

$$\pounds_V I \subset I$$

that is, for all $\omega\varepsilon I$ we also have $\pounds_V\omega\varepsilon I$. $\pounds_V$ is the Lie derivative operator with respect to V. This implies that I is invariant under a one-parameter family of mappings $\psi^s: M\to M$ where $\psi^s = \exp(sV)$ is generated by the orbits of V. Obviously the components of an isovector field correspond to infinitesimal generators of the symmetry (Lie) group of the field equations. If ϕ is a solving map the family of maps $\phi_V = \psi^s o\phi$ are also solving maps.

Hence the condition for V to be an isovector field of $I(\omega_k, \sigma_k, d\sigma_k)$ can be expressed as

$$\pounds_V\sigma_k = \lambda_{k\ell}\sigma_\ell, \quad \pounds_V\omega_k = \mu_{k\ell}\omega_\ell + \nu_{k\ell}d\sigma_\ell + \gamma_{k\ell}\wedge\sigma_\ell \tag{8}$$

for some $\lambda_{k\ell}, \mu_{k\ell}, \nu_{k\ell}\varepsilon\Lambda^o(M), \gamma_{k\ell}\varepsilon\Lambda^1(M)$. Since

$$\pounds_V d\sigma_k = d\,\pounds_V\sigma_k = d\,\lambda_{k\ell}\wedge\sigma_\ell + \lambda_{k\ell}d\sigma_\ell$$

the forms $\pounds_V d\sigma_k$ remain in the ideal if functions $\lambda_{k\ell}$ are properly determined. We know that for all $\omega\varepsilon\Lambda(M)$

$$\omega = d(V\lrcorner\omega) = d(V\lrcorner\omega) - V\lrcorner d\omega$$

where $\lrcorner$ denotes the interior product operator[4]. Therefore it follows from $(8)_1$ with

$$V\lrcorner\sigma_k = U_k - p_k\Phi - v_k\Psi = F_k, \quad V\lrcorner d\sigma_k = -P_k dX + \Phi dp_k - V_k dt + \Psi dv_k$$

that

$$\left(\frac{\partial F_k}{\partial X} - P_k\right)dX + \left(\frac{\partial F_k}{\partial t} - V_k\right)dt + \frac{\partial F_k}{\partial u_\ell}du_\ell + \left(\frac{\partial F_k}{\partial p_\ell} + \Phi\delta_{k\ell}\right)dp_\ell + \left(\frac{\partial F_k}{\partial v_\ell} + \Psi\delta_{k\ell}\right)dv_\ell$$

$$= \lambda_{k\ell}(du_\ell - p_\ell dX - v_\ell dt)$$

which yields $\lambda_{k\ell} = \partial F_k/\partial u_\ell$ and

$$P_k = \frac{\partial F_k}{\partial X} + \frac{\partial F_k}{\partial u_\ell}P_\ell, \quad V_k = \frac{\partial F_k}{\partial t} + \frac{\partial F_k}{\partial u_\ell}v_\ell, \quad U_k = F_k + \Phi p_k + \Psi v_k$$

$$\frac{\partial F_k}{\partial p_\ell} = -\Phi\delta_{k\ell}, \quad \frac{\partial F}{\partial v_\ell} = -\Psi\delta_{k\ell} \tag{9}$$

Differentiating $(9)_4$ with respect to p_m we obtain

$$\frac{\partial\Phi}{\partial p_m}\delta_{k\ell} = \frac{\partial\Phi}{\partial p_\ell}\delta_{km}$$

By contracting the indices k and ℓ we find

$$\frac{\partial \Phi}{\partial p_m} = 0$$

We similarly deduce that

$$\frac{\partial \Psi}{\partial v_m} = 0$$

On the other hand $(9)_4$ and $(9)_5$ implies that

$$\frac{\partial \Psi}{\partial p_m} \delta_{k\ell} = \frac{\partial \Phi}{\partial v_\ell} \delta_{km}$$

Contracting on k, ℓ and again on k,m we deduce immediately

$$\frac{\partial \Phi}{\partial v_m} = 0, \quad \frac{\partial \Psi}{\partial p_m} = 0$$

Hence F_k has the representation

$$F_k = -\phi(X,t,u_m)p_k - \psi(X,t,u_m)v_m + \omega_k(X,t,u_m) \tag{10}$$

and the components of the isovector fields take form

$$\Phi = \phi\ ,\ \Psi = \psi\ ,\ U_k = \omega_k\ ,\ P_k = \frac{\partial F_k}{\partial X} + \frac{\partial F_k}{\partial u_\ell} p_\ell\ ,\quad V_k = \frac{\partial F_k}{\partial t} + \frac{\partial F_k}{\partial u_\ell} v_\ell \tag{11}$$

Now writing

$$\gamma_{k\ell} = \gamma^1_{k\ell} dX + \gamma^2_{k\ell} dt + \gamma^1{}_{k\ell m} du_m + \gamma^2{}_{k\ell m} dp_m + \gamma^3{}_{k\ell m} dv_m$$

and comparing both sides of $(8)_2$, we obtain

$$\mu_{k\ell} = \frac{\partial V_k}{\partial v_\ell} + (\frac{\partial \phi}{\partial X} + \frac{\partial \phi}{\partial u_m} p_m)\delta_{k\ell}\ ,\quad \nu_{k\ell} = -\frac{\partial V_k}{\partial p_\ell} - (\frac{\partial \psi}{\partial X} + \frac{\partial \psi}{\partial u_m} p_m)C_{k\ell}$$

$$\gamma^1_{k\ell} = -\frac{\partial V_k}{\partial u_\ell} - \gamma^1_{km\ell} p_m, \quad \gamma^2{}_{k\ell} = -\frac{\partial Q_k}{\partial u_\ell} - \gamma^1_{km\ell} v_m, \quad \gamma^1_{k\ell m} = \gamma^1_{km\ell}$$

$$\gamma^2_{k\ell m} = \frac{\partial \psi}{\partial u_\ell} C_{km}, \quad \gamma^3_{k\ell m} = \frac{\partial \phi}{\partial u_\ell} \delta_{km}$$

and the relations

$$\frac{\partial Q_k}{\partial X} + \frac{\partial Q_k}{\partial u_\ell} p_\ell + \frac{\partial V_k}{\partial t} + \frac{\partial V_k}{\partial u_\ell} v_\ell = 0, \quad Q_k = C_{k\ell} P_\ell$$

$$\frac{\partial Q_k}{\partial p_\ell} - \frac{\partial V_k}{\partial v_m} C_{m\ell} + (\frac{\partial \psi}{\partial t} + \frac{\partial \psi}{\partial u_m} v_m - \frac{\partial \phi}{\partial X} - \frac{\partial \phi}{\partial u_m} p_m) C_{k\ell} = 0$$

$$\frac{\partial Q_k}{\partial v_\ell} - \frac{\partial V_k}{\partial p_\ell} - (\frac{\partial \psi}{\partial X} + \frac{\partial \psi}{\partial u_m} p_m) C_{k\ell} + (\frac{\partial \phi}{\partial t} + \frac{\partial \phi}{\partial u_m} v_m)\delta_{k\ell} = 0 \tag{12}$$

Obviously $\gamma^1_{k\ell m}$ is arbitrary subject only to the condition $\gamma^1_{k[\ell m]} = 0$
However, $\gamma_{k\ell}$ can now be put into the form

$$\gamma_{k\ell} = -\frac{\partial V_k}{\partial u_\ell}\,dX - \frac{\partial Q_k}{\partial u_\ell}\,dt + \frac{\partial\psi}{\partial u_\ell}\,C_{km}dp_m + \frac{\partial\phi}{\partial u_\ell}\,dv_k + \gamma^1_{k(\ell m)}\,\sigma_m$$

Hence in the expression $\gamma_{k\ell}\wedge\sigma_\ell$ in (8) the terms $\gamma^1_{k(\ell m)}\,\sigma_m\wedge\sigma_\ell$ vanish.

Since

$$\frac{\partial Q_k}{\partial v_\ell} = C_{km}\frac{\partial P_m}{\partial v_\ell} = -C_{k\ell}\left(\frac{\partial\psi}{\partial X} + \frac{\partial\psi}{\partial u_m}p_m\right),\quad \frac{\partial V_k}{\partial p_\ell} = -\left(\frac{\partial\phi}{\partial t} + \frac{\partial\phi}{\partial u_m}v_m\right)\delta_{k\ell}$$

it follows from $(12)_3$ that

$$\left(\frac{\partial\psi}{\partial X} + \frac{\partial\psi}{\partial u_m}p_m\right)C_{k\ell} = \left(\frac{\partial\phi}{\partial t} + \frac{\partial\phi}{\partial u_m}v_m\right)\delta_{k\ell}\ ; \qquad C_k \neq C\delta_{k\ell}$$

and

$$\frac{\partial\psi}{\partial X} + \frac{\partial\psi}{\partial u_m}p_m = \frac{\partial\phi}{\partial t} + \frac{\partial\phi}{\partial u_m}v_m = 0 \quad\text{or}\quad \frac{\partial\psi}{\partial X} = \frac{\partial\phi}{\partial t} = 0\ ,\quad \frac{\partial\psi}{\partial u_m} = \frac{\partial\phi}{\partial u_m} = 0$$

We thus obtain

$$\phi = \phi(X)\ ,\qquad \psi = \psi\,(t) \tag{13}$$

and

$$P_k = \frac{\partial\omega_k}{\partial X} - \left(\phi'\delta_{k\ell} - \frac{\partial\omega_k}{\partial u_\ell}\right)p_\ell,\quad V_k = \frac{\partial\omega_k}{\partial t} - \left(\dot\psi\,\delta_{k\ell} - \frac{\partial\omega_k}{\partial u_\ell}\right)v_\ell$$

Then $(12)_{1,2}$ yield

$$-\frac{\partial^2\omega_k}{\partial u_m\,\partial u_\ell}\,v_m v_\ell - \left(2\,\frac{\partial^2\omega_k}{\partial u_\ell\partial t} - \dot\psi\,\delta_{k\ell}\right)v_\ell - C_{k\ell}\left[-\frac{\partial^2\omega_\ell}{\partial X^2}\right] - \frac{\partial^2\omega_k}{\partial t^2} = 0 \tag{14}$$

$$C_{k\ell m}\left[\frac{\partial\omega_m}{\partial X} + \left(\frac{\partial\omega_m}{\partial u_n} - \phi'\delta_{mn}\right)p_n\right] - 2(\dot\psi - \phi')C_{k\ell} - C_{km}\frac{\partial\omega_m}{\partial u_\ell} - C_{\ell m}\frac{\partial\omega_k}{\partial u_m} = 0$$

$$C_{k\ell m} = \frac{\partial^3\Sigma}{\partial p_k\partial p_\ell\partial p_m}$$

The antisymmetric part of $(14)_2$ with respect to k and ℓ give

$$C_{km}\left(\frac{\partial\omega_m}{\partial u_\ell} + \frac{\partial\omega_\ell}{\partial u_m}\right) = \left(\frac{\partial\omega_k}{\partial u_m} + \frac{\partial\omega_m}{\partial u_k}\right)C_{m\ell} \tag{15}$$

Now we deduce from $(14)_1$ that

$$\frac{\partial^2\omega_k}{\partial u_m\partial u_\ell} = 0,\quad \frac{\partial^2\omega_k}{\partial u_\ell\partial t} = \frac{1}{2}\ddot\psi\,(t)\delta_{k\ell}$$

from which we obtain

$$\omega_k = \Lambda_{k\ell}(X,t)u_\ell + \Lambda_k(X,t),\quad \Lambda_{k\ell}(X,t) = \frac{1}{2}\dot\psi\,\delta_{k\ell} + \lambda_{k\ell}(X) \tag{16}$$

Substituting (16) into (15) we find

$$C_{km}\Lambda_{(m\ell)} = \Lambda_{(km)}C_{m\ell}$$

Hence the symmetric part of the matrix $\Lambda_{k\ell}$ commutes with symmetric arbitrary matrices $C_{k\ell}$. Therefore $\Lambda_{(k\ell)}$ and $\lambda_{(k\ell)}$ have the forms

$$\Lambda_{(k\ell)} = \Lambda_o(X,t)\delta_{k\ell}, \quad \lambda_{(k\ell)}(X) = (\Lambda_o(X,t) - \tfrac{1}{2}\dot{\psi})\delta_{k\ell} = \lambda_o(X)\delta_{k\ell}$$

The antisymmetric part of $\lambda_{k\ell}$ is represented by $\lambda_{[k\ell]} = e_{k\ell m}\,\alpha_m(X)$ where $e_{k\ell m}$ is the alternating symbol. Consequently we have

$$\omega_k = (\lambda_o(X) + \tfrac{1}{2}\dot{\psi}(t))u_k - e_{k\ell m}\alpha_m(X)u_\ell - \Lambda_k(X,t) \tag{17}$$

Introducing (17) into $(14)_{1,2}$ and equating the coefficients of u_n to zero we obtain at once

$$\lambda_o'\delta_{mn} + e_{mnr}\alpha_r' = 0, \quad \dddot{\psi} = 0$$

or

$$\lambda_o(X) = a_o, \quad \alpha_k(X) = a_k, \quad \psi(t) = b_1t^2 + 2b_2t + b_3 \tag{18}$$

Thus $(14)_1$ and $(14)_2$ reduce to

$$(-\phi''p_\ell + \frac{\partial^2\Lambda_\ell}{\partial X^2})C_{k\ell} - \frac{\partial^2\Lambda_k}{\partial t^2} = 0 \tag{19}$$

$$\{[(-\phi' + b_1t + b_2 + a_o)\delta_{mn} + e_{mnr}a_r]\,p_n + \frac{\partial\Lambda_m}{\partial X}\}\,C_{k\ell m} + 2(-\phi' + 2b_1t + 2b_2)C_{k\ell}$$

$$-(e_{kmr}C_{\ell m} + e_{\ell mr}C_{km})a_r = 0$$

Noting that

$$p_\ell C_{k\ell} = \frac{\partial}{\partial p_k}(p_\ell\frac{\partial\Sigma}{\partial p_\ell}) - \frac{\partial\Sigma}{\partial p_k}, \quad p_n C_{k\ell m} = \frac{\partial^2}{\partial p_k\partial p_\ell}(p_n\frac{\partial\Sigma}{\partial p_m}) - \delta_{nk}C_{m\ell} - \delta_{n\ell}C_{km}$$

$$p_m C_{k\ell m} = \frac{\partial^2}{\partial p_k\partial p_\ell}(p_m\frac{\partial\Sigma}{\partial p_m}) - 2C_{k\ell}$$

we easily see that Eqs. (19) can be transformed respectively into

$$\frac{\partial^2}{\partial p_k\partial X}(\Lambda + \phi'\Sigma) - \frac{\partial^2\Lambda_k}{\partial t^2} = 0, \quad \frac{\partial^2\Lambda}{\partial p_k\partial p_\ell} = 0 \tag{20}$$

where we defined

$$\Lambda = \{[(-\phi' + b_1t + b_2 + a_o)\delta_{mn} + e_{mnr}a_r]\,p_n + \frac{\partial\Lambda_m}{\partial X}\}\frac{\partial\Sigma}{\partial p_m} + 2(b_1t + b_2 + a_o)\Sigma \tag{21}$$

If we differentiate $(20)_1$ with respect to p_ℓ and $(20)_2$ with respect to X we obtain

$$\frac{\partial^3\Lambda}{\partial p_k \partial p_\ell \partial X} + \phi'' C_{k\ell} = 0, \qquad \frac{\partial^3\Lambda}{\partial p_k \partial p_\ell \partial X} = 0$$

or

$$\phi'' C_{k\ell} = 0$$

whence we find

$$\phi'' = 0 \text{ or } \phi(X) = c_1 X + c_2 \tag{22}$$

On the other hand, differentiation of $(19)_1$ with respect to t once and $(19)_2$ twice lead to

$$\frac{\partial^3\Lambda_\ell}{\partial X^2 \partial t} C_{k\ell} - \frac{\partial^3\Lambda_k}{\partial t^3} = 0, \qquad \frac{\partial^3\Lambda_m}{\partial X \partial t^2} C_{k\ell m} = 0 \tag{21'}$$

For strictly nonlinear materials $C_{k\ell m} \neq 0$ and the relations (21') result in

$$\frac{\partial^3\Lambda_k}{\partial X \partial t^2} = 0, \qquad \frac{\partial^3\Lambda_k}{\partial X^2 \partial t} = 0, \qquad \frac{\partial^3\Lambda_k}{\partial t^3} = 0$$

form which we deduce that

$$\Lambda_k(X,t) = d_k t^2 + f_k X t + g_k t + h_k(X) \tag{22'}$$

where d_k, f_k and g_k are constant. Introducing (22') into $(19)_1$ we find

$$h''_\ell C_{k\ell} - 2d_k = 0$$

which yields

$$h''_k = 0 \text{ or } h_k(X) = m_k X + n_k; \; d_k = 0 \tag{22''}$$

for *nonlinear materials*. Accordingly (21) now becomes

$$\Lambda = t\Lambda_1 + \Lambda_2 = t\{(b_1 p_k + f_k)\frac{\partial\Sigma}{\partial p_k} + 2b_1 \Sigma\}$$
$$- \left[(b_2 + a_o - c_1)p_k + e_{k\ell m} a_m p_\ell + m_k\right] \frac{\partial\Sigma}{\partial p_k} + 2(b_2 - a_o)\Sigma \tag{23}$$

If we substitute (23) into (20) we see that $(20)_1$ is satisfied identically and $(20)_2$ gives

$$\frac{\partial^2\Lambda_1}{\partial p_k \partial p_\ell} = 0, \qquad \frac{\partial^2\Lambda_2}{\partial p_k \partial p_\ell} = 0$$

or

$$\Lambda_1 = \alpha_k p_k + \alpha, \quad \Lambda_2 = \beta_k p_k + \beta \tag{24}$$

where α_k, β_k, α and β are constant. However, the right hand sides of (24) contribute only linear terms in p_k to the functional structure of Σ.

Hence they can be discarded without loss of generality if the natural state of the hyperelastic material is stress free. Consequently the conditions (24) reduce to

$$(b_1 p_k + f_k)\frac{\partial\Sigma}{\partial p_k} + 2b_1\Sigma = 0$$
$$\left[(b_2 + a_o - c_1)p_k + e_{k\ell m}\, a_m\, p_\ell + m_k\right]\frac{\partial\Sigma}{\partial p_k} + 2(b_2 - a_o)\Sigma = 0 \qquad (25)$$

Finally the components of the most general isovector field for nonlinear materials may be written as

$$\Phi = c_1 X + c_2, \qquad \Psi = b_1 t^2 + 2b_2 t + b_3$$
$$U_k = \omega_k = (b_1 t + b_2 + a_o)u_k + e_{k\ell m}\, a_m\, u_\ell + f_k X t + m_k X + g_k t + n_k$$
$$P_k = (b_1 t + b_2 + a_o - c_1)p_k + e_{k\ell m}\, a_m\, p_\ell + f_k t + m_k$$
$$V_k = -(b_1 t + b_2 - a_o)v_k + e_{k\ell m}\, a_m v_\ell + b_1 u_k + f_k X + g_k \qquad (26)$$

For an isovector field corresponding to an arbitrary material, (25) should be satisfield for all strain energy functions. This is possible if and only if we choose

$$b_1 = b_2 - a_o = b_2 + a_o - c_1 = 0, \quad f_k = a_k = m_k = 0 \qquad (27)$$

Then relations (26) reduce to

$$\Phi = c_1 X + c_2,\ \Psi = c_1 t + b_3,\ U_k = c_1 u_k + g_k t + n_k,\ P_k = 0,\ V_k = -c_1 v_k + g_k \qquad (28)$$

For a different set of parameters from those in (27), Eqs.(25) impose severe restrictions on the allowable forms of the strain energy function for which the field equations admit the given symmetry group.

It is clear from (26) that the parameters c_2 and b_3 represent, respectively, spatial and temporal homogeneities and the set (g_k, n_k) the galilean invariance of the field equations.

Let us note that (25) and (26) correspond to a general anisotropic material. For isotropic material $\Sigma = \Sigma\,(p,q)$ where $p = p_1$ and $q^2 = p_2^2 + p_3^2$ and we have

$$\frac{\partial\Sigma}{\partial p_1} = \frac{\partial\Sigma}{\partial p}, \qquad \frac{\partial\Sigma}{\partial p_\alpha} = \frac{p_\alpha}{q}\frac{\partial\Sigma}{\partial q} \qquad \alpha = 2,3$$

We can easily see that (25) is then simplified to

$$(b_1 p + f_1) \frac{\partial\Sigma}{\partial p} + b_1 q \frac{\partial\Sigma}{\partial q} + 2b_1\Sigma = 0; \quad f_\alpha = 0 \text{ or } \frac{\partial\Sigma}{\partial q} = 0,$$

$$(b_2 + a_o - c_1) \frac{\partial\Sigma}{\partial p} + (b_2 + a_o - c_1) q \frac{\partial\Sigma}{\partial q} + 2(b_2 - a_o)\Sigma = 0; \quad m_\alpha = a_\alpha = 0$$

$$\text{or } \frac{m_\alpha}{q} \frac{\partial\Sigma}{\partial q} + e_{\alpha\beta} a_\beta \left(\frac{\partial\Sigma}{\partial p} - \frac{p}{q}\frac{\partial\Sigma}{\partial q}\right) = 0 \qquad (28')$$

where $e_{\alpha\beta}$ is the two-dimensional alternating symbol.

SIMILARITY SOLUTIONS

Similarity solutions corresponding to a *particular* isovector field V are generated by imposing on the solving map ϕ of the ideal $I(\omega_k, \sigma_k d\sigma_k)$ the additional constraints

$$\phi^*(V \rfloor \sigma_k) = 0, \quad \phi^*(V \rfloor d\sigma_k) = 0, \quad \phi^*(V \rfloor \omega_k) = 0 \qquad (29)$$

Obviously $(29)_1$ leads to the linear partial differential equation

$$\phi(X) \frac{\partial u_k}{\partial X} + \psi(t) \frac{\partial u_k}{\partial t} - \omega_k(X, t, u_m) = 0 \qquad (30)$$

It is then a simple exercise to show that the relation (30) together with the fact that $\phi: S \to M$ is a solving map secures the satisfaction of both $(21)_2$ and $(29)_3$. The solution of (30) is obtainable through the method of characteristics. To this end we have to solve the ordinary differential equations

$$\frac{dX}{\phi(X)} = \frac{dt}{\psi(t)} = \frac{du_k}{\omega_k(X, t, u_m)} \quad \text{(no sum on k)} \qquad (31)$$

which yields first the similarity variable

$$\xi = \frac{\Phi(X)}{\Psi(t)}, \quad \Phi(X) = \exp\int \frac{dX}{\phi} = (c_1 X + c_2)^{1/c_1}, \quad \Psi(t) = \exp\int \frac{dt}{\psi(t)} \qquad (32)$$

Since ω_k is linear in u_m the functions u_k are expressible as

$$u_k = A_{k\ell}(X, t) U_\ell(\xi) + B_k(X, t) \qquad (33)$$

where $A_{k\ell}$ and B_k are specified functions of their arguments. Introducing (33) into the field equations we can obtain three nonlinear ordinary differential equations in ξ to determine U_k.

For an arbitrary strain energy function we can write from (28) with

$$c_2 = b_3 = 0, \quad g_k = n_k = 0$$

$$u_k = tU_k(\xi), \quad \xi = \frac{X}{t}$$

and U_k satisfy

$$\left(\frac{\partial^2\Sigma(U'_m)}{\partial U'_k \partial U'_\ell} - \xi^2\delta_{k\ell}\right)U''_\ell = 0$$

As a more complicated example, let us consider the case

$a_k = 0$, $m_k = (b_2 + a_o - c_1)f_k/b_1$, $b_2 + a_o - c_1 = b_2 - a_o$ - or $2a_o = c_1$

in which (25) is reduced to a simgle equation which can be written as

$$(p_k + h_k)\frac{\partial\Sigma}{\partial p_k} + 2\Sigma = 0, \quad f_k = b_1 h_k \tag{34}$$

whose solution is found easily as

$$\Sigma = (p_1 + h_1)^{-2}\bar{\Sigma}(P,Q), P = \frac{p_2 + h_2}{p_1 + h_1}, \quad Q = \frac{p_3 + h_3}{p_1 + h_1}$$

We can immediately see that (31) now takes the form

$$\frac{dX}{\phi(X)} = \frac{dt}{\psi(t)} = \frac{du_k}{\frac{1}{2}(\dot\psi + \phi')(u_k + h_k X) - h_k\phi} \tag{35}$$

if we also choose $g_k = 0$, $n_k = -c_2 h_k$. Let us further assume that

$$\psi(t) = b_1(t - t_1)(t - t_2), \quad t_1, t_2 \in \mathbb{R}$$

Then we can readily obtain from (35) the similarity variable

$$\xi = \frac{t - t_2}{t - t_1}(X + \alpha)^{\beta(t_1 - t_2)}, \quad \alpha = \frac{c_2}{c_1}, \qquad \beta = \frac{b_1}{c_1} \tag{36}$$

On a characteristic, by differentiating with respect to X we have

$$\psi = \phi t' \Rightarrow \quad \dot\psi t' = \phi t'' + \phi' t' \quad \Rightarrow \quad \dot\psi = \phi' + \phi\frac{t''}{t'}$$

and (35) gives the relation

$$(u_k + h_k X)' = \left(\frac{\phi'}{\phi} + \frac{1}{2}\frac{t''}{t'}\right)(u_k + h_k X)$$

which can be integrated to

$$u_k + h_k X = (\phi\psi)^{\frac{1}{2}}\, U_k(\xi) \tag{37}$$

Noting that

$$\frac{\partial\xi}{\partial X} = \frac{\xi}{\phi} \ , \ \frac{\partial\xi}{\partial t} = -\frac{\xi}{\psi}$$

we can easily find out that

$$p_k + h_k = \left(\frac{\psi}{\phi}\right)^{\frac{1}{2}}(\xi U_k' + \frac{1}{2} c_1 U_k), \quad v_k = -\left(\frac{\phi}{\psi}\right)^{\frac{1}{2}}(\xi U_k' - (2b_1 t + b_2) U_k)$$

$$\frac{\partial v_k}{\partial t} = \frac{1}{\psi}\left(\frac{\phi}{\psi}\right)^{\frac{1}{2}}(\xi^2 U_k'' + \xi U_k' + (b_1 b_3 - b_2^2) U_k) \tag{38}$$

where primes denote differentiation with respect to the similarity variable ξ. In a similar fashion we obtain

$$\Sigma = \frac{\phi}{\psi}(\xi U_1' + \frac{1}{2} c_1 U_1)\bar{\Sigma}(P,Q), \quad P(\xi) = \frac{\xi U_2' + \frac{1}{2} c_1 U_2}{\xi U_1' + \frac{1}{2} c_1 U_1}, \quad Q(\xi) = \frac{\xi U_3' + \frac{1}{2} c_1 U_3}{\xi U_1' + \frac{1}{2} c_1 U_1}$$

$$\frac{\partial\Sigma}{\partial p_k} = \left(\frac{\phi}{\psi}\right)^{3/2} A_k(\xi), \quad A_1(\xi) = -(\xi U_1' + \frac{1}{2} c_1 U_1)^{-3}(2\bar{\Sigma} + P\frac{\partial\bar{\Sigma}}{\partial P} + Q\frac{\partial\bar{\Sigma}}{\partial Q}) \tag{39}$$

$$A_2(\xi) = (\xi U_1' + \frac{1}{2} c_1 U_1)^{-3}\frac{\partial\bar{\Sigma}}{\partial P}, \quad A_3(\xi) = (\xi U_1' + \frac{1}{2} c_1 U_1)^{-3}\frac{\partial\bar{\Sigma}}{\partial Q}$$

Introducing (38) and (39) into the field equations we arrive at the following three ordinary differential equations in ξ

$$\xi A_k' + \frac{3}{2} c_1 A_k - \xi^2 U_k'' - \xi U_k' - (b_1 b_3 - b_2^2) U_k = 0 \tag{40}$$

REFERENCES

1)

Eringen, A.C. and Şuhubi, E.S., " Elastodynamics, Vol.I-Finite Motions", Academic Press, New York (1974).

2)

Cartan, E., " Les Systèmes Différentiels Extérieurs et Leurs Applications Géometriques", Hermann, Paris (1945).

3)

Harrison, B.K. and Estabrook, F.B., " Geometric Approach to Invariance Groups and Solution of Partial Differential Systems", J.Math.Phys.12, 653-666 (1971).

4) Edelen,D.G.B., " Applied Exterior Calculus", John Wiley and Sons, New York (1985).

THEORETICAL FEATURES OF GUIDING CENTER KINETIC THEORY

Massimo Tessarotto
Dipartimento di Scienze Matematiche
Università degli Studi di Trieste
TRIESTE, Italy

ABSTRACT

A general strategy for guiding–center kinetic theory is formulated adopting non–canonical phase–space variables. The approach, which is useful for developing asymptotic schemes based either on canonical or non–canonical perturbation theories, can – in particular – be used to investigate previous guiding–center theories, in order to ascertain their agreement with such a strategy, as well as their *consistency*, i.e. whether conservation laws are preserved by such theories, and their *accuracy*, i.e. the degree of agreement between the solutions of the asymptotic theories and those of the exact hamiltonian particle dynamics.

1 – INTRODUCTION

A common feature of existing particle–dynamics and kinetic theories for magnetoplasmas is that they rely on asymptotic theories of hamiltonian systems with all solution nearly periodic or at least conditionally periodic[1]. Two wellknown examples are given respectively by guiding–center (or gyro–)[2–10] and quiver–mechanics and kinetic theory [11–12]. Thus guiding–center mechanics is obtained from (non–relativistic or relativistic) mechanics – in a fashion analogous to the time–honoured Chew, Golberger and Low technique [2] – assuming that a charged point particle is immersed in a "strong" magnetic field B, which exibits time variations weak in an appropriate sense, i.e. requiring that the Larmor frequency $\Omega = ZeB/Mc$, with B the magnetic field, be much larger than all remaining "characteristic" frequencies which may be associates to the dynamical system (see below) and by introducing an asymptotic expansion in an infinitesimal nondimensional parameter ϵ – to be identified with the largest ratio between such frequencies and Ω, i.e. usually equivalent to the ratio between the Larmor radius and a characteristic macroscopic scale length L, i.e. $\epsilon = r_L/L$. Similarly quiver–mechanics is obtained by assuming the electromagnetic field, and hence the Lorentz force, are themselves periodic in time, with frequency ω assumed much larger that all remaining "characteristic" frequencies, or, at most, comparable with the largest of the remaining frequencies.

A basic aspect of customary approaches, besides the aforementioned expansion technique, is the adoption of a suitable averaging. Thus, in guiding center mechanics the average is performed with respect to the Larmor gyrophase, related to the fast degree of freedom corresponding to Larmor rotation. As typical of asymptotic theories, the expansion in terms of ϵ is truncated at prescribed order in ϵ, yielding a closed set of equations which correspond to possible models of guiding–center mechanics (qui- ver–mechanics) and respectively of gyro– (quiver–) kinetics. A problem is, how- ever, that such a truncation may lead to the loss of exact conservation laws and symmetries; for example, it may result, in particular, in the violation of Galilei invariance or the non conservation of Poincarè invariants (assuring in turn the conservation of phase–space measure).

Various approaches have been proposed in order satisfy, at least in part, such requirements and, at the same time, furnish a rational basis for a guiding–center mechanics, as well as guiding–center kinetic theories, and similarly for quiver mechanics and quiver–kinetics. In particular, in the case of guiding center theories, such approaches – apart the meaninful example given by Mynick [12)] – share the common feature of being based on non–canonical perturbative schemes, i.e. for which use is made of a set of non–canonical phase space variables. Thus Boozer [6)], considering the particular case of time–independent fields, proposed an "ad hoc" modification of the drift kinetic equation,in such a way to make possible the fulfillment of approximate conservation laws (i.e., the gyrophase–average of the canonical momentum conjugate to an ignorable spatial coordinate in the case of exact spatial symmetries). Littlejohn, instead,[5,7)],while neglecting kinetic theory, developed a perturbative scheme for guiding–center mechanics applying to time dependent fields and large drift based on Lie–transform methods and the adoption of so called quasi–canonical phase–space variables. He showed that, using Lie , transform methods to generate a sequence of suitable phase–space diffeomeorfisms it is possible to preserve at least some conservation theorems and, at the same time, take into account higher-order corrections to guiding-center mechanics. The approach of Mynick [8], which concerns, as the previous one of Littlejohn,a perturbative scheme for guiding center mechanics, is based instead on a canonical perturbation scheme developed adopting both Lie transform methods and mixed variables generating functions.

Finally, in another independent approach, Bernstein and Catto [9], by making use of non-canonical variables, developed an approach able to a yield a guiding center kinetic theory accurate to second order in the Larmor ra- dius. Their approach is based on a phase-space , so-called gyrokinetic, transfor- mation leading to a set of non canonical variables exibiting time derivatives with gyrophase dependence demoted in an suitable sense and which are close to the phase space variables of the guiding center. Upon inversion of such a transformation by power series expansion in ϵ and by expressing in terms of them the kinetic equation, an approximate kinetic (i.e., gyrokinetic) equation is obtained which results authomatically correct through second order in the Larmor radius.

Purpose of this paper is to define a general strategy for guiding-center theory, adopting a set of generally non-canonical phase-space variables. The scheme here proposed, to which previous guiding center theories should be compared, is particularly suited for the investigation of their accuracy, i.e. their agreement with the hamiltonian particle dynamics and moreover of conservation laws, including the case of adiabatic invariants, and symmetries fulfilled by previous asymptotic theories.

The fulfillment of exact conservation laws seems particularly relevant for certain classes of problems in kinetic theory, especially when dealing with nonlinear dynamics and long time behaviour. In any case, as a general rule, it seems obviously preferable, when possible, to adopt a (kinetic) theory fulfilling exact conservation laws than theories in which such laws are, in some sense, only approximate.

2 – A STRATEGY FOR GUIDING CENTER MECHANICS AND KINETIC THEORY

Let us firstly recall that all guiding-center theories (for particle mechanics and kinetic theory) are based on the construction of a hierarchy of phase space transformations, which are obtained by (power series) asymptotic expansions in ϵ.The starting point of such theories is usually based on the introduction of a preliminary non-canonical (or canonical [8]) phase-space transformation (diffeomorfism) $\gamma:\underline{x}\rightarrow\rightarrow\underline{x}'=(\underline{x}_o,\varphi)$, yielding a transformed distribution $F(\underline{x}')= f(\underline{x}(\underline{x}'),t)$, where the set of transformed variables $\underline{x}'$includes,in particular, a gyrophase φ describing, in a suitable and generally non inertial reference frame, the rapid motion of gyration undergone by the perpendicular component of the particle velocity (i.e. in a plane perpendicular to the magnetic field $\underline{B}$).

As a starting point let us therefore express the Fokker-Planck (F-P) kinetic equation in a set of extended canonical variables $\underline{x}=(\underline{q},\underline{p},t,-h)$,with extended hamiltonian $H(\underline{q},\underline{p},t,-h)=\frac{1}{2m}(\underline{p}-\frac{q}{c}\underline{A})^2+ q\Phi - h$, where $\underline{A}$,and Φ denote the electromagnetic potentials. The Fokker-Planck equation for the distribution function $f(\underline{x})$ expressed in terms of such canonical variables reads then:

$$(1)\qquad [\underline{x},H].\frac{\partial}{\partial\underline{x}}f(\underline{x}) = K(f|f)$$

where [A,B] is the extended Poisson bracket (i.e. $[A,B]=\frac{\partial A}{\partial\underline{q}}\cdot\frac{\partial B}{\partial\underline{p}}-\frac{\partial B}{\partial\underline{q}}\cdot\frac{\partial A}{\partial\underline{p}}-\frac{\partial A}{\partial t}\cdot\frac{\partial B}{\partial h}+\frac{\partial B}{\partial t}\cdot\frac{\partial A}{\partial h}$); furthermore $[x,H].\frac{\partial}{\partial\underline{x}}$ denotes the Vlasov operator and $K(f|f)=\frac{\partial}{\partial\underline{v}}.\underline{\Gamma}$ represents the Fokker-Planck collision operator.

Consider now the action of an arbitrary, generally non-canonical and time dependent, $(C^{\infty}-)$diffeomorfism $\gamma_1:\underline{x}=(\underline{q},\underline{p},t,-h)\rightarrow\underline{Z}=(\underline{z},\varphi)$ in the (extended) phase space of the representative particle of the s-th species, where $\underline{Z}$ denotes a 8-vector spanning the phase space, $\underline{z}$ an appropriate 7-vector belonging to a (restricted) phase space and φ an angle-like generalized coordinated to be later indentified with some gyrophase. Denoting by $F(\underline{Z},t)=f(\underline{r}(\underline{Z}),\underline{v}(\underline{Z}),t)$ the transformed distribution function and by $J(\underline{x};\underline{Z})=\frac{\partial(\underline{x})}{\partial(\underline{Z})}$ the Jacobian of the transformation γ_1, the F-P equation in the new variables reads,

$$(2)\qquad \frac{\partial H}{\partial Z^i}\sigma^{ij}\frac{\partial}{\partial Z^j}F(\underline{Z})=K_{(\underline{Z})}(F|F)$$

Similarly, usińg Liouville's Theorem:

(3) $\frac{\partial}{\partial Z^i}\{\sigma^{ij}\frac{\partial H}{\partial Z^j}J(\underline{x};\underline{Z})\} = 0$

there results the analogous so–called conservation form of the F–P equation:

(4) $\frac{\partial}{\partial Z^i}\{\sigma^{ij}\frac{\partial H}{\partial Z^j}J(\underline{x};\underline{Z})F(\underline{Z}) - J(\underline{x};\underline{Z})\underline{\Gamma}.\frac{\partial}{\partial \underline{v}}Z^i\}=0$

where $\sigma^{ij}(\underline{Z})$ is the Poisson tensor $\sigma^{ij}=[Z^i,Z^j]$ and the the time derivatives $\frac{d}{dt}Z^i$ read $\frac{d}{dt}Z^i = \sigma^{ij}\frac{\partial H}{\partial Z^j}$.

The aim of guiding center kinetic theory may then be defined as consisting in constructing a further phase–space transformation $\gamma_2:\underline{Z}=(\underline{z},\varphi)\to\overline{\underline{Z}}=(\overline{\underline{z}},\overline{\varphi})$ *to a new set of phase-space, to be termed here **guiding–center variables**, such that their time derivatives* $\frac{d}{dt}\overline{Z}^i = \sigma^{ij}\frac{\partial \overline{H}}{\partial \overline{Z}^j}(\overline{z})$ *are independent of* $\overline{\varphi}$. *It then follows that by introducing the transformed distribution function* $\overline{F}(\overline{\underline{Z}})=F(\underline{Z}(\overline{\underline{Z}}))=f(\underline{x}(\overline{\underline{Z}}))$ *and by taking the gyrophase–average* (in terms of $\overline{\varphi}$) *of the F–P equation, it takes the simple form* (see e.g. Eq.(3)):

(5) $\frac{\partial \overline{H}_1}{\partial \overline{z}_1^i}\overline{w}^{ij}\frac{\partial}{\partial \overline{z}_1^j}<F(\overline{\underline{Z}}_1)> = <K_{(\overline{\underline{Z}}_1)}(F|F)>$

*to be called as **guiding–center F–P equation**; where* $[\overline{z}^i,\overline{z}^j] = \overline{w}^{ij}$ *and the brackets* "<.>" *denote the gyrophase average.i.e.:* $<A(\overline{\underline{z}},\overline{\varphi},t)> = \frac{1}{2\pi}\int_0^{2\pi} d\overline{\varphi}\, A(\overline{\underline{z}},\overline{\varphi},t)$. *Furthermore, upon using Eqs.*(3) *and* (4), *there results analogously the **guiding–center F–P equation in conservation form:***

(6) $\frac{\partial}{\partial \overline{z}_1^j}\{\frac{\partial \overline{H}_1}{\partial \overline{z}_1^i}\overline{w}^{ij}<J(\underline{x};\underline{Z})><F(\underline{Z})>\} - <\frac{\partial}{\partial Z^i}\{J(\underline{x};\underline{Z})\underline{\Gamma}.\frac{\partial}{\partial \underline{v}}Z^i\}>=0$

Thus, in guiding center theories two steps can be distinguished, which are required for constructing respectively the two phase–space transformations $\gamma_1:\underline{x}\to\underline{Z}=(\underline{z},\varphi)$ and $\gamma_2:\underline{Z}=(\underline{z},\varphi,)\to\overline{\underline{Z}}=(\overline{\underline{z}},\overline{\varphi})$. The first step can be (non uniquely) accomplished by a non–canonical transformation displaying the gyrophase dependence in the kinetic equation. As recently pointed out by the author [10], such transformations can be defined in such a way to yield authomatically a Galilei invariant guiding center kinetic theory. In fact, γ_1 can be constructed by introducing a set of velocity space coordinates which are themselves Galilei invariant, i.e. defined with respect to a non–inertial reference frame moving with the velocity of the fluid element. For the subsequent step, required for constructing the transformation γ_1 which defines the *guiding center* phase–space variables, one of the several methods known in the literature [7–9] can in principle be used. We notice that this scheme is by no means unique, since, in particular the product transformation $\Gamma=\gamma_1 \circ \gamma_2$ can, of course, be nonuniquely defined (and hence also the product diffeomorfisms γ_1 and γ_2).

3 – CONCLUSIONS

A strategy for developing guiding center theory has been pointed out, which adopts an extended, and generally non–canonical, phase–space description. The scheme can in principle be adopted for developing a whole new class of asymptotic perturbative theories. Possibile application include, in particular, the development of asymptotic theories fulfilling all basic conservation laws and the investigation of previous guiding–center teories, in order to ascertain their consistency and accuracy.

ACKNOWLEDGMENTS

The author is indebted to Prof.Ira B.Bernstein (Yale University, New Haven, Conn.,USA) and Dr.Peter J.Catto (Lodestar Research Corporation. Boulder, Colorado, USA) for clarifying discussions. Work performed in the framework of the MPI 40% Research Program "Modelli e problemi non lineari in Fisica Matematica" of the Italian "Ministero della Pubblica Istruzione".

REFERENCES

1 – M.Kruskal, J.Math.Phys.,3,806 (1962).
2 – G.F.Chew,M.L.Goldberger,F.E.Low,Proc.Roy.Soc.(London),**A236**,112 (1956).
3 – T.G.Northrop, J.A.Rome, Phys.Fluids,21,384 (1978)
4 – P.J.Catto,Plasma Phys.20,719 (1978).
5 – R.G.Littlejohn, J.Math..Phys.,20 5 (1979)
6 – A.H.Boozer,Phys.Fluids,29,904 (1980).
7 – R.G.Littlejohn, Phys.Fluids,24,1730 (1981).
8 – H.J.Mynick, Phys.Fluids,23,1888 (1981).
9 – I.B.Bernstein, P.J.Catto, Phys.Fluids,28,1342 (1985).
10 – M.Tessarotto, Proc. 16th International Symposium on Rarefied Gas Dynamics (July 1988,Pasadena, California, USA); Proc.XVI European Conference on Controlled Fusion and Plasma Physics (March 1989, Italy).
11 – J.Cary and A.Kaufman, Phys.Fluids, 24,1238 (1981).
12 – P.J.Catto,J.M.Myra,Phys.Fluids B,1,1193 (1989)

Forces in Tchebychev Nets

William O. Williams
Department of Mathematics
Carnegie Mellon University
Pittsburgh, PA

ABSTRACT: A derivation is given for representation of forces in Tchebychev nets based on a functional representation for deformation of nets.

1. Introduction

Tchebychev nets are models for the behavior of cloth. The cloth is represented as a continuum of inextensible cross fibres which can be deformed so that the angle between fibres can change but the fibres do not slide along one another. These have been studied by Tchebychev [T], Servant [S], and Voss [V] in the last century; the study was revived by Rivlin [R] and his coworkers in the fifties, and recently have been intensely studied by Pipkin [P1–7]. Pipkin has approached the problem of determining static equilibria of such nets by introducing a variational formulation which I follow here.

One interesting aspect of Tchebychev nets is that they may naturally be expected to carry both distributed and concentrated loads, loads that in either case may be modeled as reactions to the constraint of inextensibility, and it is this aspect of their behavior which I discuss here. My aim is to find a functional setting in which these reactions can be precisely characterized and their behavior delimited. The only significant mathematical tool involved in this is the Riesz representation theorem. The resulting representations involve stresses, concentrated loads and singular loads and I examine the balance equations for each sort. Of particular interest is the question of the mode of transition between distributed and concentrated loads, which is found to occur in a natural and predictable way. Such force types may be expected to occur in all models of continuum mechanics of course, and the same questions of the relationshops of the various balance equations and force–transitions are of interest; the theory of Tchebychev nets offers a particularly straightforward and convenient first case to begin a study of these effects.

2. Formulation; the representations

We will treat a net based on $[0,1]^2$, that is, each fibre is a strand represented by x=constant or y=constant, and each is of unit length. Deformations of the net can be specified by giving a continuous map $f: [0,1]^2 \to \mathbb{R}^3$ which leaves each strand of unit length. In fact, we can deduce more smoothness: since clearly

$$|f(x,y) - f(r,s)| \leq |x-r| + |y-s|, \tag{2.1}$$

we see that f is Lipschitz–continuous in $[0,1]^2$ and hence is absolutely continuous. In particular, the partial derivatives of f exist almost everywhere, or more precisely, for each y, $f,_1(\cdot,y)$ exists a.e. and for each x, $f,_2(x,\cdot)$ exists a.e. Of course the length constraint ensures that each is a unit vector, and the condition of absolute continuity ensures that the displacement is obtainable from integration:

$$\begin{aligned} f(x,y) &= f(0,0) + \int_0^x f,_1(\cdot,0) + \int_0^y f,_2(x,\cdot) \\ &= f(0,0) + \int_0^x f,_1(\cdot,y) + \int_0^y f,_2(0,\cdot). \end{aligned} \tag{2.2}$$

Clearly, f then can be obtained by integration along any path starting with a value at any point. We also find it necessary to assume that the mapping which carries x into $f,_2(x,\cdot)$ is continuous, and likewise for the y mapping[1]. Thus we describe the space of functions with which we will work as

$\mathfrak{H} := \{ f:[0,1]^2 \to \mathbb{R}^3 \mid$ a) f is absolutely continuous (obeys (2.2)

b) $y \longmapsto (f,_1(\cdot,y)):[0,1] \to L_1([0,1], \mathbb{R}^3)$ is continuous,

c) $x \longmapsto (f,_2(x,\cdot)):[0,1] \to L_1([0,1],\mathbb{R}^3)$ is continuous.$\}$

[1]It seems reasonable that this should follow from the assumtpions, but I can neither prove that it does nor find a counterexample.

and we describe the set of deformations as

$$\mathfrak{F} = \mathfrak{H} \cap \{ f:[0,1]^2 \to \mathbb{R}^3 \mid \text{d) } \forall x \in [0,1] \ |f_{,2}(x,\cdot)|=1, \text{ a.e.},$$
$$\text{e) } \forall y \in [0,1[\ |f_{,1}(\cdot,y)|=1, \text{ a.e.}\}.$$

It is not difficult to verify the following

2.1 **Lemma.** $\mathfrak{H}$, *with the norm*

$$\|f\| = |f(0,0)| + \sup_x \|f_{,2}(x,\cdot)\|_1 + \sup_y \|f_{,1}(\cdot,y)\|_1 \tag{2.3}$$

is a Banach space.

It should be noted that the norm given is equivalent to the norm of the same form with $|f(0,0)|$ replaced by $\sup |f|$, due to the integrability condition (2.2).

Now let us examine the boundary conditions which we will impose on the space of displacements. We assume first that we are given measurable complementary subsets of the boundary of $[0,1]^2$: B_p, on which the **boundary displacement,** $f_p \in \mathscr{C}(B_p, \mathbb{R}^3)$ is specified, and B_t, with the **boundary force,** an $\mathbb{R}^3$–valued Radon measure α supported on B_t. We shall suppose that B_p is not empty, and require that

$$f(x,y) = f_p(x,y) \text{ if } (x,y) \in B_p; \tag{2.4}$$

and thus introduce compatibility conditions on f_p, which we shall not detail.[2] α will enter in the variational formulation below. Then we also anticipate displacement and traction being specified on parts of $(0,1)^2$ as well. We suppose given measurable sets S_p, on which is specified the **surface displacement** $f_p \in \mathscr{C}(S_p, \mathbb{R}^3)$, and S_t with the **surface force,** a Radon measure β supported on S_t. Of course, we will require, that

$$f(x,y) = f_p(x,y) \text{ if } (x,y) \in S_p, \tag{2.5}$$

and β will appear in the variational formulation below. Note that an area–continuous part of β would represent a body force such as gravity.

---

[2]Some boundary conditions of place may produce extreme restrictions, due to the constraints. Zero displacement on the entire boundary ensures that the admissible manifold reduce to a single point, $f = 0$.

Each of the kinematical constraints gives rise to a reaction: for use in deducing the form of these reactions, we introduce the constraint mappings H_1, H_2, H_b, and H_s, as

$$\begin{aligned}
&H_1:\mathfrak{H} \longrightarrow \mathscr{C}([0,1],L_1[0,1]) \quad \{H_1(f)\}(y) = |f,_1(\cdot,y)| \\
(2.6)\qquad &H_2:\mathfrak{H} \longrightarrow \mathscr{C}([0,1],L_1[0,1]) \quad \{H_2(f)\}(x) = |f,_2(x,\cdot)| \\
&H_b:\mathfrak{H} \longrightarrow \mathscr{C}(B_p,\mathbb{R}^3) \quad H_b(f) = f|_{B_p} \\
&H_s:\mathfrak{H} \longrightarrow \mathscr{C}(S_p,\mathbb{R}^3) \quad H_s(f) = f|_{S_p}.
\end{aligned}$$

2.2 **Lemma:** *Each of the constraint mappings is differentiable on* $\mathfrak{F}$.

This is trivially true for the bounded linear mappings H_s and H_p. We note that H_1 is the composition of a partial derivative operator and the length operator

$$(2.7)\qquad |\cdot|:\mathfrak{H} \longrightarrow (L_1([0,1],\mathbb{R}^3) \longrightarrow L_1[0,1])$$

and thus H_1 is differentiable, in particular, at functions in $\mathfrak{F}$. The derivative is given by

$$(2.8)\qquad \{\nabla H_1(f)[h]\}(y) = f,_1(\cdot,y)\cdot h,_1(\cdot,y).$$

A similar calculation applies to the mapping H_2.

The variational formulation of the problem: *Given the above data maximize the functional*

$$G(f) = \int_{B_t} f\cdot d\alpha + \int_{S_t} f\cdot d\beta$$

for $f\in\mathfrak{F}$ *and obeying the boundary conditions* (2.4) *and* (2.5).

As remarked by Pipkin [P5], this is an unusual form for a variational inequality, and reflects the fact that the system does not posses any internal energy: the maximization is simply the minimizing of the work done by the external loads. This contributes to the extreme ill–posedness of many problems.

It is easy to re–express the constraint and boundary conditions in terms of the mappings introduced in (2.6). Our main representation result utilizes the following.

Lemma: *Given Banach spaces* $\mathfrak{H}$ *and* $\mathfrak{f}$, *smooth mappings*

$$G:\mathfrak{H} \longrightarrow \mathbb{R},\ H:\mathfrak{H} \longrightarrow \mathfrak{f},$$

and $h \in \mathfrak{f}$, *a necessary condition that* G *take a maximum value at* w *in the manifold* $\mathcal{M}$,

$$\mathcal{M} := \{ z \in \mathfrak{H} \mid H(z) = h \}$$

is that there exist a linear mapping λ_w *in the dual of* $\mathfrak{f}$ *such that*

$$\nabla G(w) = \lambda_w \circ \nabla H(w).$$

To complete our representation, we need to establish that these forms are continuous, to determine a representation for each and then to identify them physically. It can be shown that continuity does not obtain for all choices of **w**; one can establish this property for a rather general but not universal class of functions. For our current purposes, let us simply assume that **w** is in this class so that all of λ_1, λ_2, λ_s, and λ_b are continuous. Given this continuity, we can use standard representation results to find the form of the λ_k. To dispense first with the boundary terms, we note that by the Riesz representation theorem, there exist $\mathbb{R}^3$–valued Radon measures γ and ϵ such that

$$\lambda_b(g) = \int_{B_p} g \cdot d\gamma, \qquad \lambda_s(g) = \int_{S_p} g \cdot d\epsilon.$$

The values of these measures γ and ϵ are the **reaction forces** associated to the kinematic constraints.

Of more interest are the reactions introduced by the length constraints of the filaments of the net. We apply a generalized Riesz representation (see Federer [F], p.102) to deduce the existence of a Radon measure μ and of a $L^\infty[0,1]$–valued μ–measurable function $y \longmapsto t_y$ such that

$$\lambda_1(f)[g] = \int_0^1 \int_0^1 t_y(x) f_{,1}(x,y) \cdot g_{,1}(x,y) dx \, d\mu(y).$$

and similarly that

$$\lambda_2(f)[g] = \int_0^1 \int_0^1 t_x(y) f_{,2}(x,y) \cdot g_{,2}(x,y) dy \, d\nu(x).$$

In the representation formula in [F], the functions t_x and t_y are unit–valued. For our purposes, it is more natural to transfer the magnitude of μ and ν to these densities; since μ is identifiable with a function of bounded variation, we write $\mu = \mu_c + \mu_a + \mu_s$, where

μ_c is absolutely continuous, the saltus function μ_a has jumps on the countable set $\mathcal{J}$ and the singular function μ_s has support on the ℓ–null set $\mathcal{S}$. We define

$$t(x,y) := \begin{cases} t_y(x)\mu_c'(y) & \text{for } y \in [0,1]\backslash(\mathcal{J} \cup \mathcal{S}) \\ t_y(x)\mu_a(y) & \text{for } y \in \mathcal{J} \\ t_y(x) & \text{for } y \in \mathcal{S} \end{cases}$$

and similarly define $s(x,y)$ using t_x and ν. We redefine the measures μ and ν in the obvious way. Now we can identify $t(x,y)$ and $s(x,y)$ as the **tensions** in the x– and y–fibres respectively. If we identify

$$\mathbf{a}(x,y) := \mathbf{f}_{,1}(x,y),\ \mathbf{b}(x,y) := \mathbf{f}_{,2}(x,y)$$

s the **unit tangent vectors** to the x– and y–fibres respectively, we can define the **vector–tensions**, as

$$\mathbf{t}(x,y) := t(x,y)\mathbf{a}(x,y) \quad \mathbf{s}(x,y) := s(x,y)\mathbf{b}(x,y),$$

and rewrite the equations as

$$\lambda_1(\mathbf{f})[\mathbf{u}] = \int_0^1 \int_0^1 \mathbf{t}(x,y)\cdot\mathbf{u}_{,1}(x,y)dxd\mu(y),$$

$$\lambda_2(\mathbf{f})[\mathbf{u}] = \int_0^1 \int_0^1 \mathbf{s}(x,y)\cdot\mathbf{u}_{,2}(x,y)dyd\nu(x).$$

We will see below that the integrals can be reversed so that we can identify the integral

$$\int_a^b \mathbf{t}(c,y)dy,$$

for example,as a **stress** resultant force exerted across the line $x=c$, $a \le y \le b$, while

$$\mathbf{t}(c,Y)$$

represents a concentrated force along the fibre $y=Y$ at a point $x=c$. We will call such a fibre a **loaded fibre**. The integral

$$\int_a^b \mathbf{t}(c,y)d\mu_s(y)$$

is a sort of "noise force" exerted across the line.

LEMMA: *The function* $\mathbf{t}$ *is integrable with respect to* $\ell\times\mu$, *and the function* $\mathbf{s}$ *is integrable with respect to* $\nu\times\ell$

It is important to notice that the density μ, the y–concentration of the force is independent of x and likewise that the x–concentration ν is independent of y.

3.Local Balance Equations; Boundary Conditions

In this paper we will assume that the applied loading and displacement occur only on the boundary of the region, with only a body force in the interior. (The case in which these also occur in the interior is more complicated but not more difficult.) Thus we have

$$G(f) = \int_0^1 f(0,\cdot)\cdot d\alpha_0 + \int_0^1 f(1,\cdot).d\alpha_1$$

(3.1) $$+\int_0^1 f(\cdot,0)\cdot d\alpha^0 + \int_0^1 f(\cdot,1).d\alpha^1 + \iint_{[0,1]^2} g\, d\mathcal{L}^2,$$

using α_0 to represent the external force α restricted to the y–axis, and α_1 its restriction to x=1, etc. The support of each of these measures in general is not all of [0,1]; on the complementary part of the interval the displacement, rather than the force, is specified. The representation theorem then assures the existence of reaction forces (measures) on this complement. For the purposes of this section, we choose to incorporate (the negative of) these measures into the α. Thus, for example, the measure α_0 will henceforth represent the applied force on the appropriate part of the line x=0 and the negative of the reaction force on the remainder of that line.

The constraint equation then tells us that at an extreme point $w\in\mathfrak{F}$ for all $f\in\mathfrak{H}$,

(3.2) $$G(f) = \iint_{[0,1]^2} t(x,y)\cdot f_{,1}(x,y)\, dx\, d\mu(y) + \iint_{[0,1]^2} s(x,y)\cdot f_{,2}(x,y)\, d\nu(x)\, dy.$$

The fundamental assumption we shall make in this section then is

ASSUMPTION: *For μ–almost every y the function $t(\cdot,y)$ is of bounded variation; for ν–almost every x the function $s(x,\cdot)$ is of bounded variation.*

In fact, for those points y which are atoms of μ it is easily provable that $t(\cdot,y)$ is of bounded variation, but we require it more generally. We choose to let each of these bounded–variation–functions be left–continuous at 1 and right–continuous at 0.

Now we can rewrite equation (3.2) as

$$G(f) = \int_0^1 \{t(1,y)\cdot f(1,y) - t(0,y)\cdot f(0,y)\}d\mu(y)$$

$$+ \int_0^1 \{s(x,1)\cdot f(x,1) - s(x,0)\cdot f(x,0)\}d\mu(x)$$

$$(3.3) \qquad - \int\int_{[0,1]^2} f(x,y)\cdot d_1 t(x,y)d\mu(y) - \int\int_{[0,1]^2} f(x,y)\cdot d_2 s(x,y)d\nu(x),$$

and note that all terms in the equation represent Radon measures acting on the absolutely continuous function f. By the uniqueness of the representation of linear functionals we conclude that the measures are equal. To save burdensome notation, let us agree to write $g\eta$ for the term $\int g(p)d\eta(p)$, and if we equate the measures, after combining (3.1) and (3.3), we have

$$\delta_o \times \boldsymbol{\alpha}_o + \delta_1 \times \alpha_1 + \alpha^{o} \times \delta_o + \alpha^1 \times \delta_1 + g\ell^2 =$$

$$(3.4) \qquad = t(\delta_1 \times \mu) - t(\delta_o \times \mu) - d_1 t d\mu$$

$$+ s(\nu \times \delta_1) - s(\nu \times \delta_o) - d_2 s d\nu,$$

where δ_o and δ_1 represent the dirac delta measures at 0 and 1, respectively.

The fact that these measures have differing supports leads to several different "pointwise" consequences of this equation. The remainder of this section will consist in a systematic listing of these equations, with remarks on the interpretation of the most significant of them.

1. **Area supported part**. The part of (3.4) continuous with respect to ℓ^2 is

$$(3.5) \qquad g\ell^2 = -t,_1 \ell^2 - s,_2 \ell^2$$

or, in terms of the densities,

$$(3.6) \qquad t,_1 + s,_2 + g = 0 \qquad \ell^2\text{–a.e.}$$

which is the usual **equation of balance of forces for stresses**. The boundary conditions for the stresses are included in the analysis of the load on an end fibre: consider, for example, x=0. The measures which are continuous with respect to $\delta_o \times \ell$ are

$$\delta_o \times \boldsymbol{\alpha}_o = - t(\delta_o \times \mu) - d_2 s d\nu,$$

yielding the equation

(3.7) $$\frac{d\alpha}{d\ell} = -\mathbf{t}(0,\cdot) - \mathbf{s},_2(0,\cdot)\nu(\{0\}) \quad \ell\text{-a.e.}$$

Thus, *the endvalues of the stress* **t** *are equal to the corresponding applied stress unless the endfibre is loaded, in which case the value is modified by the derivative of the corresponding tension.* Of course there are corresponding equations at the other endfibres.

2. **Balance along fibres.** We have considered the endfibres above; let us choose $X\in(0,1)$ to be an **atom of** ν and consider the measures continuous with respect to $\delta_X\times\ell$

(3.7′) $$0 = -d_1\mathbf{t}d\mu - d_2\mathbf{s}d\nu$$

so that the densities are

(3.8) $$0 = [[\mathbf{t}]]_1(X,\cdot) + \mathbf{s},_2(X,\cdot)\nu(\{X\}) \quad \ell\text{-a.e.}$$

where we have used the notation $[[\cdot]]_1$ to denote the jump of the function with respect to x as y is held constant. This says that *the stress* **t** *is discontinuous at a point where it crosses a loaded fibre, with the discontinuity proportional to the derivative of the tension along the loaded fibre.* To introduce boundary values for the tension **s** we look at the measures continuous with respect to $\delta_X\times\delta_o$. One has

(3.9) $$\alpha_o\times\delta_o = -d_1\mathbf{t}d\mu - \mathbf{s}(\nu\times\delta_o)$$

or, in terms of densities,

(3.10) $$\alpha o(\{X\}) = -\nu(\{X\})\mathbf{s}(X,0) - \mu(\{0\})[[\mathbf{t}]]_1(X,0)$$

This says that *the endvalue of the tension in a loaded fibre is balanced by a corresponding concentrated force in the loading or by a discontinuity in the (necessarily loaded) endfibre* or by a combination of the two.

3. **Point Balance.** Given points X amd Y in (0,1), we consider measures continuous with respect to $\delta_X\times\delta_Y$:

(3.11) $$0 = -d_1\mathbf{t}d\mu - d_2\mathbf{s}d\nu,$$

or

(3.12) $$0 = \mu(\{Y\})[[\mathbf{t}]]_1(X,Y) + \nu(\{X\})[[\mathbf{s}]]_2(X,Y).$$

Thus *discontinuities of the tensions in loaded fibres occur at crossings with other loaded fibres and the two discontinuities are equal and opposite.* Of course this does not apply to endfibres, where we introduce also the concentrated forces due to the loading.

4.Singular Stresses. We now consider the singular parts of the force system. This type of force is the least physical of the three types, and would not occur in the applied forces. On the other hand, as a reaction force it cannot be ruled out *a priori*, and could conceivably be important in some minimization problems. Letting the singular part of each measure γ be denoted by γ^S, the equation reduces to

$$\begin{aligned}\delta_o\times\alpha_o^S + \delta_1\times\alpha_1^S + \delta_o\times\alpha o^S + \delta_1\times\alpha^{1S} &= \mathbf{t}(\delta_1\times\mu^S) - \mathbf{t}(\delta_o\times\mu^S) - \mathrm{d}_1^S\mathbf{t}\mathrm{d}\mu \\ &+ \mathbf{s}(\nu^S\times\delta_1) - \mathbf{s}(\nu^S\times\delta_o) - \mathrm{d}_2^S\mathbf{s}\mathrm{d}\nu.\end{aligned} \tag{3.13}$$

The equation decomposes exactly as the previous ones, with doubly singular terms

$$\mathrm{d}_1^S\mathbf{t}\mathrm{d}\mu^S + \mathrm{d}_2^S\mathbf{s}\mathrm{d}\nu^S = 0, \tag{3.14}$$

and fibre by singular terms such as

$$\mathbf{t},_1(\mathrm{x},\cdot)\mathrm{d}\mu^S + \mathrm{d}_2^S\mathbf{s}(\mathrm{x},\cdot) = 0 \qquad \measuredangle\text{-a.e. } \mathrm{x}\in(0,1). \tag{3.15}$$

The latter shows that *the singular part of the variation of the stress along a fibre is* $\measuredangle$*-almost everywhere absolutely continuous with respect to the singular part of the force concentration*, and thus, in particular, is zero whenever the latter is. Likewise, for the atomic by singular parts, one has equations such as

$$[[\mathbf{t}]]_1(\mathrm{X},\cdot)\mathrm{d}\mu^S + \nu(\{\mathrm{X}\})\mathrm{d}_2^S\mathbf{s}(\mathrm{X},\cdot) = 0 \tag{3.16}$$

saying that *the singular part of the variation of the tension along a loaded fibre similarly is absolutely continuous with respect to the singular part of the force concentration.*

Finally, we look at boundary conditions: at $\mathrm{x}=0$, for example,

$$\alpha_o^S = -\mathbf{t}(0,\cdot)\mu^S - \nu(\{0\})\mathrm{d}_2^S\mathbf{s}(0.\cdot), \tag{3.17}$$

so that *the boundary loads have singular parts absolutely continuous with respect to the singular part of the force concentration provided the endfibre is not loaded.*

References

[F] Federer, H., Geometric Measure Theory, Springer Verlag, New York, 1969.

[G] Goldstine, H. H., A multiplier rule in abstract spaces, *Bull. Am. Math. Soc.* 44:388–394, 1938.

[DS] Dunford, N. and J. T. Schwartz, Linear Operators, Part I: General Theory, Interscience, New York, 1966.

[P1] Pipkin, A. C., Some developments in the theory of inextensible networks, *Quart. Appl. Math.*38:343–355, 1980.

[P2] Pipkin, A. C., Inextensible networks with slack, *Quart. Appl. Math.* 40:63–71, 1982.

[P3] Pipkin, A. C., Inextensible networks deformed as translation surfaces, *IMA J. Appl. Math.*31:261–267, 1983.

[P4] Pipkin, A. C., Energy minimization for nets with slack, *Quart. Appl. Math.* 44:249–253, 1986.

[P5] Pipkin, A. C., Some examples of crinkles, in *Homogenization and Effective Moduli*, ed. ,Springer Verlag, New York, 1988.

[P6] Pipkin, A. C., Equilibrium of Tchebychev nets, *Arch. Rational Mech. Anal.*,

[P7] Pipkin, A. C., Continuously distributed wrinkles in fabrics, *Arch. Rational Mech. Anal.*

[R] Rivlin, R. S., The deformation of a membrane formed by inextensible cords, *Arch. Rational Mech. Anal.* 2:447–476, 1959.

[S] Servant, M. Sur l'habillage des surfaces, *C.R.Acad. Sci.*135:575–577, 1902.

[T] Tchebychev, P. L., Sur ls coupe des vêtements, *Oeuvres II*, 708, Acad. Sci. St. Petersbourg, 1907.

[V] Voss, A. Über ein neues princip der abbildung krummer oberflächen, *Math. Ann.* 19:1–26, 1882.

ASYMPTOTIC STABILITY OF A PERTURBED LOTKA-VOLTERRA MODEL

Fergola P., Rionero S. and Tenneriello C.

Dipartimento di Matematica e Applicazioni, Università di Napoli, Via Mezzocannone, 8 - 80134 Napoli, Italy.

ABSTRACT. The stability features of a positive equilibrium to a perturbed non-autonomous Lotka-Volterra system are investigated. By means of the Lyapunov direct method, conditions on the perturbation terms are given in order to ensure the exponential stability of this equilibrium.

1. INTRODUCTION

The perturbed Lotka-Volterra predator-prey system has been widely analyzed by many authors (see, for instance, [3],[4],[5],[7],[8],[9], [10],[12],[13],[14]). Perturbation terms have been introduced in order to account both for (small) variations of the classic Lotka-Volterra model or to put controls on the growth of both predator and the prey. So, the stability of the positive equilibrium point of the classic Lotka-Volterra model as well as the existence of periodic solutions and perturbed critical points have been tested under the influence of small autonomous perturbations. The multiple scale method has been applied in [6] in order to study the stability of the equilibria in presence of time dipendent perturbation terms, depending on a small parameter.

In this paper we consider the same type of non-autonomous perturbation terms as in [6], by giving up the character of the "smallness" of these perturbations. The considered terms are given by the product between a very general function of time and the difference between the population densities of the prey and predator. The same kind of perturbation terms have been already considered in [1] and [2], where they were related to "diffusion" effects. Then by using the Lyapunov direct method (see, for instance, [11]) conditions on the perturbation terms are given in order to ensure the nonlinear asymptotic exponential stability of a positive equilibrium.

2. PRELIMINARIES

It is well known that the classic Lotka-Volterra model which describes the prey-predator interaction

$$(1) \qquad \begin{cases} \dot{x} = a x - b x y \\ \dot{y} = - c y + d x y \end{cases}$$

$x, y \in R$ and a,b,c,d positive constants, admits the biologically meaningful positive equilibrium

$$(2) \qquad x^* = \frac{c}{d} \quad , \quad y^* = \frac{a}{b}$$

which is a center. Now we want to study the influence of time-varying perturbation terms on the stability of the equilibrium (2). Precisely we consider the perturbed model

$$(3) \qquad \begin{cases} \dot{x} = a x - b x y + D(t)\,(y - x) \\ \dot{y} = - c y + d x y + D(t)\,(x - y) \end{cases} ,$$

where $D: R^+ \to R$ is $L^1_{loc}(R^+)$. We suppose that

$$(4) \qquad E = y^* - x^* = 0$$

so that (2) is an equilibrium for the system (3) too.

Performing in (3) the change of variables

$$(5) \qquad x = x^* + u \quad , \quad y = y^* + v$$

because of (2) and (4), we obtain

$$(6)\qquad \begin{cases} \dot{u} = -D(t)\,u + m(t)\,v - b\,u\,v \\ \dot{v} = n(t)\,u - D(t)\,v + d\,u\,v \end{cases}$$

where

$$(7)\qquad m(t) = D(t) - a \quad , \quad n(t) = D(t) + c > 0$$

Now we are able to start for the stability analysis which will be carried out by using the Lyapunov direct method (see, for instance, [11]). We introduce the new variables

$$(8)\qquad X = u\,e^{\int_0^t D(\tau)d\tau} \quad , \quad Y = v\,e^{\int_0^t D(\tau)d\tau}$$

and system (6) can be written as

$$(9)\qquad \begin{cases} \dot{X} = m(t)\,Y - b\,e^{-\int_0^t D(\tau)d\tau}\,X\,Y \\ \dot{Y} = n(t)\,X + d\,e^{-\int_0^t D(\tau)d\tau}\,X\,Y \end{cases}$$

Standard calculation allows to obtain from the system (9) the following equation

$$(10)\qquad \frac{1}{2}\frac{d}{dt}(X^2 + Y^2) = (m+n)\,X\,Y + (-b\,X^2Y + d\,X\,Y^2)\,e^{-\int_0^t D(\tau)d\tau}$$

It is useful in the sequel to set

$$(11)\qquad \ell = \frac{1}{2}(a - c)$$

3. MAIN RESULTS

In this Section we suppose that the parameters a and c which appear in the equ. (1) are such that

$$(12)\qquad c < a$$

and therefore the constant ℓ defined by (11) is positive.

Now we shall state two theorems ensuring the exponential asymptotic stability of the equilibrium (2) correspondingly to the cases $D(t) \geq \ell \;\; \forall t \in R^+$ and $D(t) \in]\ell/2, \ell[\;\; \forall t \in R^+$.

<u>THEOREM 1</u> . - <u>Let us suppose that</u>

i) $a/b = c/d$

ii) $c < a$

iii) $D(t) \in L^1_{loc}(R^+)$, $D(t) \geq \ell \quad \forall t \in R^+$.

<u>Then the positive equilibrium</u> $x^* = c/d$, $y^* = a/b$ <u>of system (3) is asymptotically exponentially stable and the attractivity region A is</u>

$$A = \left\{ (x_\circ, y_\circ) \in R^{+^2} : \sqrt{(x_\circ - x^*) + (y_\circ - y^*)} < 2\ell/k \ , \ k = 2(d + \frac{2}{3\sqrt{3}} b) \right\}.$$

PROOF. Let us consider first the case

(13) $$D(t) > \ell \qquad \forall t \in R^+ \ .$$

Because of (7) the hypotheses imply that

(14) $$m + n > 0 \ .$$

Then, using the Cauchy inequality, from (10) we obtain

(15) $$\frac{1}{2}\frac{d}{dt}(X^2+Y^2) \leq \frac{m+n}{2}(X^2+Y^2) + (dXY^2 - bX^2Y)\, e^{-\int_0^t D(\tau)d\tau}$$

Now we observe that the following inequality holds

(16) $$\frac{d\,X\,Y^2 - b\,X^2 Y}{(X^2+Y^2)^{3/2}} < \frac{k}{2}$$

where

(17) $$\frac{k}{2} = d + \frac{2}{3\sqrt{3}}\, b$$

Therefore from (15) it follows

(18) $$\frac{d}{dt}(X^2+Y^2) - (m+n)(X^2+Y^2) < k\,(X^2+Y^2)^{\frac{3}{2}}\, e^{-\int_0^t D(\tau)d\tau}$$

which gives

(19) $$\frac{d}{dt}\left[(X^2+Y^2)\, e^{-\int_0^t (m+n)d\tau}\right] < k\,(X^2+Y^2)^{3/2}\, e^{-\int_0^t (m+n+D)d\tau} \ .$$

Then, by putting

(20) $$H = (X^2+Y^2)\, e^{-\int_0^t (m+n)d\tau}$$

we obtain

(21) $$\frac{dH}{dt} < k\,H^{\frac{3}{2}}\, e^{\frac{1}{2}\int_0^t (m+n-2D)d\tau}$$

Hence, because of (7), (4) and (11), it follows

(22) $$H^{-\frac{1}{2}}(t) > H^{-\frac{1}{2}}(0) - \frac{k}{2\ell} .$$

Then, if we suppose that

(23) $$H^{\frac{1}{2}}(0) < \frac{2\ell}{k} ,$$

it follows

(24) $$H(t) < \frac{1}{\left[H^{-\frac{1}{2}}(0) - \frac{k}{2\ell}\right]^2} .$$

By means of (20), going back to the variables u,v through (8) and using again (7), (4), (11) we can write (24) in the form

(25) $$u^2 + v^2 < \frac{e^{-2\ell t}}{\left[H^{-\frac{1}{2}}(0) - \frac{k}{2\ell}\right]^2}$$

which ensures the asymptotic exponential stability of the solution u = = v = 0 of system (6). This implies, because of (5), that the positive equilibrium (2) of system (3) is asymptotically exponentially stable with an attractivity region defined through (23).

In the case

(26) $$D(t) = \ell \qquad \forall t \in R^+$$

because of (i), (ii) and (7) we have

(27) $$m + n = 0$$

and, reasoning as before, the proof can be easily obtained.

If the function D(t) is bounded from above by the constant ℓ given in (11), the following theorem can be proved by similar arguments.

<u>THEOREM 2</u>.- <u>Let us assume that</u>

(i) $a/b = c/d$,

(ii) $c < a$,

(iii) $\exists h \in \left]0, \frac{\ell}{2}\right[$ such that $D(t) \in \left[\frac{\ell}{2} + h, \ell\right[\quad \forall t \in R^+$.

Then the positive equilibrium $x^* = c/d$, $y^* = a/b$ of system (3) is asymptotically exponentially stable and the attractivity region B is

$$B = \left\{ (x_o, y_o) \in R^{+^2} : \sqrt{(x_o - x^*)^2 + (y_o - y^*)^2} < \frac{4h}{k} \right\} .$$

REFERENCES

[1] Beretta E., Fergola P. and Tenneriello C.: Ultimate boundedness for non-autonomous diffusive Lotka-Volterra patches. Math. Biosci. 92 (1988), 29-53

[2] Beretta E., Fergola P. and Tenneriello C.: Partial ultimate boundedness for completely non-autonomous diffusive Lotka-Volterra patches. R. Trappl (ed.), Cybernetics and Systems '88. Kluwer Academic Publishers, 1988, 399-406

[3] Comins H.N. and Blatt D.W.E.: Prey-predator models in spatially heterogeneous environments. J. Theor. Biol., 48 (1974), 75-83

[4] Dutt R. and Glosh P.K.: Nonlinear correction to Lotka-Volterra oscillations in a prey-predator system. Math. Biosci. 27 (1975), 9-16

[5] Dutt R., Glosh P.K. and Karmaker B.B.: Application of perturbation theory to the nonlinear Volterra-Gause-Witt model for prey predator interaction. Bull. Math. Biol. 37 (1975), 139-146

[6] Fergola P. and Nayfeh A.H.: Application of the method of multiple scales to perturbed Lotka-Volterra models. Proceedings of V International Conference "Waves and Stability in Continuous Media" (Taormina, Italy, Sept. 1987), S.Giambò editor, Editel, Cosenza

[7] Freedman H.I.: Deterministic mathematical models in population ecology. Marcel Dekker, Inc., New York, 1980

[8] Freedman H.I. and Waltman P.: Perturbation of two dimensional predator-prey equations. SIAM J. Appl. Math., 28 (1975), 1-10

[9] Freedman H.I. and Waltman P.: Periodic solutions of perturbed Lotka-Volterra Systems. International Conference on Differential Equations, H.A. Antosiewicz editor, Academic Press, New York, 1975, 312-316

[10] Freedman H.I. and Waltman P.: Perturbation of two dimensional predator-prey equations with an unperturbed critical point. SIAM J. Appl. Math., 29 (1975), 719-733

[11] Hahn W.: Stability of motion. Springer-Verlag, Berlin, 1967

[12] Levin S.A.: Dispersion and population interactions. Amer.Nat., 108 (1974), 207-228

[13] Samuelson P.A.: Generalized predator-prey oscillations in ecological and economic equilibrium. Proc.Nat. Acad. Sci. USA, 68(1971), 980-983

[14] Walter C.: The global asymptotic stability of prey-predator systems with second order dissipation. Bull.Math.Biol., 36(1974),215-217

ACKNOWLEDGMENTS. This research has been performed under the auspices of the G.N.F.M. of the C.N.R. and has been partially supported by the M.U.R.S.T. under 40% and 60% contracts and by C.N.R.-S.P.A.I.M.

AUTHORS INDEX

Series on Advances in Mathematics for Applied Sciences

Aims and Scope

This Series reports on new developments in mathematical research relating to methods, qualitative analysis, interaction with computer science and mathematical modelling in the applied and the technological sciences.

This Series includes Books, Lecture Notes, Proceedings, Collections of research and review papers. Proceedings and Monographic Collections will generally have a Guest Editorial Board.

High quality, novelty of the content and potential for the applications to modern problems in applied science will be the guidelines for the selection of the content of this series.

Instructions for Authors

Submission of proposals should be addressed to the Editor-in-Charge or to any member of the Editorial Board. Acceptance of books and Lecture Notes will generally be based upon the description of the general content and scope of the work as well as upon samples (about one third) of the book or Lecture Notes including the parts judged more significantly by the Authors themselves.

Acceptance of Proceedings will be based upon relevance of the topics and of the lecturers contributing to the volume.

Authors are urged, in order to avoid re-typing, not to begin the final preparation of the text before having received the publisher's guidelines. They will receive from World Scientific the instructions to prepare the final manuscript in a camera-ready form.